Characterization of Metal and Polymer Surfaces

VOLUME 2   *Polymer Surfaces*

ACADEMIC PRESS RAPID MANUSCRIPT REPRODUCTION

Characterization of Metal and Polymer Surfaces

Volume 2 Polymer Surfaces

EDITED BY

LIENG-HUANG LEE

Xerox Corporation
Rochester, New York

ACADEMIC PRESS, INC. New York San Francisco London 1977
A Subsidiary of Harcourt Brace Jovanovich, Publishers

COPYRIGHT © 1977, BY ACADEMIC PRESS, INC.
ALL RIGHTS RESERVED.
NO PART OF THIS PUBLICATION MAY BE REPRODUCED OR
TRANSMITTED IN ANY FORM OR BY ANY MEANS, ELECTRONIC
OR MECHANICAL, INCLUDING PHOTOCOPY, RECORDING, OR ANY
INFORMATION STORAGE AND RETRIEVAL SYSTEM, WITHOUT
PERMISSION IN WRITING FROM THE PUBLISHER.

ACADEMIC PRESS, INC.
111 Fifth Avenue, New York, New York 10003

United Kingdom Edition published by
ACADEMIC PRESS, INC. (LONDON) LTD.
24/28 Oval Road, London NW1

Library of Congress Cataloging in Publication Data

Symposium on Advances in Characterization of Metal
    and Polymer Surfaces, New York, 1976.
    Characterization of metal and polymer surfaces.

    Sponsored by the Division of Organic Coatings
and Plastics Chemistry, the Division of Analytical
Chemistry, and Cellulose, Paper, and Textile
Division of the American Chemical Society.
    Includes bibliographical references and indexes.
    CONTENTS: v. 1.   Metal surfaces.
    1.   Metallic surfaces–Congresses.   2.   Polymer
surfaces–Congresses.   3.   Spectrum analysis–
Congresses.   I.   Lee, Lieng–Huang,   Date
II.   American Chemical Society.   Division of
Organic Coatings and Plastics Chemistry.
III.   American Chemical Society.   Division of
Analytical Chemistry.   IV.   American Chemical
Society.   Cellulose, Paper, and Textile Division.
V.   Title
QD506.A1S95   1976   541'.3453   77-2255
ISBN 0–12–442102–4 (v. 2)

PRINTED IN THE UNITED STATES OF AMERICA

Contents

Contributors	ix
Preface	xi
Contents of Volume 1	xiii

PART I: Electron Spectroscopy for Chemical Analysis

Introductory Remarks D. M. HERCULES	3
Plenary Lecture: Application of ESCA to Structure and Bonding in Polymers D. T. CLARK	5
Sputter-Induced Compositional Change During ESCA/Sputtering of Polymers DWIGHT E. WILLIAMS AND LAWRENCE E. DAVIS	53
Characterization of Chemically Modified Cottons by ESCA DONALD M. SOIGNET	73
Surface Analysis of Plasma Treated Wool Fibers by X-Ray Photoelectron Spectroscopy MERLE M. MILLARD	86
Plasma Modification of Polymers Studied by Means of ESCA D. T. CLARK AND A. DILKS	101
XPS Studies of Polymer Surfaces for Biomedical Applications JOSEPH D. ANDRADE, GARY K. IWAMOTO, AND BONNIE McNEILL	133
Discussion	143

PART II: Infrared and Laser Raman Spectroscopy

Introductory Remarks: Surface Characterization of Polymers by Infrared and Raman Spectroscopy 147
    LIENG-HUANG LEE

Plenary Lecture: Transmission and Reflection Spectroscopy, Nature of the Spectra 153
    N. J. HARRICK

The Study of Thin Polymer Films on Metal Surfaces Using Reflection-Absorption Spectroscopy — Oxidation of Poly(1-Butene) on Gold and Copper 193
    D. L. ALLARA

Use of Laser Raman Techniques in the Study of Polymers 207
    R. D. ANDREWS AND T. R. HART

Discussion 241

PART III: Microscopy for Polymers

Introductory Remarks 249
    L. H. PRINCEN

Microscopical Analysis of Chemically Modified Textile Fibers 251
    WILTON R. GOYNES AND JARRELL H. CARRA

Laboratory Study of Fiber Fracture Using the Scanning Electron Microscope 267
    ALFREDO G. CAUSA

The Investigation of Poly(tetrafluoroethylene) Wetting Behavior by Scanning Electron Microscopy 289
    N. E. WEEKS, G. M. KOHLMAYR, AND E. P. OTOCKA

Fluoropolymer Surface Studies, II 313
    DAVID W. DWIGHT

Structural Characterization of Poly(N-vinylcarbazole) 333
    C. H. GRIFFITHS

Discussion 359

PART IV: Surface-Chemical and Radiation Analyses

Characterization of Latexes by Ion Exchange and Conductometric Titration   365
    J. W. VANDERHOFF

Surface Area of Polymer Latexes by Angular Light Scattering   397
    ROBERT L. ROWELL AND RAYMOND S. FARINATO

Evaporative Rate Analysis: Its First Decade   409
    JOHN LYNDE ANDERSON

Radiation Absorption for Polymers   429
    JOHN R. HALLMAN, J. REED WELKER, AND C. M. SLIEPCEVICH

About Authors   *447*
Author Index   *449*
Subject Index   *457*

Contributors

D. L. ALLARA   Bell Laboratories, Inc.
JOHN LYNDE ANDERSON   ERA Systems, Inc.
JOSEPH D. ANDRADE   University of Utah
R. D. ANDREWS   Stevens Institute of Technology
JARRELL H. CARRA   Southern Regional Research Center, USDA
ALFREDO G. CAUSA   Goodyear Tire and Rubber Co.
D. T. CLARK   Durham University, England
LAWRENCE E. DAVIS   Physical Electronics Industries
A. DILKS   Durham University, England
DAVID W. DWIGHT   Virginia Polytechnic Institute and State University
RAYMOND S. FARINATO   University of Massachusetts
WILTON R. GOYNES   Southern Regional Research Center, USDA
C. H. GRIFFITHS   Xerox Corporation
JOHN R. HALLMAN   Nashville State Technical Institute
N. J. HARRICK   Harrick Scientific Corporation
T. R. HART   Stevens Institute of Technology
D. M. HERCULES   University of Georgia
GARY K. IWAMOTO   University of Utah
G. M. KOHLMAYR   Pratt & Whitney Aircraft
LIENG-HUANG LEE   Xerox Corporation
BONNIE McNEILL   University of Utah
MERLE M. MILLARD   Western Regional Research Center, USDA
E. P. OCTOCKA   Pratt & Whitney Aircraft
L. H. PRINCEN   Northern Regional Research Center, USDA
ROBERT L. ROWELL   University of Massachusetts
C. M. SLIEPCEVICH   University of Oklahoma
DONALD M. SOIGNET   Southern Regional Research Center, USDA
J. W. VANDERHOFF   Lehigh University
N. E. WEEKS   Pratt & Whitney Aircraft
J. REED WELKER   University of Oklahoma
DWIGHT E. WILLIAMS   Dow Corning Corporation

Preface

1976 was the Centennial year of the American Chemical Society. During this Centennial Meeting, the Symposium on Advances in Characterization of Metal and Polymer Surfaces was held to mark the achievements of surface science. Because of its broad appeal, the Symposium was jointly sponsored by the Division of Organic Coatings and Plastics Chemistry, the Division of Analytical Chemistry, the Division of Colloid and Surface Chemistry, and Cellulose, Paper, and Textile Division. A total of eight sessions took place between April 5 and 8 in the Statler Hilton Hotel, New York City.

We are grateful to have had as many physicists as chemists speak on various subjects of surface science. As a result, some sessions had lively discussions, and it was not easy to recapture all those words in writing. Nonetheless, we have attempted, to our best ability, to assemble the reviewed papers and most of the discussions in these two volumes of Proceedings. It is probable that one hundred years from now in some corner of the world, these two volumes may still be preserved in book form or on microtapes to mark the historic event of this conference.

Not all papers in these two volumes follow the original order of presentation owing to regrouping under the subtitles *Metal Surfaces* for the first volume and *Polymer Surfaces* for the second. Volume 1 is further subdivided into five parts, Volume 2 into four.

Volume 1: METAL SURFACES

    I. Atom-Probe and Mössbauer Spectroscopy
    II. Auger Electron Spectroscopy and Electron Microprobe
    III. Low Energy Electron Diffraction
    IV. Secondary Ion Mass Spectroscopy
    V. Photoelectron and Electron Tunneling Spectroscopy

Volume 2: POLYMER SURFACES

    I. Electron Spectroscopy for Chemical Analysis
    II. Infrared and Laser Raman Spectroscopy
    III. Microscopy for Polymers
    IV. Surface-Chemical and Radiation Analyses

For each part, there is at least one paper to provide an authoritative survey of the subject matter. Other contributed papers then present recent research results related to the same theme. Two post-conference contributions have been included, while several original papers were published elsewhere.

We would like to thank Session Chairpersons and all contributors to these two volumes. We would like to acknowledge the Petroleum Research Fund, administered by the American Chemical Society, for assisting our speakers from overseas. We sincerely appreciate the editorial assistance of Robert M. S. Lee of Princeton University.

Lieng-Huang Lee

Contents of Volume 1

Centennial Tribute: Surface Science and Polymer Technology
  Lieng-Huang Lee

PART I: ATOM-PROBE AND MÖSSBAUER SPECTROSCOPY

Plenary Lecture: Surface Analysis at Atomic Level Using the Atom-Probe
  Erwin W. Müller and S. V. Krishnaswamy

Applications of Mössbauer Spectroscopy to the Study of Corrosion
  G. W. Simmons and H. Leidheiser, Jr.

Characterization of Bulk and Surface Properties of Heterogeneous Ruthenium Catalysts by Mössbauer and ESCA Techniques
  C. A. Clausen, III, and M. L. Good

Discussion

PART II: AUGER ELECTRON SPECTROSCOPY AND ELECTRON MICROPROBE

Introductory Remarks
  M. L. Good

Plenary Lecture: Low Energy Electrons as a Probe of Solid Surfaces
  Robert L. Park, Marten den Boer, and Yasuo Fukuda

Surface Characterization by Electron Microprobe
  Ian M. Stewart

Auger Electron Spectroscopy of Solid Surfaces
  J. T. Grant

A Study of Passive Film Using Auger Electron Spectroscopy
  C. E. Locke, J. H. Peavey, O. Rincon, and M. Afzal

Discussion

PART III: LOW ENERGY ELECTRON DIFFRACTION

Introductory Remarks
D. L. Allara

Plenary Lecture: LEED Studies of Surface Layers
Peder J. Estrup

The Use of Direct Methods in the Analysis of LEED
David L. Adams and Uzi Landman

Surface Structure by Analysis of 'LEED' Intensity Measurements
P. M. Marcus

Computation Methods of LEED Intensity Spectra
N. Stoner, M. A. Van Hove, and S. Y. Tong

Discussion

PART IV: SECONDARY ION MASS SPECTROMETRY

Plenary Lecture: Ion Microscopy and Surface Analysis
G. H. Morrison

Surface Characterization by Ion Microprobe Analyzer
Ian M. Stewart

Study of Adhesive Bonding and Bond Failure Surface Using ISS-SIMS
W. L. Baun

Discussion

PART V: PHOTOELECTRON AND ELECTRON TUNNELING SPECTROSCOPY

Introductory Remarks
Ruth Rogan Benerito

Plenary Lecture: Surface Characterization Using Electron Spectroscopy (ESCA)
David M. Hercules

Photoemission Study of Chemisorption on Metals
Thor Rhodin and Charles Brucker

The Study of Organic Reactions on the Surface of Magnetic Pigments by X-Ray Photoelectron Spectroscopy (ESCA)
Robert S. Haines

Molecular Spectroscopy by Inelastic Electron Tunneling
Kenneth P. Roenker and William L. Baun

Discussion

About Authors, Author Index, Subject Index

PART I

Electron Spectroscopy for Chemical Analysis

Introductory Remarks:

D.M. Hercules

*Department of Chemistry
University of Georgia
Athens, Georgia   30602*

The variety of papers presented in this section is characteristic of the scope of research important to ESCA or to which ESCA has been applied. It is important in a symposium such as this to consider a variety of topics, as a challenge for new applications of ESCA as well as to explain phenomena intrinsic to this form of spectroscopy.

The measurement of binding energies and chemical shifts remain an important aspect of electron spectroscopy both for organic and inorganic compounds. These measurements are important in defining substituent effects for organic compounds as well as characterizing species on surfaces. Two topics related to binding energy measurements are of interest. One is deconvolution of spectra. Deconvolution is particularly important for the measurement of binding energy shifts in the absence of an x-ray monochromator. Use of deconvolution permits more precise definition of peaks and reveals subtlety of structure. A second area is studying satellite structure produced by processes like shake-up and shake-off. These phenomena can be diagnostically useful for elucidation of structure.

Intensity measurements in ESCA are important to the future development of the technique. It is imperative that we understand the factors which contribute to quantitative measurements. ESCA intensity measurements must be put on a quantitative basis for both relative and absolute intensity measurements. Intensity measurements can be valuable to several applications. Ratioing various functional groups in organic molecules has been done for some time. Contamination on surfaces can be quantitated as can changes in composition induced by a variety of chemical or physical treatments. Of particular interest is the difference between bulk and surface analyses which rely entirely on valid measurements of intensity.

The application of ESCA to chemical problems is always interesting. One of the most important tools used along with ESCA is ion etching and it is particularly important to establish the nature of changes which ion etching can produce in the samples. Modification of materials by treatment with plasmas is another topic which can be studied by ESCA. Escape depths for inorganic materials have been well defined, but escape depth measurements for structured organic films are not common. Another application of ESCA is its use for end group analysis. Adsorption of organic materials on the surface of inorganic species has also been demonstrated. A particularly intriguing result is the use of ESCA for the characterization of plasmas.

In summary, ESCA is becoming an important spectroscopic tool both qualitatively and quantitatively. The papers in this session point out clearly new advances in the qualitative and quantitative aspects of ESCA and indicate a variety of interesting, novel problems to which it can be applied.

Plenary Lecture: Application of ESCA to Structure and Bonding in Polymers

D. T. Clark

*Department of Chemistry
University of Durham
South Road, Durham City, England*

The continued development of ESCA as a major spectroscopic tool for investigating aspects of structure and bonding pertaining to surface, subsurface and bulk polymer systems is described.

INTRODUCTION

The unique capability of ESCA as a spectroscopic tool of enabling features of structure and bonding in surface, subsurface and bulk regions of polymer systems to be elaborated is now well established as is the relevance of such information to many fields of industrial, academic and technological importance[1,2]. The investigations which have been published to-date have largely pertained to fluoro polymer systems where advantage has been taken of the relatively large span in binding energy range consequent upon the large substituent effect of fluorine on core level binding energies. With a background of information on well characterized homogeneous polymer systems, it becomes possible to elaborate theoretical models for the quantitative interpretation of data pertaining to both absolute and relative binding energies[1,2]. By making use of core levels of differing escape depth dependencies, it becomes possible to quantify data with respect to inhomogeneous films allowing the separation of features of structure and bonding for the surface, subsurface and bulk of appropriate systems[3]. With this background data, it is now possible to extend the scope of the technique to polymer systems where the range of shifts and kinetic energy dependencies of mean free paths for the various core levels is somewhat smaller than for the systems previously investigated. For many polymer systems, however, the fact remains that the primary sources of ESCA data; namely, absolute and relative binding energies and relative peak intensities are insufficient of themselves to provide unambiguous information with regard to structure bonding and surface morphology. A particular class of materials which fit into this category are hydrocarbon based polymers. Since the factors which determine both absolute and relative binding energies in ESCA are relatively short range in nature and sensitive only to rather large perturbations caused by particularly electronegative (or electropositive) substituents, carbons in formally sp^3, sp^2 or sp hybridization have binding energies for their core levels which are closely similar. Thus both theoretical calculations and experimental data indicate that in the series ethane, ethylene, and acetylene the C_{1s} levels span a range of less than 0.4 eV[4,5]. With a conventional x-ray source the typical FWHM for C_{1s} levels of a given environment might be ∼1.1 eV. With such an adverse shift to line width range, it is difficult to extract information from the overall band profile for C_{1s} levels in an unambiguous fashion[6]. The advent of efficient x-ray monochromatization schemes, particularly for $Al_{K\alpha_{1,2}}$ (hυ 1486.6 eV) by employing dispersion from spherically bent quartz disks, offers the possibility however of substantially reducing overall line widths by reducing the contribution to the overall linewidth arising from the inherent width of the x-ray photon source[7]. In addition, removal of the bremmstrahlung improves overall signal to background and hence signal to noise ratios resulting in improved counting statistics. Unfortunately with both of the currently available commercial designs based on either slit filtering[8a] or dispersion compensation[8b], care must be taken in studying thick insulating samples to obviate charging which can be orders of magnitude larger than with a conventional spectrometer employing a slitted design and unmonochromatized x-ray source. By providing a low energy electron source either with a flood gun[8b] or by general illumination of the spectrometer sample chamber by means of a low power u.v. lamp[9], it is possible to maintain overall electroneutrality of the sample so that energy referencing may be accomplished. The advent of multiple collector assemblies will undoubtedly improve overall rates of data acquisition by an order of magnitude compared with the current situation for instruments based on a slit filtering design; and since such designs are capable of providing an x-ray band pass of 0.2 eV it can be envisaged that in appropriate cases composite line widths for C_{1s} levels may be reduced to ∼0.5 eV[10]. When such a situation obtains the analysis of band profiles even for systems in which the shift range is small will be practicable at a level which will leave little room for ambiguity. For the time being, however, spectra of adequate counting statistics may be obtained for uniform polymer films with an x-ray band pass of 0.4 eV (for $Al_{K\alpha_{1,2}}$) which gives composite line widths for C_{1s} levels of ∼0.8 eV. To illustrate the improvement in resolution thus obtainable, Fig. 1 shows the C_{1s} levels of polyethylene terephthalate (PET) obtained with either unmonochromatized $Mg_{K\alpha_{1,2}}$ or monochromatized $Al_{K\alpha_{1,2}}$ photon sources[11].

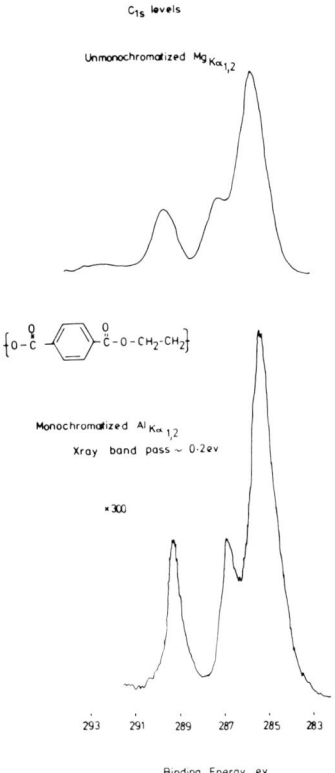

Fig. 1. C_{1s} levels for polyethylene terephthalate measured with unmonochromatized $Mg_{K\alpha_{1,2}}$ and monochromatized $Al_{K\alpha_{1,2}}$ x-ray sources.

The situation with regard to systems for which the chemical shift ranges for core levels is small is transformed, however, when information derived from the study of the excited states of the core hole states are considered. As will become apparent, this represents one of the most important growth areas in ESCA applied to studies of structure and bonding in polymeric systems and considerably broadens the scope of the technique[12-14].

The hierarchy of information levels available in ESCA are shown in Table 1. It is clear that in many areas of study relating to the chemical, physical, electrical, mechanical, etc. properties of polymer coatings, the information derived from ESCA studies is unique whilst in other areas the technique complements the information derived from more established spectroscopic tools[1,2]. In many areas of application, ESCA provides data at a somewhat 'coarser' level than most other spectroscopic tools, however this is more often than not outweighed by the great range of information levels available as will become apparent in this review. The principal advantages of the technique in relation to the study of polymers are set out in Table 2. It is the purpose of this review to outline some of the recent progress in the application of ESCA to studies of structure and bonding in polymers which illustrate the unique advantages of the technique in this area accruing from the large range of available information levels.

Table 1

Hierarchy of Information Levels Available from ESCA Studies

Core Levels

1) Absolute binding energies, relative peak intensities, shifts in binding energies. Element mapping, analytical depth profiling, identification of structural features, etc. Short range effects directly, longer range indirectly.

2) Shake up - shake off satellites. Monopole excited states energy separation wrt direct photoionization peaks and relative intensities, 'singlet and triplet' components short and longer range effects directly.

(Analogue of UV)

3) Multiplet effects.

For paramagnetic systems, spin state, distribution of unpaired electrons.

(Analogue of ESR)

4) Angular and photon energy dependence of (1) - (3), symmetries of levels.

Valence Levels

1) Absolute and relative binding energies and relative peak intensities. Valence bands of insulators, densities of states for metals. Core-like valence levels for analytical depth profiling.

Table I (cont'd).

2) Angular and photon energy dependence. Symmetries of levels.

3) Shake up phenomena (less useful than for core levels).

Table 2

Principal Advantages of ESCA as a Spectroscopic Tool
in the Study of Polymers

---

1) Technique non destructive.

2) Study in situ in working environment, minimum of preparation.

3) Large number of information levels available from single experiment.

4) Capability of studying surface, subsurface and bulk depth profiling.

5) Information levels such that 'ab initio' investigations are feasible.

6) Theoretical basis well understood results quantifiable.

7) Data often complementary to those obtained by other techniques. Unique capabilities central to the development of a number of very important fields.

Applications of ESCA to Studies of Structure and Bonding
from Measurements of Absolute and Relative Binding Energies
and Relative Peak Intensities

1. <u>Mass Transfer Between Contacting Surfaces</u>. The detection and semi-quantitative estimation of surface contamination of polymers is of importance in a number of fields. We have previously shown that ESCA has a unique capability for the detection of the initial stages of oxidation of polymers initiated at the surface and of adsorbed water and hydrocarbon contaminants[15]. One area where knowledge of the precise nature of the surface is of critical importance is in delineating models for the interpretation of data relating to triboelectric phenomena. Contacting polymer films from opposite ends of the so-called triboelectric series results in

charge transfer such that there is a considerable buildup of static charge on each component of the contacting pair. Such charge transfer could conceivably occur via electron transfer from the material of low to that of higher work function thus equalizing the fermi levels or alternatively by mass transfer (transfer of ions) between the two components of the contacting pair. Even more likely is that both processes are of importance, however the two possible mechanisms are not entirely separable since the propensity for adsorption of ions at the surface of a given polymer will undoubtedly be a subtle function of its electronic structure as will the work function. It is known that triboelectric phenomena in general are explicable in terms of a charge density of the order of 1 in 10^4 of surface sites which is probably an order of magnitude lower than can currently be detected by ESCA[16]. Nonetheless, it is of interest to see if in contacting polymer samples there is mass transfer or not between the components. If mass transfer is observed, it certainly does not resolve the problem of how charging occurs but is certainly allows one to say that mass transfer cannot be ruled out as a possible mechanism.

We have, therefore, studied the surfaces of a variety of polymer films both before and after contacting events. The great advantage of such an investigation by means of ESCA is the ability to look at both halves of a contacting pair. As one example, Fig. 2 shows the core level spectra for PTFE and PET films, which are highly characteristic for each polymer. In addition, there is also shown the core level spectra for the polyethylene terephthalate component after lightly contacting the PTFE film. Since the polymers are from the opposite ends of the triboelectric series, the films show a strong tendency to adhere to one another, even on a light contact.

The observation of the high binding energy component in the C_{1s} spectrum and the observation of the F_{1s} levels shows that some PTFE transfers to the PET surface. It may be estimated that this represents fractional monolayer coverage. Of particular interest is the fact that in the F_{1s} peak in addition to the major high binding energy component associated with covalent CF_2 linkages, there is a lower binding energy peak attributable to fluoride ion thus providing evidence for bond cleavage accompanying the mass transfer[6]. Examination of the other half of the component; namely PTFE, shows the presence of PET as evidenced by the characteristic O_{1s} and C_{1s} levels. These simple experiments illustrate the utility of ESCA in this area.

2. **Substituent Effects on Core and Valence Levels.** Previous studies of substituent effects on core levels in simple monomeric systems have shown that these are highly characteristic for a given substituent and follow simple additivity models[17]. The results may be quantified by detailed non-empirical calculations, and this forms

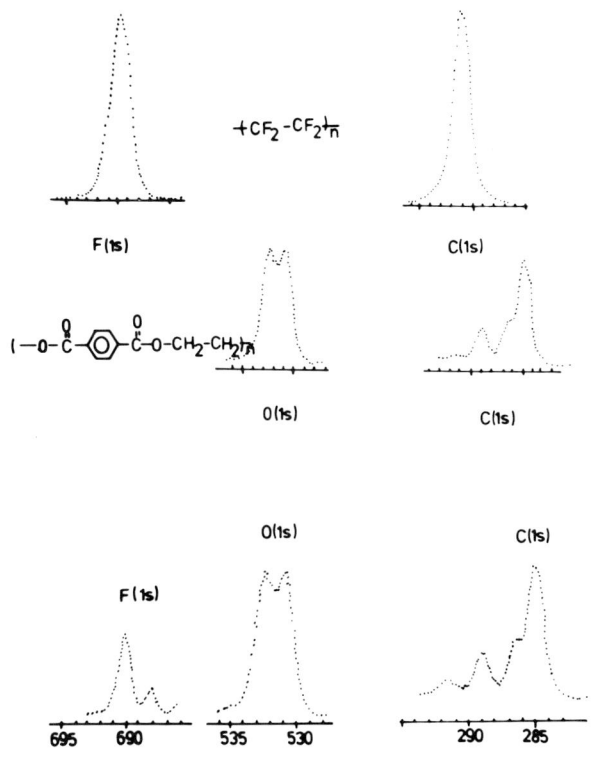

Fig. 2. Core level spectra for PTFE and PET and for the PET component after lightly contacting the two polymer films.

a sound basis for understanding the electronic factors determining both absolute and relative binding energies[18]. This has enabled computationally inexpensive models based on an all valence electron CNDO/2 SCF MO formalism to be developed which can be extended to quantitatively describe polymers[19]. A large amount of data has previously been reviewed which relates to fluorocarbon based polymers[1,2]. However, a systematic study of a large number of homopolymers of simple vinyl monomers provides a compilation of substituent effects on C_{1s}, N_{1s}, O_{1s}, F_{1s}, Si_{2p}, P_{2p}, S_{2p} and Cl_{2p} levels[20]. Fig. 3, for example, shows some of the data pertaining to substituent effects on C_{1s} levels in polymers.

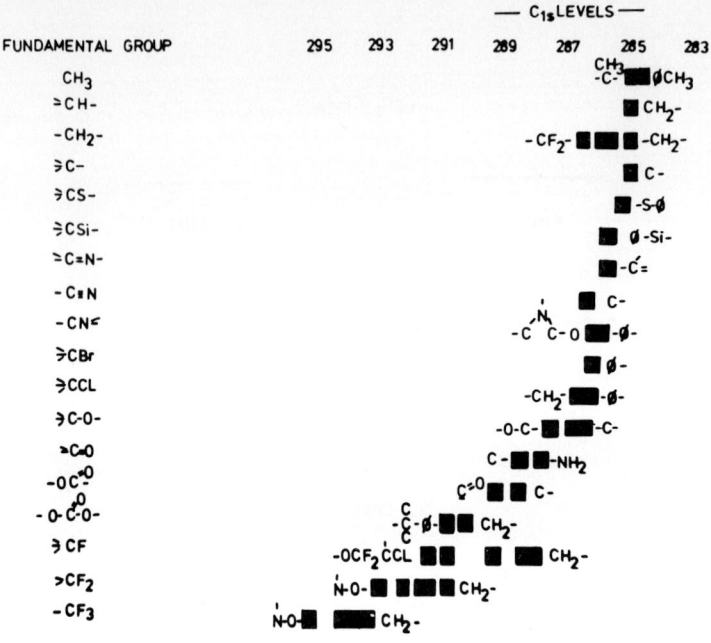

Fig. 3. Correlation diagram for C_{1s} levels in polymeric system as a function of electronic environment. (The horizontal scale for each block is taken to indicate the range of binding energies found for a given structural type.)

The characteristic nature of many substituent effects can now be used as a fingerprint much in the same manner as one might use infrared or NMR data. The fingerprint nature of the core levels of many polymer systems has previously been alluded to. Further examples are provided in Fig. 4 which shows the core level spectra for PVC and polyisopropyl acrylate. In each case, the identification of core levels taken in conjunction with their characteristic fine structure leads to an unambiguous assignment of gross structure.

It is sometimes the case that isomeric species have core level spectra which are virtually identical. Although the spectra may therefore be used to identify the gross structure, they may not allow a distinction to be drawn between isomeric species. As a simple example, we might consider the isomeric polybutyl acrylates.

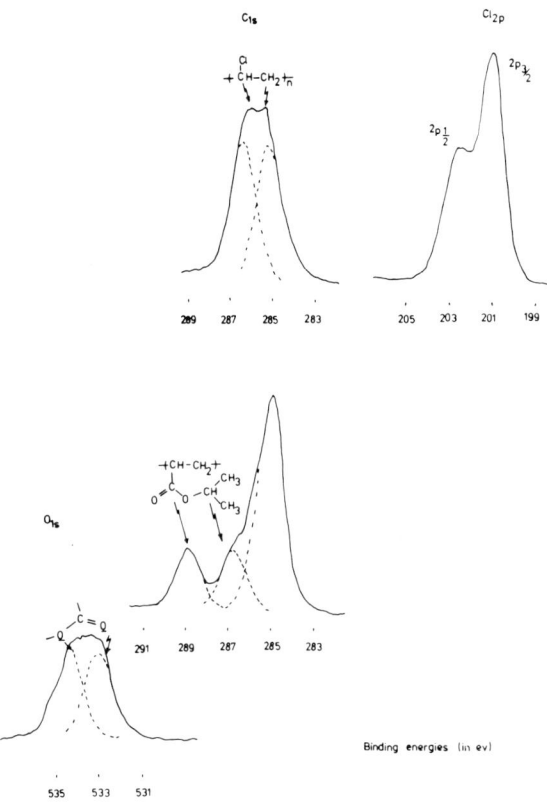

Fig. 4. Core level spectra for polyvinylchloride (upper curves) and poly (isopropylacrylate) (lower curves).

The O_{1s} and C_{1s} core level spectra for samples of these polymers are indistinguishable. By contrast, the valence levels are highly characteristic of the side chain structure since this forms an appreciable part of the whole. This is clearly apparent from the valence levels shown in Fig. 5; moreover, comparison with appropriate model systems allows an unambiguous assignment of particular structural isomers[21].

3. <u>End Group Analysis</u>. It is often the case that particular structural features may be characteristic of the end groups of a given polymer system. The direct detection of such end groups by

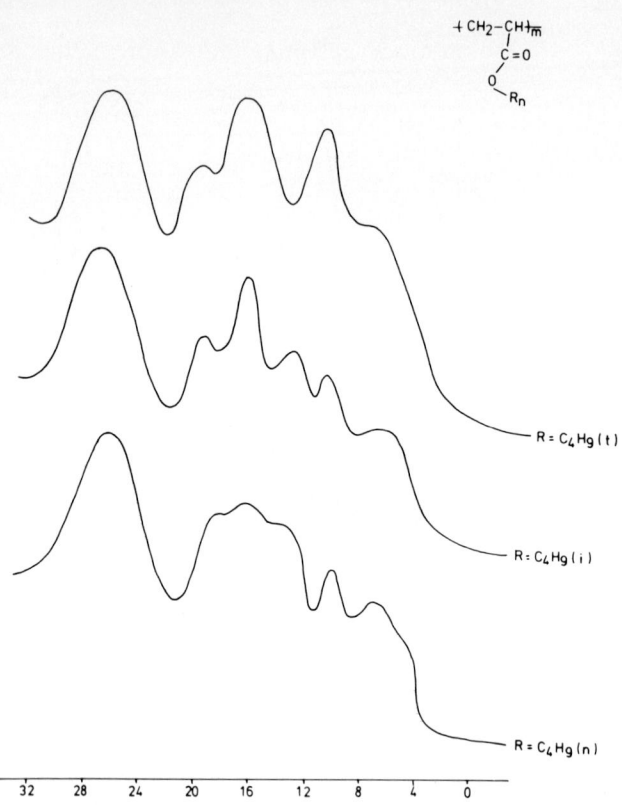

Fig. 5. Valence level spectra of isomeric polybutyl acrylates excited by $Mg_{K\alpha_{1,2}}$ radiation.

means of their characteristic binding energies provides a convenient means of establishing DP's in relatively low molecular weight material. A particularly favorable situation arises for systems for which the terminal groups involve $\underline{CF_3}$ residues. If due care is taken to ensure that ESCA statistically samples the repeat unit (by for example considering the relative intensities of differing levels of the same element with differing escape depth dependencies), then the comparison of area ratios for chemically shifted components of a given core level may be used to straightforwardly estimate DP's. For example, in a series of fluorocarbonate polymers of the general

Structure and Bonding in Polymers

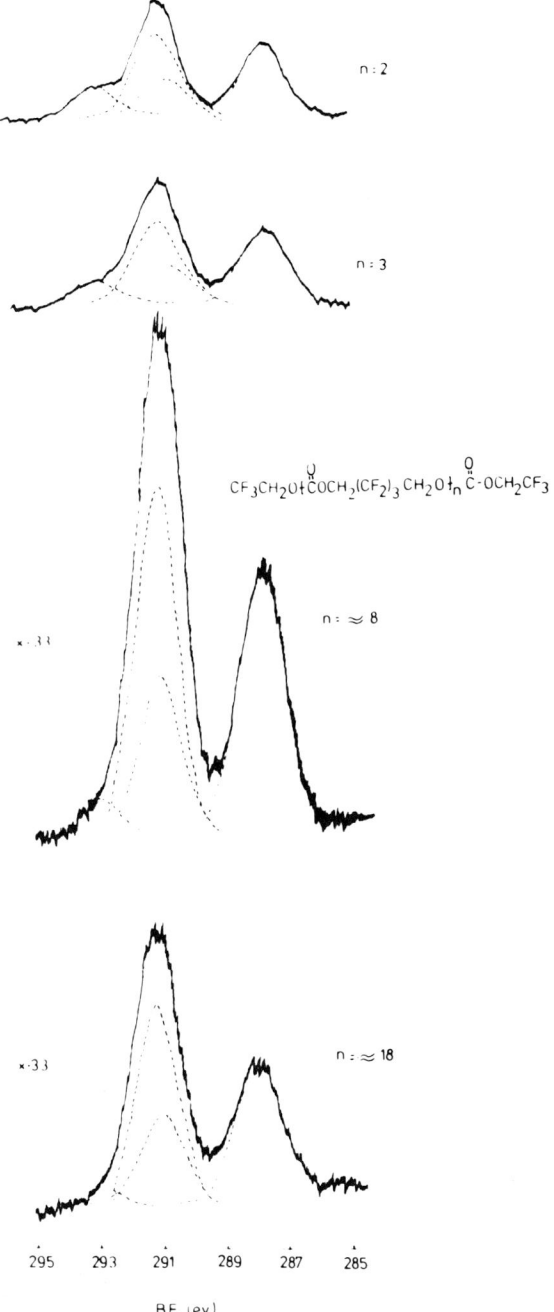

Fig. 6. C_{1s} core level spectra for a series of fluorocarbonate polymers.

formulae shown in Fig. 6, it may readily be shown that the carbon 1s levels appropriate to carbonate $-O-\overset{\overset{O}{\|}}{\underline{C}}-O$ and $\underline{C}F_2$ environments occur at approximately the same binding energy[22]. The C_{1s} levels for the series of low molecular weight materials shown in Fig. 6 fall into three distinct regions, and with appropriate calibration of linewidths and lineshapes for individual components from the study of model compounds, the lineshape analysis produces the components indicated by the dotted curves. From the relative areas of the $\underline{C}F_3$ carbons to the composite of CF_2 and carbonate carbon peaks and from the low binding energy $\underline{C}H_2$ components, the DP's may be elaborated as indicated in the figure. The technique may, therefore, be used to obtain the gross chemical structure from the C_{1s} and F_{1s} area ratios in a manner akin to that previously described for simple homo- and co-polymers[1,2,19]. From a consideration of the fine structure of the C_{1s} levels, details of the structure and of the chain length may then be elaborated.

4. <u>Comparisons of ESCA Derived Data on Structure and Bonding with that Pertaining to the Bulk</u>. Although there are well developed techniques for studying chemical compositions and features of structure and bonding pertaining to the bulk of polymer samples, until the advent of ESCA, information with regard to the surface composition could only be inferred rather indirectly by, for example, measurements of surface free energies. Since any solid communicates with the rest of the world primarily by way of its surfaces, such information is important in many areas. As we have previously demonstrated in the particular area of fluorocarbon based polymers, the question "Is the surface composition typical of the bulk?" may conveniently be answered by ESCA[15,19].

As examples of how this type of analysis can be extended to systems for which the shift range is somewhat less favorable, we may consider briefly work on a series of polyalkylacrylates[21].

The core level spectra for a series of polyalkylacrylates are shown in Fig. 7. The O_{1s} levels show a doublet structure (somewhat obscured in the case of polyacrylic acid because of hydrogen bonding effects); the binding energies for the two components being characteristic for an ester group (cf. polyethylene terephthalate). The C_{1s} levels in each case show a high binding energy component attributable to $-\underline{C}\underset{\diagdown O}{\diagup O}$ type environments shifted ~4 eV from the main peak which in each case arises from $\underline{C}H_3$, $\underline{C}H_2$ and $\underline{C}H$ type environments from the backbone carbons and carbons of the alkyl group not attached to oxygen. In going from polyacrylic acid to the polymethylacrylate, a shoulder to the high binding energy side of the low binding energy component develops, shifted by ~1.6 eV and attributable to the carbon attached to the ester oxygen. This shoulder gradually decreases in relative intensity with respect to

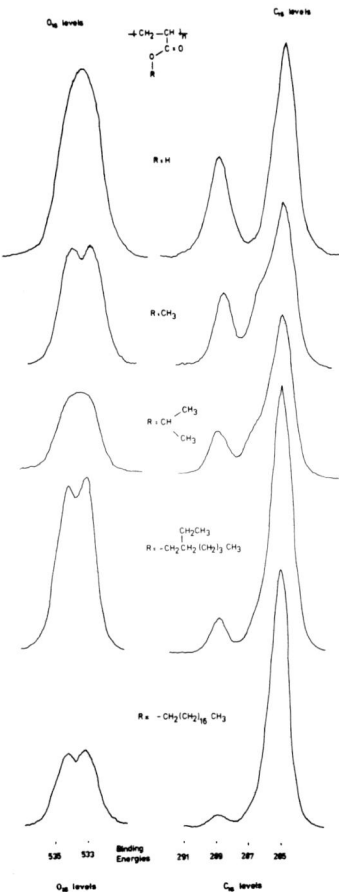

Fig. 7. Core level spectra for a series of polyalkylacrylates.

the main peak as the chain length of the alkyl group increases. The assignment of core levels is readily confirmed by reference to simple model compounds for which detailed theoretical analyses have been performed[21].

In previous studies on polymeric systems, we have shown that the substantial differences in escape depth dependence for deep lying valence levels which are core-like in character (viz. F_{2s}) with respect to tightly bound core levels (e.g. F_{1s}) may usefully be employed for analytical depth profiling[3]. A study of the valence levels for simple model compounds reveals that in esters the O_{2s}

levels are well separated from the remainder of the valence band and are essentially core-like in nature. The approximate kinetic energies pertaining to photoemitted electrons from O_{1s} and O_{2s} levels using a $Mg_{K\alpha_{1,2}}$ photon source are $\sim$720 eV and $\sim$1227 eV respectively and from a consideration of the generalized curve of escape depth versus kinetic energy these should correspond to significant differences in electron mean free paths[1,3]. By studying a series of simple oxygen containing organic molecules as condensed films such that ESCA statistically samples the molecules, it can be shown that with the particular experimental arrangement peculiar to our spectrometer the apparent O_{1s} to O_{2s} area ratio is 11 ± 1[21]. By measuring the area ratios of the O_{1s} to C_{1s} peaks versus the stoichiometric ratios for these molecules, it is also possible to derive the instrumentally dependent apparent sensitivity ratios for oxygen with respect to carbon for unit stoichiometry. Typical data are shown in Fig. 8.

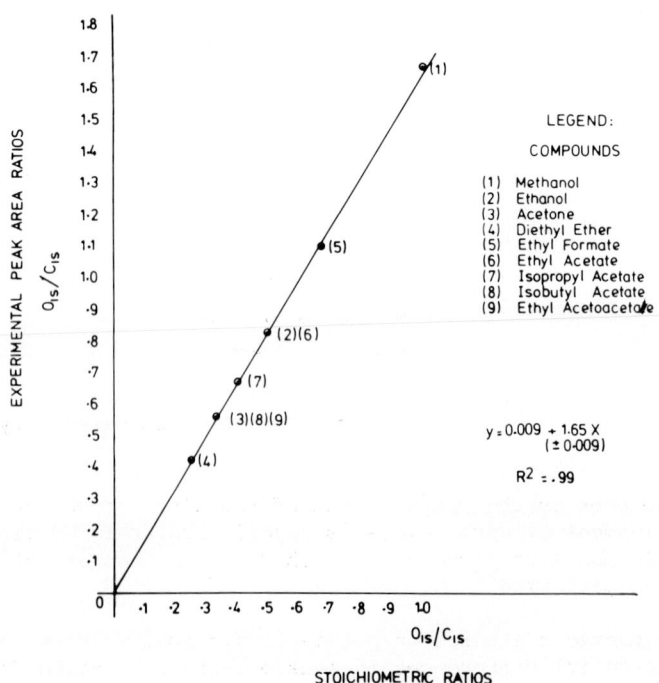

Fig. 8. Plot of measured O_{1s}/C_{1s} area ratios versus stoichiometric ratios for model compounds.

As an initial test of the homogeneity of the polymer samples on the ESCA depth profiling scale, we may compare the measured O_{1s}/O_{2s} area ratios for these polymers with those obtained for the homogeneous thin films of the model systems, since as we indicated the escape depth dependencies for the two levels are significantly different. It is fortunate, in this respect, that the O_{2s} levels for the polymers are sufficiently core-like to be readily identifiable, and also that the cross section for photoionization is large enough such that even for the long chain alkyl systems the signals arising from these levels have adequate intensity to be detected. The measured area ratios for the polymers fall within a narrow range (12 ± 1) which is within experimental error the same as that for the model compounds. As we have previously noted, there are two essentially independent means of establishing the polymer compositions from the ESCA data: Firstly, from the C_{1s}/O_{1s} area ratios employing the instrumentally dependent sensitivity factors for the core levels established from a study on the model systems. Secondly, from the area ratios for the relevant component peaks of the C_{1s} levels. The data are shown in Fig. 9.

For polyacrylic acid, the measured areas for the various structural features for the O_{1s} levels and C_{1s} levels and the overall ratios for the C_{1s} to O_{1s} levels (corrected for differing sensitivity factors) are 1.0, 2.0 and 1.6 respectively, in excellent agreement with the theoretical values of 1.0, 2.0 and 1.5 based on a statistical sampling of the polymer repeat unit. The area ratios for the individual components for the C_{1s} levels show an excellent correlation with the number of carbon atoms in the alkyl groups*. (Slope; Experimental 0.99, Theoretical 1.0). By contrast, the plot of total C_{1s}/O_{1s} area ratios (corrected) against the chain length for the alkyl group fall on a smooth curve; however replotting the data in a different form, as shown in Fig. 10, reveals the underlying linear correlation with the appropriately derived theoretical parameter.

Within experimental limits, the slope is unity as required by theory, if the ESCA experiment statistically samples the repeat

---

* It is convenient in this correlation to plot the area ratios for the two best resolved peaks, namely the carbonyl carbon which, although of low relative intensity for the longer chain systems, nonetheless, is well removed from the main component arising from the backgone and carbons not directly attached to oxygen. This obviates any error due to deconvoluting the signal arising from the other carbons directly attached to oxygen (viz. of the ester group), since for the long chain systems it is obviously preferable to have a small error in a large rather than a small quantity. Since the area ratio does not therefore include the carbons of the ester group which are directly bonded to oxygen, this leads to an obvious break in the curve for polyacrylic acid.

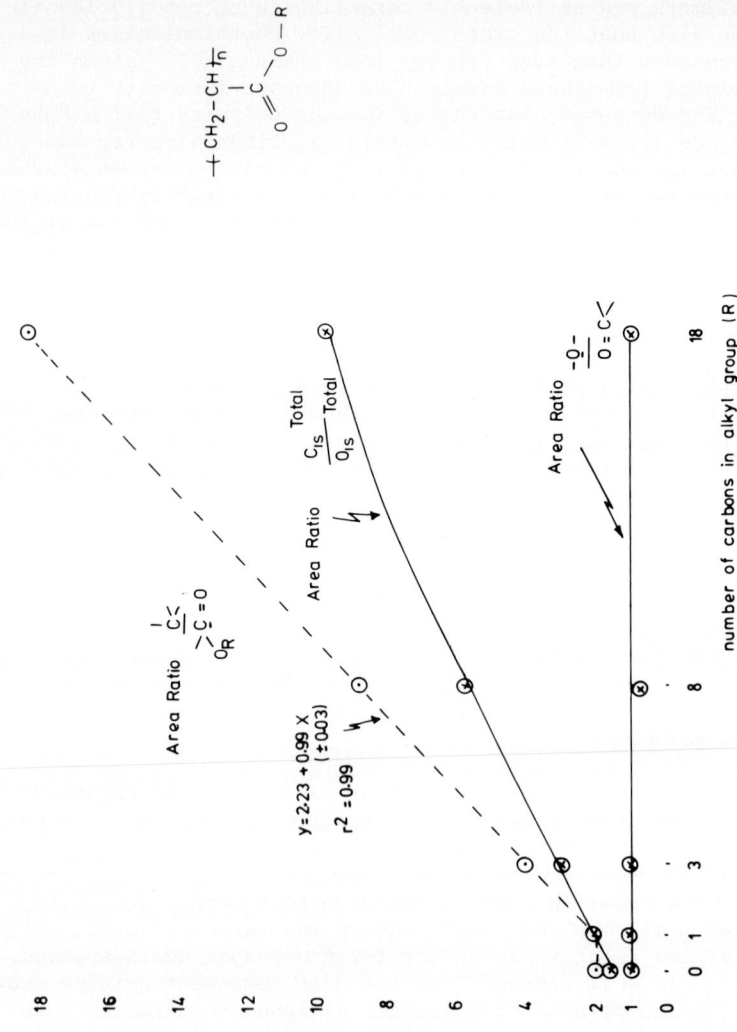

Fig. 9. Plot of area ratios for the core levels versus number of carbons on the alkyl group of a series of polyalkyl acrylates. (These have been corrected for differences in cross section and instrumental sensitivity (see text)).

units of the polymers. In sum total, therefore, the ESCA data shows that for these systems the outermost few tens of Angstroms of the samples are representative of the bulk and that compositions, integrity of the immediate surface, and homogeneities may routinely be established. The analysis also strongly suggests that there are no specific orientation effects of side chain alkyl groups at the surface.

The theoretical models previously developed to quantitatively describe absolute and relative core binding energies for fluoropolymers based on the charge potential model may readily be extended to the polyalkylacrylates, once appropriate values for the charge potential parameters k and E° are established for each core level. This is readily accomplished by studying model compounds as shown in Fig. 11.

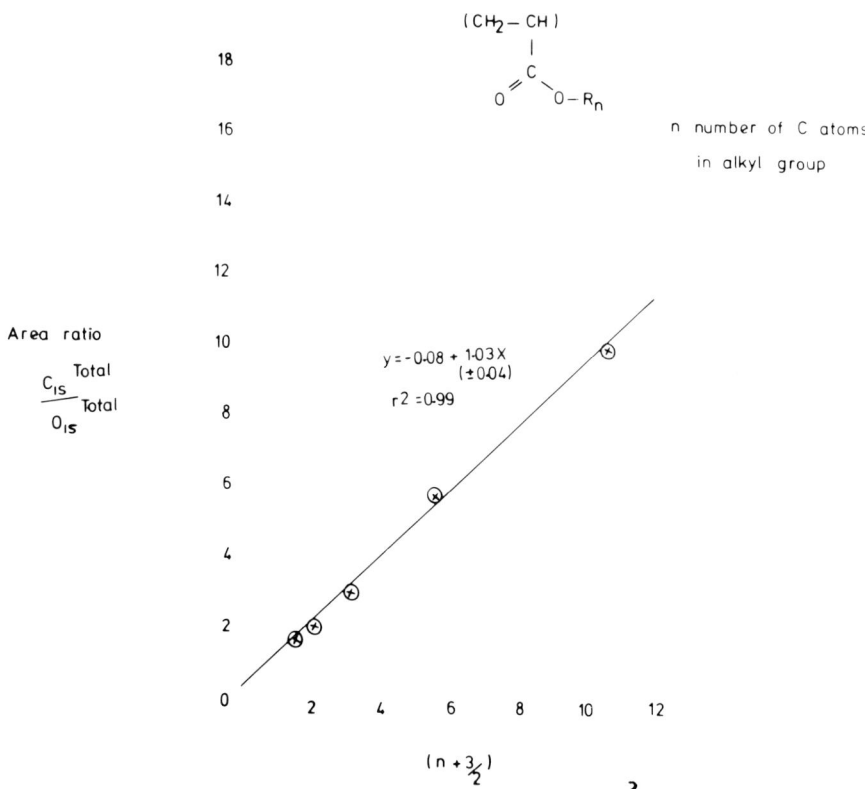

Fig. 10. Plot of area ratios for the C_{1s} and O_{1s} levels of a series of polyalkyl acrylates as a function of chain lengths of the alkyl group.

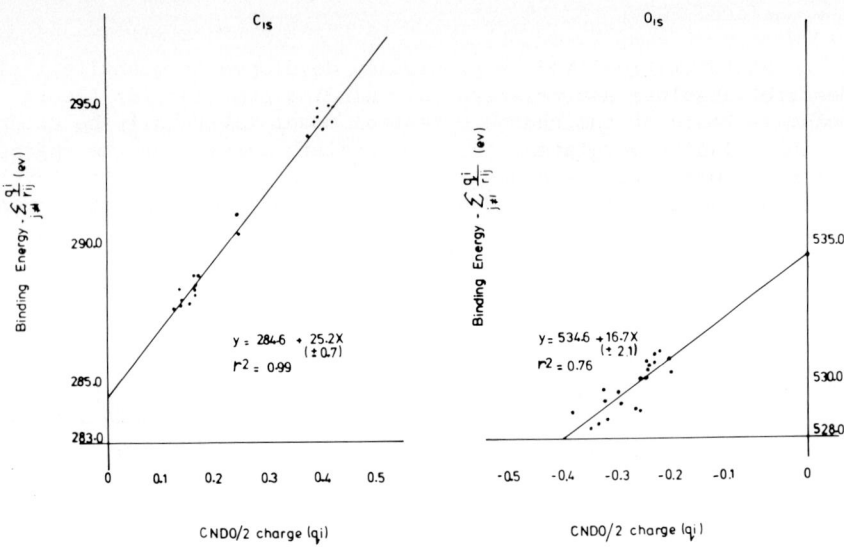

Fig. 11. Charge potential correlations for the C_{1s} and O_{1s} levels of model compounds shown in Fig. 8.

The correlation for the C_{1s} and O_{1s} levels and least squares analysis of the data yields values of 284.6 eV and 25.2 for $E^°$ and k respectively for the C_{1s} levels, the correlation coefficient being 0.99. These values are in excellent agreement with those previously determined for simple fluorocarbon systems[19a]. The relatively small range of binding energies leads to a considerable scatter in the data for the O_{1s} levels, and the correlation coefficient (0.76) is hence somewhat lower. Nonetheless, the concomitant error limits still lead to an unambiguous assignment of the O_{1s} core levels and this has been checked by carrying out detailed non-empirical LCAO SCF MO calculations of absolute binding energies from computations on the neutral molecules and the relevant hole states[21]. Having established the charge potential parameters k and $E°$ for the C_{1s} and O_{1s} levels of these model systems, appropriate models of the polymer systems may be investigated. Such an investigation has two primary objectives. Firstly, with an unsymmetrical vinyl monomer the possibility exists for structural isomerism (viz. head-to-tail vs. head-to-head and tail-to-tail bonding). Secondly, for a given structural isomer the relative conformation of the pendant groups are of some interest. In previous reviews[1,2,3,6,15] we have shown that the factors which determine differences in binding energies are relatively short range in nature and may therefore be quantitatively described by calculation on model systems incorporating a small number of monomer units such that all of the important short range interactions are quantified. In particular cases (e.g. nitroso rubbers)[19b], we have shown that structural isomerisms may be investigated directly by ESCA; however in general for homopolymers based on simple unsymmetrical vinyl monomers both theory and experiment agree that both absolute and relative binding energies are virtually the same for regular and irregular structures[19c].

Even at the semi-empirical all valence electron SCF MO CNDO/2 level, computations on model systems for the polymers studied in this work are extremely time consuming. Computing limitations therefore dictated that the central monomer unit of the model chains were linked to a single monomer unit at each end. Calculations previously reported show that such a model incorporates all the important short range interactions with respect to the central unit. The model systems chosen for study were polyacrylic acid and polymethyl acrylate. In all cases, standard bond angles and lengths were employed for the various structural features[23]. The model systems studied are shown in Figs. 12 and 13 and in each case the carbonyl group was taken to eclipse the CH bond of the backbone, and this we refer to as the π eclipsed arrangement. In particular cases, calculations were also carried out on staggered conformers in which the carbonyl group was rotated through an angle of 60° (with respect to the eclipsed configuration) about an axis through the carbon-carbon bond linking the pendant group and backbone. Without exception, such conformers were calculated to be significantly higher in energy than for those involving a π eclipsed arrangement. For a head-to-head (irregular) arrangement (HH-HT linkages), the side chains were again taken in the eclipsed pi conformations with the pendant groups either all being on the same side of a plane drawn through the backbone or in an alternating arrangement. These are analogous to the 'isotactic' and 'syndiotactic' arrangements in the regular head-to-tail model system. Similar calculations were carried out on a model of polymethyl acrylate as prototype for the polyalkyl acrylates. In this case, however, computing limitations dictated that on one end of the model system the adjacent backbone carbons were simulated by taking a hydrogen atom rather than a methyl group. Since the calculated binding energies for such a model are taken with respect to the central unit, this further approximation has virtually no effect,

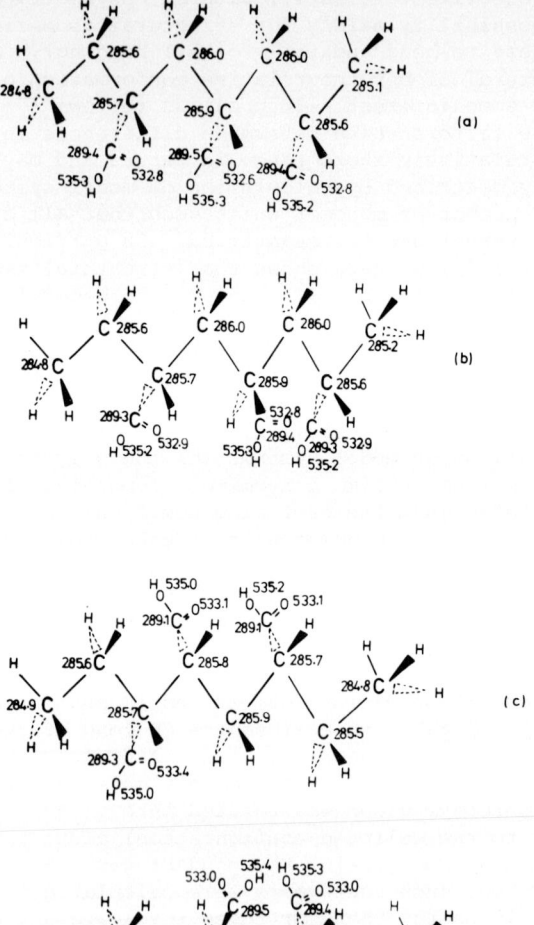

Fig. 12. Conformational models of polyacrylic acid with calculated binding energies from the charge potential model shown for (a) HT-HT 'isotactic' model, (b) HT-HT 'syndiotactic' model, (c) HH-HT 'isotactic' model and (d) HH-HT 'syndiotactic' model.

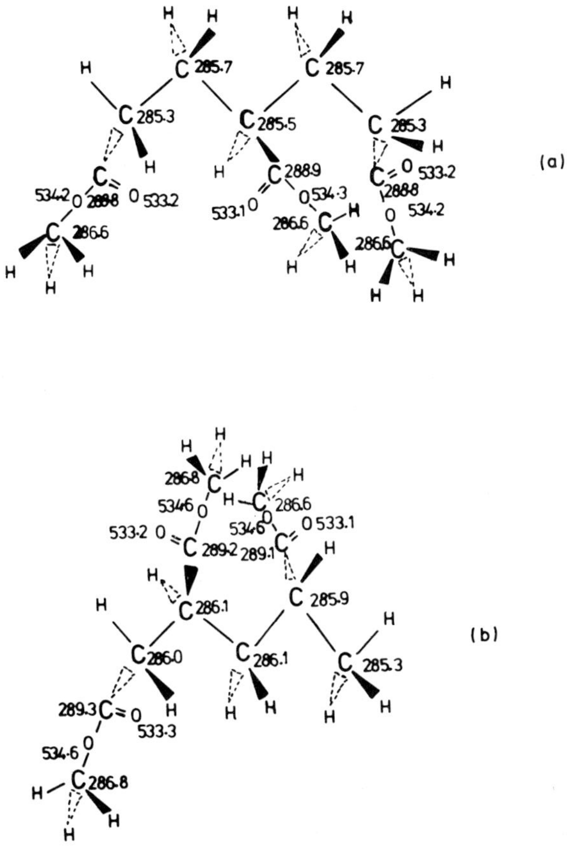

Fig. 13. Conformational models of polymethylacrylate with calculated binding energies from the charge potential model shown for (a) HT-HT 'syndiotactic' model and (b) HH-HT 'syndiotactic' model.

as will become apparent in the discussion of the results.

The results for the models of polyacrylic acid are displayed in Fig. 12 where the absolute binding energies have been computed using the charge potential parameters derived from the study of simple model systems as described in a previous section. Considering firstly the regular head-to-tail arrangement ((a) and (b) Fig. 12), it is clear that the factors determining both the absolute and relative binding energies are insensitive to the overall stereochemistry of the system. Comparison with the corresponding data for the staggered (with respect ot the carbon-oxygen double bond) conformers reinforces this conclusion since the calculated binding energies are in exact correspondence with those for the eclipsed conformers shown in Fig. 12. A comparison of the central monomer units with the adjacent units, effectively demonstrates the short range nature of the factors determining absolute and relative binding energies in these systems. A comparison with the experimental data reveals the overall adequacy of the theoretical model in respect of both absolute and relative binding energies for both the O_{1s} and C_{1s} levels. For the regular models, the second methyl-group in the chain linked directly to the methyl group provides an indication of the likely binding energy for methylene groups appropriate to a tail-to-tail structural arrangement. This is also apparent from the corresponding irregular HH-HT models ((c) and (d) Fig. 12). For the two possibilities considered, it is evident that the calculated binding energies show a small dependence on stereochemistry arising from the significant interaction between carbonyl groups oriented cis to one another on adjacent carbon atoms. For both models, the backbone carbons are predicted to have closely similar binding energies, a feature common to the regular models alluded to previously. The shift in binding energy for the O_{1s} levels range from 2.7 eV for the HT-HT 'isotactic' model to 1.9 eV for the HH-HT 'isotactic' model. This compares with the experimentally determined value of 1.3 eV. The discrepancy is largely accounted for by the effect of inter and intra chain hydrogen bonding which has somewhat of a levelling effect on the relative charge distribution about the two types of oxygen. This effect has previously been noted and is manifested in a distinctly increased linewidth for the individual components of the O_{1s} levels[6]. In going from polyacrylic acid to polymethyl acrylate, such interactions disappear with a concomitant decrease in linewidth for the individual components of the O_{1s} levels. The two models chosen for this system are shown in Fig. 13. The calculated absolute and relative binding energies for the O_{1s} levels are seen to be in excellent agreement and bracket the experimental data. The calculations suggest that whereas the O_{1s} levels and C_{1s} levels for the -$\underline{C}$=O and -$\underline{C}$-O structural features should fall within a narrow range (±0.2 eV) for an atactic material the C_{1s} levels for the backbone carbons should span a much larger range (±0.4 eV) with the extremes being represented by tail-to-tail and head-to-head arrangements. All of the polymer samples studied in this work were synthesized by free radical processes at relatively high temperature and should therefore be predominantly atactic.

The theoretical models suggest that this would be manifest in the ESCA data by a somewhat larger linewidth for the C_{1s} signal arising from the backbone carbons than from the pendant groups. Since the signal arising from the carbons of the alkyl group (other than that directly attached to oxygen) falls on the same region as that for the backbone carbons, one corollary of the theoretical analysis is that as the alkyl chain becomes longer and therefore proportionally provides a larger contribution to the low binding energy region of the C_{1s} spectra, the FWHM should approach that appropriate to the alkyl chain and hence be largely independent of the tacticity of the polymer system. Analysis of the experimental data provides evidence for this. Thus, whilst the FWHM for the deconvoluted components of the C_{1s} and O_{1s} levels for polymethyl acrylate are 1.55 eV, 1.4 eV, and 1.55 eV for the $\underline{C}H$, $-\underline{C}=O$ and $C=\underline{O}$ structural features, for polyoctadecyl acrylate the corresponding figures are 1.35 eV, 1.35 eV and 1.5 eV.

Whilst for the polymer samples discussed above, structure and bonding in the outermost few tens of Angstroms sampled by ESCA corresponds to that in the bulk this is not always the case. For example, Fig. 14 shows the ESCA spectra for a further series of polyalkyl acrylates. A distinctive feature clearly evident in all of the spectra is the obvious inequality in intensity of the two component peaks of the O_{1s} levels. A similar analysis to that presented in a previous section provides the following information. Fig. 15, for example, shows a plot of the ratio of intensities for the individual components of the O_{1s} levels and also the total O_{1s}/O_{2s} ratios.

For comparison purposes, the dotted lines indicate the correlations expected for samples which on the ESCA depth profiling scale correspond to a statistical sampling of the appropriate repeat unit in the polymer. It is clear that there are considerable deviations from such correlations in a direction which overall suggests that the samples are oxidized. If we consider the polydecyl acrylate for example, the O_{1s}/O_{2s} ratio is significantly higher than for the reference compounds suggesting that since the mean free path for the O_{1s} levels is considerably shorter than for the O_{2s} level that the oxidation is largely confined to the surface. The absolute binding energies in each case for the O_{1s} component levels which have apparently increased in intensity corresponds to $\underline{C}=O$ structural features, as is apparent from a comparison with data for the model systems. It is interesting to note that high resolution infrared studies revealed no major distinction of the type clearly evident from the ESCA spectra, and the carbonyl region for all of the samples showed only a single peak in the range 1734 ± 6 cm^{-1} consistent with $-C\genfrac{}{}{0pt}{}{\nearrow O}{\searrow O-R}$ structural features[24]. This is readily understandable since the infrared data pertains essentially to the bulk. Further evidence for the oxidized nature of the poly-n-decyl acrylate surface is provided by the greatly increased wettability with respect to water compared with polyisopropyl acrylate as a representative example of the unoxidized samples. This was immediately apparent from the relative contact angles assessed from the photographs for the two samples. A comparison was also made with poly-2-ethylhexyl and polyoctadecyl acrylates; the latter had a contact angle closely similar to that of polyisopropyl acrylate whilst the former showed

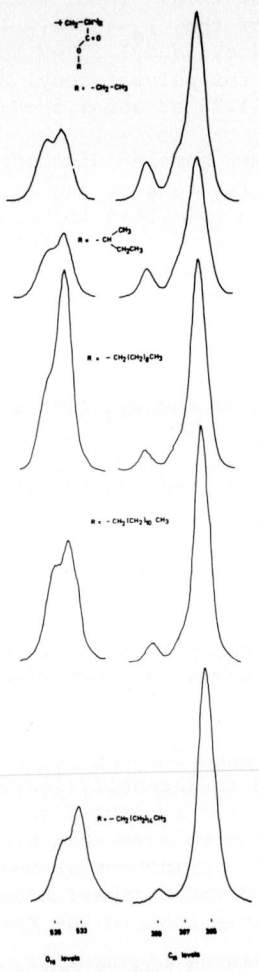

Fig. 14. Spectra for O_{1s} and C_{1s} core levels for a series of surface oxidized polyalkyl acrylates.

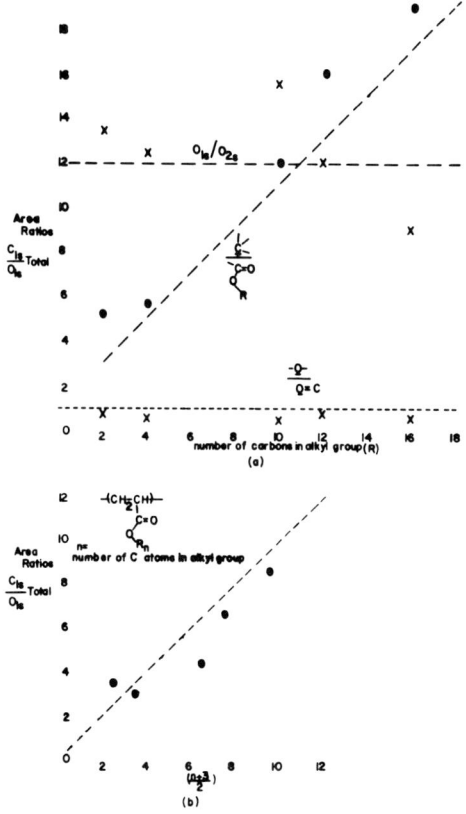

Fig. 15. (a) Plot of the intensity ratios for the individual components of the O_{1s} levels and also the O_{1s}/O_{2s} ratios for a series of polyalkyl acrylates.

(b) Plot of the C_{1s} and O_{1s} area ratios versus the number of carbons in the alkyl groups.

a wettability intermediate between that of polyisopropyl acrylate and poly-n-decyl acrylate. It is interesting to note that although the data for the poly-2-ethylhexyl acrylate generally fits well into the overall analysis previously presented as is evident from Figs. 9 and 10, a close inspection of the relative intensities of the component peaks of the O_{1s} levels reveals some evidence for a small extent of oxidation (Fig. 7).

If the surface oxidation inferred from the inequality of the component peaks of the O_{1s} levels is attributable to surface carbonyl features, then this should also be manifest in the carbon 1s levels. It should, however, be emphasized that since the escape depth dependence for photoemitted electrons in the energy range considered is such that the mean free path increases with increasing kinetic energy, then any surface feature will be relatively more prominent for the more tightly bound O_{1s} levels than for the C_{1s} levels. A detailed examination of the C_{1s} spectra for the series of surface oxidized samples (Fig. 14) shows that the overall line profiles can only be quantitatively fitted with the addition of a small peak in the C_{1s} spectrum appropriate to isolated carbonyl features as might arise from oxidation. A more extensive discussion of work on these systems and the related polyalkylmethacrylates is given elsewhere[21].

5. <u>Shake Up Phenomena in Polymers</u>. The removal of a core electron (which is almost completely screening as far as the valence electrons are concerned) is accompanied by reorganization of the valence electrons in response to the effective increase in nuclear charge. This perturbation gives rise to a finite probability for photo-ionization to be accompanied by simultaneous excitation of a valence electron from an occupied to an unoccupied level (shake up) or ionization of a valence electron (shake off). These processes giving rise to satellites to the low kinetic energy side of the main photo-ionization peak, follow monopole selection rules and considerably extend the scope of ESCA as a technique as will become apparent. The relationship between direct photo-ionization, shake up and shake off are shown schematically in Fig. 16. Before considering shake up phenomena in general, it is worthwhile considering briefly some theoretical aspects of the processes involved. We have previously emphasized that the binding energy is characteristic of a given core level and varies within narrow limits (chemical shifts). Relaxation energies (associated with contraction of the valence electron cloud consequent upon core ionization) are also characteristic of a given core level and also vary within narrow limits as a function of the bonding environment of the atom on which the core level is located[18]. For C_{1s} levels for neutral systems for example, binding energies measured with respect to the Fermi level as energy reference fall in the range 285 - 295 eV whilst relaxation energies might typically fall in the range 12 ± 2 eV. The direct relationship between shake up and shake off processes and relaxation energies may be readily understood from

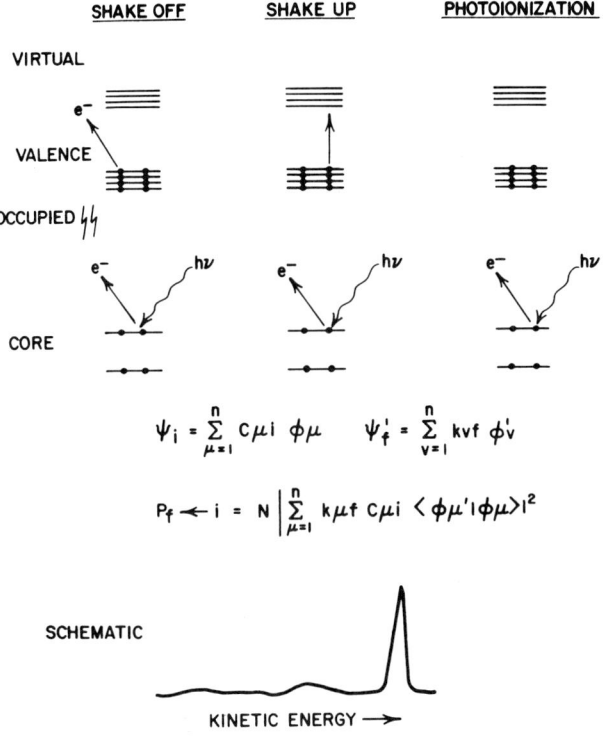

Fig. 16. Schemstic of relationship between direct photoionization, shake up and shake off processes.

theoretical relationships first established by Manne and Aberg[25]. They showed that the weighted average over the direct photo-ionization and shake up and shake off peaks corresponds to the binding energy appropriate to the unrelaxed systems and this is shown schematically in Fig. 17. Since relaxation energies fall within such a narrow range for a given core level[18], it is clear that shake up and shake off are perfectly general phenomena which are present in every system, the feature which changes from one system to another being the weighting coefficients (probabilities) for each transition. It is clear that transition probabilities for high energy shake off processes should be relatively small and that

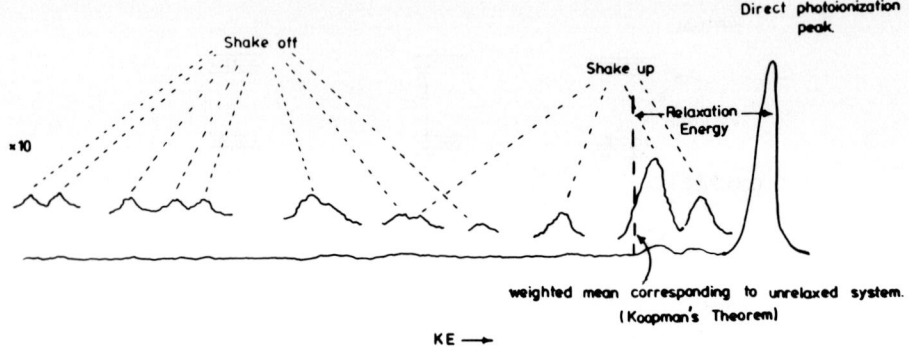

Fig. 17. Relationship between direct photoionization shake up and shake off process; relaxation energies and Koopmans' Theorem.

transitions of highest probability should fall reasonably close to the centroid. In principle, relaxation energies should be available from experiment provided all of the relevant shake up and shake off processes can be estimated in terms of energies and intensities. In practice, this is not a feasible proposition particularly for solids, since the overall situation is considerably complicated by the presence of the general inelastic tail (arising from photoemission from a given core level followed by energy loss by a variety of scattering processes) which provides a broad energy distribution usually peaking for organic systems $\sim$20 eV below the direct photo-ionization peaks. This generally obscures any underlying high energy shake up or shake off processes such that it is only for systems exhibiting relatively high intensity low energy shake up peaks that information derived from this source can conveniently be exploited. Fortunately, such a situation generally obtains for polymer systems which contain either unsaturated backbones[12] or pendant groups[13,14] since low energy $\pi \to \pi^*$ shake up transitions are available.

As a typical example, fig. 18 shows the C_{1s} spectra for typical saturated polymers-polyethylene (high density), polydimethylsiloxane, polystyrene and polydiphenylsiloxane which represent prototype systems with saturated backbones and unsaturated pendant groups.

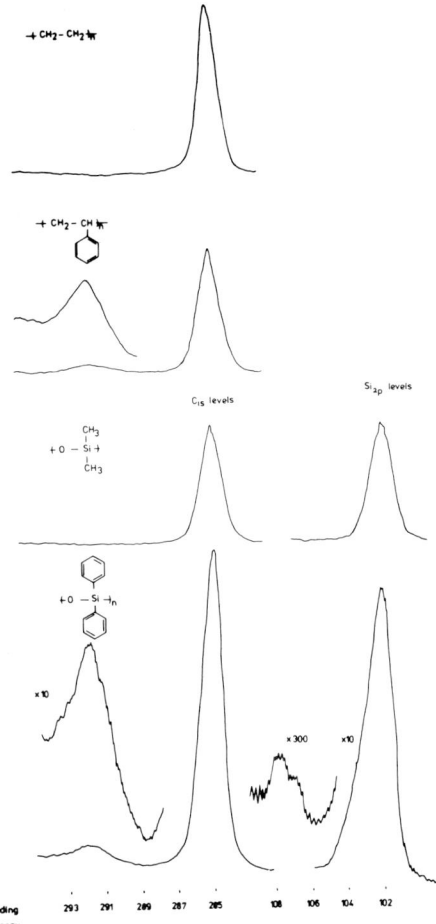

Fig. 18. Core level spectra of polyethylene, polystyrene, polydimethylsiloxane and polydiphenylsiloxane showing shake up structure.

For the latter, well developed shake up structures are apparent which clearly distinguishes them from the saturated systems although the differences in lineshape and linewidths for the main photo-ionization peaks are closely similar. (For the siloxanes of course, the relative intensities of the C_{1s} with respect to the O_{1s} and Si_{2p} levels may be used to effect a ready distinction between the two siloxanes.) It will become apparent that the transitions giving rise to the satellite structures in PS and PDPS are due to $\pi \rightarrow \pi^*$ transitions as might indeed be inferred from the much smaller shake up peak associated with the Si_{2p} levels and the lack of any low

energy shake up structure accompanying the O_{1s} levels for PDPS.
To elucidate the nature of these shake transitions as a preliminary
to utilizing such data for structural studies, we have made a systematic study of para-substituted polystyrenes the objective being
to study the transition energies and peak intensities as a function
of the electronic demand of the substituents in the classic mould
of physical organic chemistry[14,26]. To complement the experimental
studies, theoretical computations of shake up probabilities have
been made within the sudden approximation employing the equivalent
cores concept and a semi-empirical all valence electron SCF MO
formalism. Within this framework, the calculation of shake up
probabilities involves summation over weighted overlap terms involving the occupied and virtual orbitals involved in the transitions.
For the parent polystyrene, the marked asymmetry of the satellite
structure to the C_{1s} levels strongly suggests that at least two
transitions are involved. The substituents investigated included
ring nitrogen (poly-4-vinyl pyridine), para CH_3, $(CH_3)_3C$, Cl, Br,
NH_2 and OMe, and thus ranged from an extremely powerful π electron
acceptor to strong π electron donors. The effect on the sense of
asymmetry of the satellite peaks for the C_{1s} levels in this series
depends markedly on the electronic demands of the substituent, being
in an opposite sense for the two extremes of donor and acceptor
properties. The centroids for the satellite structures increase
in energy separation with respect to the direct photo-ionization
peaks as the para substituent changes from being an overall pi
electron donor to a pi electron acceptor. The most striking feature,
however, is the strong dependence (on the substituent) of the overall intensity of the shake up satellites, with respect to the direct
photo-ionization peak. Table 3 shows the relevant data. By contrast,
the satellite structure accompanying core ionization from the para
substituent is symmetrical in nature, and the centroid corresponds
to that for the lower energy component of the C_{1s} satellite structure if this is deconvoluted into two components.

The trend in intensities for the substituent shake up satellites
is in the opposite sense to that for the C_{1s} levels and this becomes
more evident on consideration of the ratio corrected for equal
numbers of atoms. (The intensity ratio for the methoxy derivative
is almost certainly a lower limit since there is some evidence
from the O_{1s} spectrum that there is a small amount of water and/or
oxidation at the surface.)

Comparison of the U.V. spectrum of polystyrene in the 2600 Å
region with that of toluene shows a close relationship in terms of
both extinction coefficients and vibronic fine structure. The
effect of para substituents is most conveniently characterized by
the shift in the band corresponding to the v_{0-0} transition. The
comparison of substituent effects on the electronic excited states
of the para-substituted polystyrenes parallels those for the
corresponding para-substituted toluenes. Such a correlation would
only be expected if the π → π* transitions were effectively localized

Table 3

Core Level Binding Energies, and Transition Energies and Intensities
for Low Energy Shake Up Structures in Poly Para-Substituted Styrenes

Substituent X	Core Level	Binding Energies[†] (eV)		Mean Shake Up[††] Energy (eV)	Total Shake Up[†††] Intensity (%)
(N)	C1s	285.0	285.9	7.1	5.8
H	C1s	285.0		6.6	8.1
tBu	C1s	285.0		6.5	7.5
Me	C1s	285.0		6.5	7.0
Cl	C1s	285.0	286.3	6.7	6.3
Br	C1s	285.0	286.2	6.6	5.9
OCH$_3$	C1s	285.0	286.7	6.7	3.7
NH$_2$	C1s	285.0	286.3	6.4	3.3
(N)	N1s	399.5		6.5	7.1 (7.3)
Cl	Cl2p	200.5	201.9	6.5	2.7 (3.0)
Br	Br3d	69.8	70.9	6.8	1.9 (2.3)
OCH$_3$	O1s	534.2		6.6	2.0 (3.8)
NH$_2$	N1s	400.4		6.4	5.8 (12.3)

† Binding energies relative to hydrocarbon at 285.0 eV.
†† Measured to centroids of asymmetric satellite peaks.
††† Intensities expressed as a percentage of the total intensity due to the core level from atoms in the ring and directly attached to the ring. Figures in brackets refer to the relative substituent/carbon shake up intensity for equal numbers of atoms.

35

within a given pendant group of the polymer system. This conclusion is reinforced by the observation that polystyrene and toluene show similar shake up structure in their ESCA C_{1s} spectra with respect to both band profiles and intensities (when due allowance has been made for the differing number of carbon atoms in the repeat unit). It is evident that para substituted toluenes are good model systems for the dipole excited states for the poly-para-substituted styrenes, and it will become apparent that this carries over to the monopole excited (shake up) states. Two parameters relating to substituent effects on the low lying $\pi \to \pi^*$ excited states of substituted benzenes are Platt's spectroscopic moment and the coulomb integral of the substituent. Platt's spectroscopic moment relates to intensity changes consequent upon introduction of a given substituent[27,28]. Whilst the coulomb integral of a given substituent derives from a localized orbital description of the dipole excited states of substituted alternants and non-alternants and is a measure of the change in potential of an electron in a $2p_z$ orbital on an adjacent carbon atom[28]. This may be derived from experimental data by analyzing the first order inductive shift in non-alternants and second order shift in alternants. Both substituent constants are, therefore, intimately related to substituent effects on the $\pi \to \pi^*$ dipole excited states.

Fig. 19 shows the correlation with the shake up intensities. The trends displayed are quite striking and leave little doubt that the satellites arise from $\pi \to \pi^*$ excitations.

In a review such as this, it is inappropriate to present a detailed theoretical interpretation of the results since this is available elsewhere[14,26]. Suffice it to say that it may readily be shown that the $\pi \to \pi^*$ shake up transitions involve the highest occupied and lowest unoccupied (virtual) orbitals of the pendant.

It is a comparatively straightforward matter to establish that on the ESCA depth scale the surface morphology of the polymers studied are such that the repeat units are statistically sampled, and it is therefore appropriate to use the corresponding para-substituted toluenes as model systems. Fig. 20 shows the four one-electron transitions involving the two highest occupied and lowest unoccupied orbitals (symmetries designated with respect to approximate level C_{2v} symmetry of the pi electron system).

Of the four transitions, those arising from the $a_{2\pi} \to b_{1\pi}$ and $b_{1\pi} \to a_{2\pi}$ excitations are formally monopole forbidden. However, the strong perturbation consequent upon removal of a core electron effectively removes the symmetry restriction except for hole states corresponding to photo-ionization from core levels associated with atoms located on C2 axes (e.g. C1, C2, C5 and X). The calculated shake up probabilities for the parent system and for a prototype system with a good pi electron donor are shown in Table 4. Two features are evident. First, the intensity for the low energy

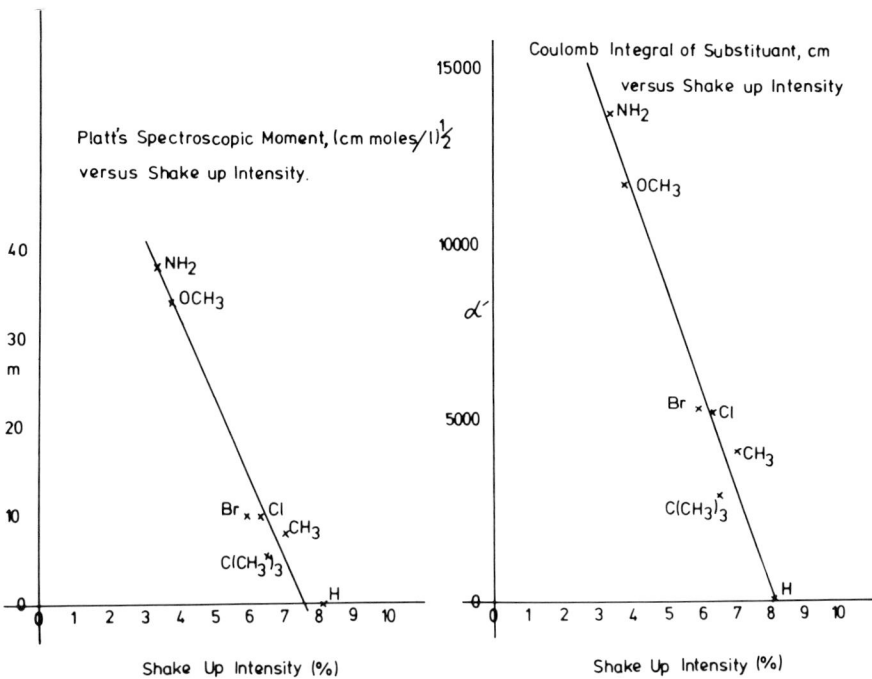

Fig. 19. Correlation of low energy shake up intensity accompanying C_{1s} photo-ionization (for a given parasubstituted polystyrene) with Platt's spectroscopic moment and the coulomb integral of the substituent.

shake up satellites to the C_{1s} spectra largely derives from two transitions ($b_{1\pi} \rightarrow b_{1\pi}$ and $a_{2\pi} \rightarrow b_{1\pi}$), and indeed the distinct asymmetry of the satellites would tend to confirm as we have already pointed out that more than one transition is involved. Secondly, when hydrogen is replaced by a pi electron donating substituent, whilst the intensity of the $b_{1\pi} \rightarrow b_{1\pi}$ transition is predicted to remain essentially the same, that for the $a_{2\pi} \rightarrow b_{1\pi}$ transition is calculated to decrease. The computed net overall decrease in shake up intensity predicted by the model is therefore qualitatively in agreement with experiment. As an interesting sidelight it might be mentioned that computations on the orthogonal conformation of the methoxy derivative ($2p_z$ lone pair on oxygen perpendicular to the ring) shows a substantial difference in shake up probabilities. This suggests that shake up structure should be sensitive to conformational preference which contrasts markedly with the behavior of the direct photo-ionization peaks.

The analysis of the shake up satellites in terms of a two

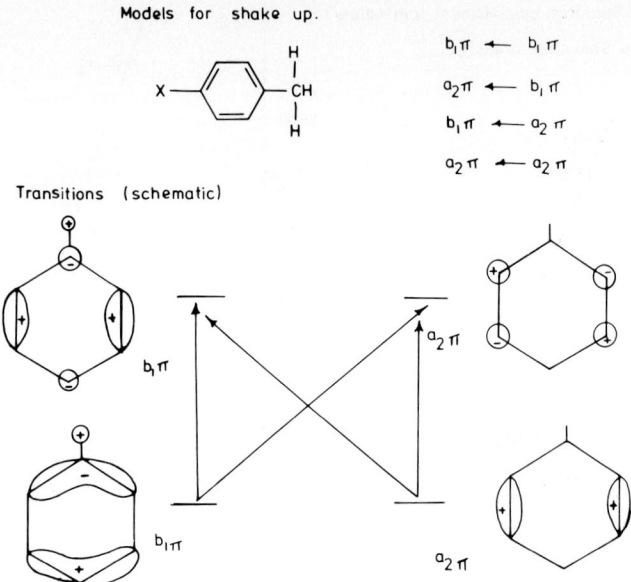

Fig. 20. The four one electron transitions involving the highest occupied and lowest unoccupied orbitals of substituted benzenes (in local C_{2v} symmetry).

component structure leads to the correlations shown in Fig. 21 where for convenience the data has been analyzed in terms of the coulomb integrals of the substituents.

It is gratifying to note that the theoretical calculations on model systems reproduce the trends shown in Fig. 21 providing strong confirmation for the overall validity of the interpretations. Low energy shake up satellite structures are often highly characteristic of the pi electronic structure of the pendant group as is clear from a comparison of Fig. 18 and Fig. 22. In each case, theoretical analysis indicates that the transitions involve the highest occupied and lowest unoccupied orbitals and this is indicated in Fig. 23.

Clearly, the distinctive nature of shake up satellites can add an extra dimension to the utility of ESCA as a structural tool. As one straight-forward example, Fig. 24 shows the C_{1s} and O_{1s} levels for poly-n-hexyl-acrylate and polyphenylacrylate.

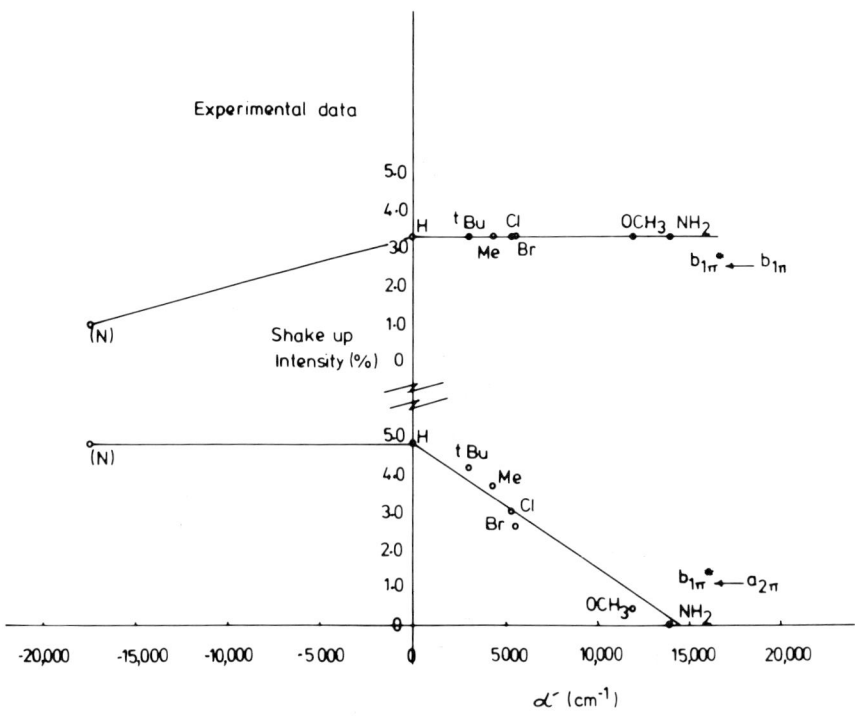

Fig. 21. Experimentally derived shake up probabilities for components contributing to the low energy shake up satellites in para-substituted polystyrenes.

The core level spectra are essentially the same; however, the distinctive nature of the shake up structure for the unsaturated pendant group allows a ready distinction to be made.

Having identified and qualitatively understood the main features of the low energy satellite structures arising predominantly from shake up processes*, we may now consider the implications with regard to enlarging the scope of the technique. Before we consider these,

---

*Discrete energy loss peaks will also contribute to the overall structure. Comparison of model systems studied in the gas and condensed phases show, however, that the low energy structures in the systems in question are dominated by shake up processes[29].

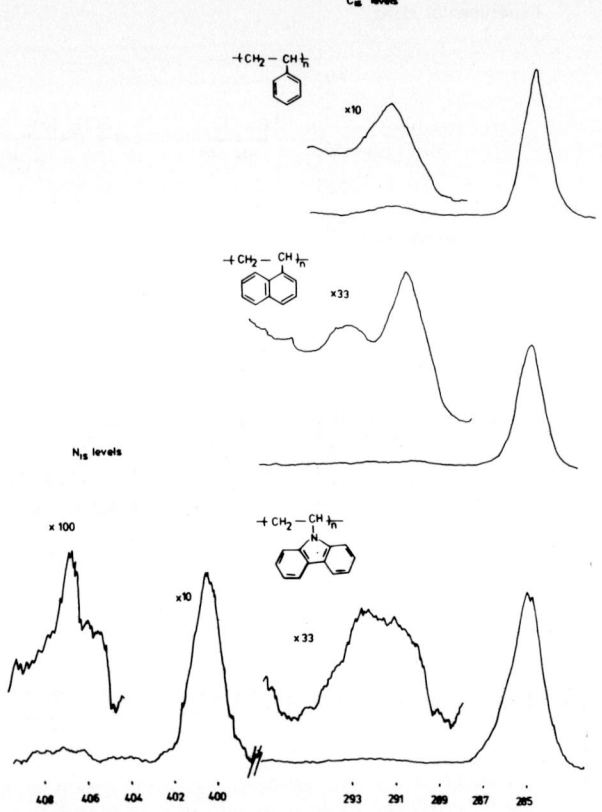

Fig. 22. Core level spectra for polystyrene, poly-1-vinylnaphthalene and polyvinylcarbazole showing low energy shake up structure.

however, it is worthwhile considering a further aspect of shake up phenomena in general. We have alluded previously to the parallel which exists between the dipole transitions to excited states of the neutral system and the monopole shake up processes accompanying core ionizations, and it is worthwhile investigating this analogy in somewhat more detail. For a closed shell system, excitation of an electron from an occupied to an unoccupied level can give rise to two states of the same excitation configuration; namely, a singlet and a triplet state, and this is illustrated schematically

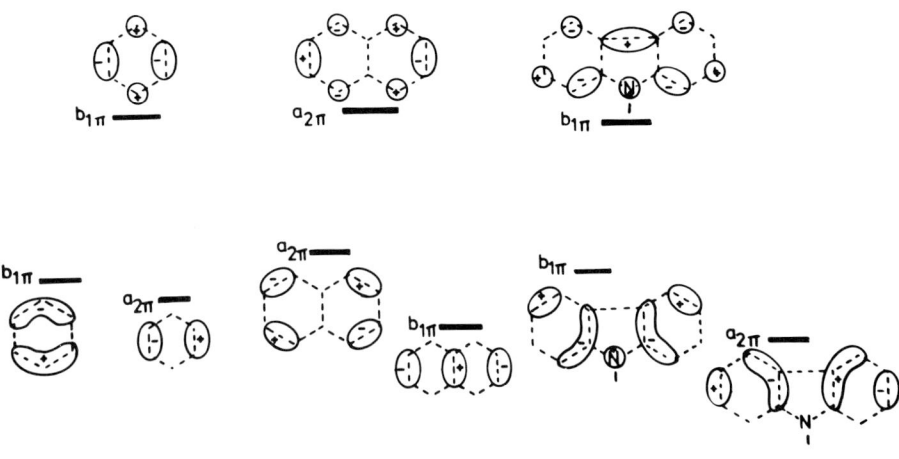

Fig. 23. Orbitals involved in the low energy shake up satellites accompanying core ionization in polystyrene, poly-1-vinylnaphthalene and polyvinylcarbazole.

in Fig. 25. In the orbital approximation, the energy difference between singlet and triplet states is 2 K where K is an exchange repulsion integral which is always positive, and therefore the triplet is always of lower energy than the singlet of the same excitation configuration[28]. Direct transitions to the triplet state from the ground state are of course spin forbidden and can only be observed under rather special circumstances. On the other hand, transitions to the singlet state are spin allowed and follow dipole selection rules. The parallel situation for the monopole excited states of a given core hole state are also shown in Fig. 25. The shake up states of the same excitation configuration are both doublet states; one of which may be thought of as deriving from 'singlet' parentage and the other of 'triplet' parentage; transitions to both states, however, are spin allowed. By analogy with the transitions of the parent system, it might naively be thought that transitions to the shake up states of 'singlet' origin should have the greater probability. Detailed theoretical studies for small poly-atomic systems show this to be the case in general[30]. High resolution gas phase studies on the core levels of small

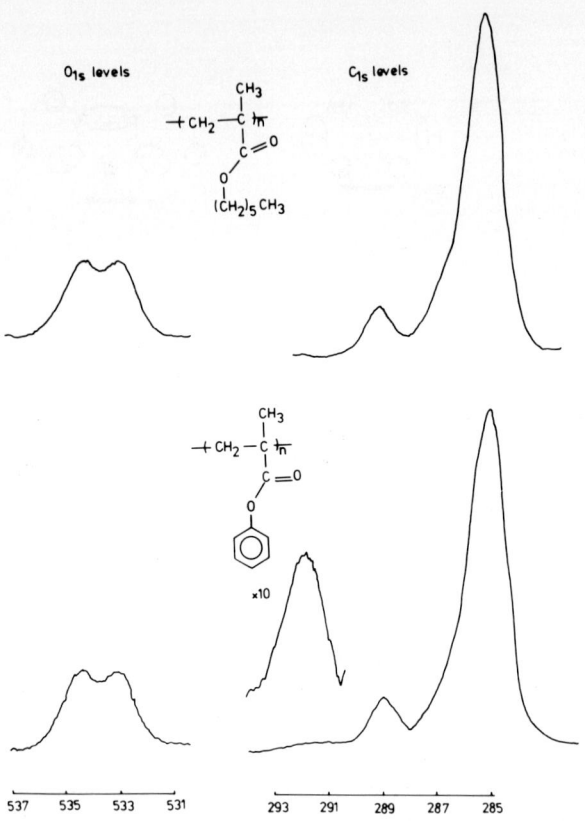

Fig. 24. Core level spectra for poly-n-hexylacrylate and poly-phenylacrylate showing the shake up structure for the latter.

molecules have indeed shown that shake up structure corresponding to transitions to states of 'singlet' and 'triplet' parentage may be observed with the latter being generally of lower intensity and at substantially lower energy[30,31]. We have of necessity painted a fairly simplistic picture of a complicated situation which for its quantitative elaboration requires detailed considerations of configuration interaction in the doublet manifold. It is clear, however, from the work carried out to date that the shake up structure which has been described for the polymer systems corresonds to

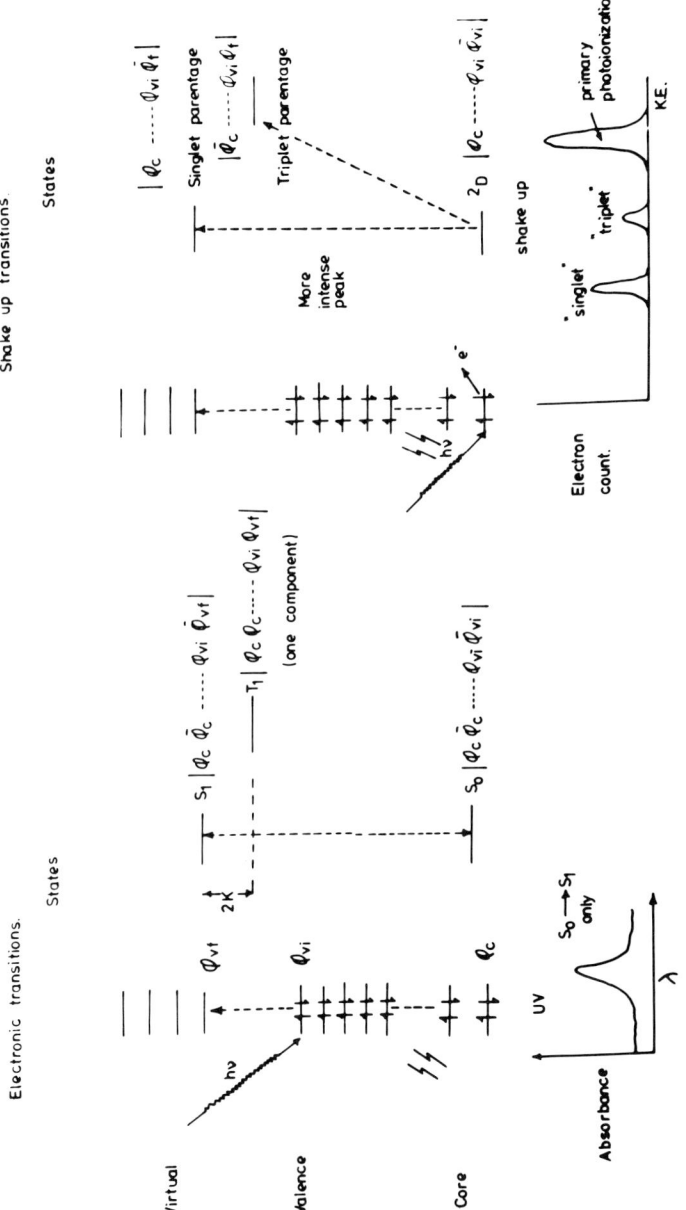

Fig. 25. Schematic illustrating the relationship between the dipole excited states of a neutral system with singlet and triplet states formally of the same excitation configuration and the monopole shake up states for a core hole state.

Table 4

Calculated Shake Up Probabilities for Low Energy $\pi \to \pi^*$ Transitions in Model Compounds[26]

Transition	Atom				% Shake Up					Total	
	1	2	3	4	5	6	7	8	9	C_{1s}	O_{1s}
$b_{1\pi}^* \leftarrow b_{1\pi}$	0.90	9.13	3.40	2.23	9.61	2.23	3.40	—	—	4.41	—
$a_{2\pi}^* \leftarrow b_{1\pi}$	0	0	0.06	0	0	0	0.06	—	—	0.02	—
$b_{1\pi}^* \leftarrow a_{2\pi}$	0	0	5.81	6.80	0	6.80	5.81	—	—	3.60	—
$a_{2\pi}^* \leftarrow a_{2\pi}$	0.11	0.01	0	0	0.02	0	0	—	—	0.02	—
$b_{1\pi}^* \leftarrow b_{1\pi}$	1.15	10.20	3.02	4.07	10.24	3.59	4.08	1.39	0.45	4.60	1.39
$a_{2\pi}^* \leftarrow b_{1\pi}$	0	0	0.06	0.10	0	0.10	0.06	0.48	0.71	0.13	0.48
$b_{1\pi}^* \leftarrow a_{2\pi}$	0	0	6.06	4.84	0	5.36	4.72	0.05	0	2.62	0.05
$a_{2\pi}^* \leftarrow a_{2\pi}$	0.10	0.04	0	0	0.07	0	0	0.03	0.10	0.04	0.03

that arising from the component of 'singlet' origin. The corresponding 'triplet' components are undoubtedly of much lower energy and intensity, and indeed recent work in these laboratories has shown that if the signal/noise ratio, general background and resolution can be improved by employing monochromatized x-rays at an appropriate band pass, then such transitions can be observed[32]. These transitions may also be characterized in terms of transition energies and intensities. Such investigations are likely to be of considerable importance in the future, and results from our current research program will be described in due course. This, therefore, adds a further refinement to the information available from ESCA from observation of the relative intensities and transition energies of components of shake up transitions.

As a particular example in this section, we outline an investigation of some alkane-styrene copolymers of general formula:

$$-[CH-CH_2CH_2-CH-(CH_2)_n]_m$$

(with phenyl groups on the CH carbons)

$n = 0, 1, 3, 5, 6, 10$

which illustrate the utility of ESCA for studying copolymer compositions in systems for which the primary sources of information are of themselves insufficient and for which the extra dimension is provided by observation of shake up satellites[14,33].

It should be evident from the previous discussion that characteristic low energy shake up structure accompanying direct photo-ionization of the C_{1s} levels should be specifically associated with the styrene component. Fig. 26 shows the measured C_{1s} levels and shake up satellites for the series of alkane-styrene copolymers, and it is evident by visual inspection that the relative intensities of the shake up satellites with respect to the main photo-ionization peaks decrease with increasing chain length of the alkane component.

The measured intensities and energy separations are given in Table 5. A clear trend exists between shake up intensity and the chain length of the alkane component and the structure of the shake up satellites and the energy separations remain essentially constant. This becomes clearer from a graphical representation of the data as shown in Fig. 27. Also shown is the correlation expected on the basis that the repeat units of the polymers are statistically sampled. It is clear that copolymer compositions may be established from the measurement of shake up intensities and the least square plot of the intensity ratio of the direct photo-ionization peak to shake up satellites versus n (the chain length of the alkane component) gives a correlation coefficient of 0.997 the slope being 1.91 and intercept 12.91. The latter may be compared with the measured

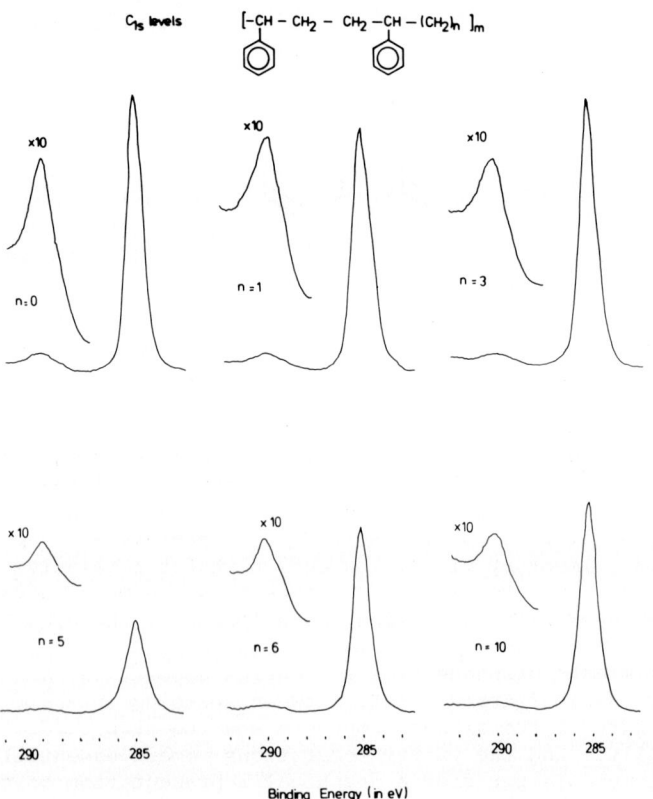

Figure 26 C_{1s} levels for a series of alkane-styrene copolymers showing shake up structure.

Table 5

ESCA Data for the Alkane-Styrene Copolymers

n	C_{1s}	C_{1s}^S	Δ	Area Ratios (C 1s/C_{1s}^S)†
	Binding Energies/eV*			
0	285.0	291.6	6.6	13.4
1	285.0	291.6	6.6	14.5
3	285.0	291.6	6.6	18.5
5	285.0	291.6	6.6	22.7
6	285.0	291.6	6.6	23.8
10	285.0	291.6	6.6	32.2

* Relative to C 1s at 285.0 eV, Δ given with respect to centroid of asymmetric shake-up peak.

† Ratio of main photo-ionization peak to shake-up peak.

value of 13.4 for the parent system (i.e. polystyrene). The calculated slope assuming statistical sampling of the repeat unit is 0.90. The fact that an additive model applies to the experimental data but with a much larger dependence of intensity on n than predicted theoretically would strongly suggest that there are specific orientation effects of the polymer chains in the surface regions sampled by ESCA. An alternative possibility is that the samples are contaminated with hydrocarbon. Such contamination would contribute to the C_{1s} peak at 285 eV but not to the low energy satellite structure. Even if the extent of contamination were such as to produce the linear correlation found experimentally (which is most unlikely on the basis of the method of preparation)[34], on the basis of the likely escape depth dependence on kinetic energy for C_{1s} levels employing $Mg_{K\alpha_{1,2}}$ radiation such an explanation is untenable[3].
What is clearly required is a model in which the repeat unit is not statistically sampled in such a sense that the phenyl groups are discriminated against. As n becomes large, we might reasonably expect a structure based on the folded chain structure of polyethylene[35]. With this in mind and with the aid of models, we have considered possible structures which would lead to the results illustrated in Fig. 27. Two such models which exhibit an increased gradient with respect to that expected if the data correspond to statistically sampling the repeat unit are shown in Fig. 28 for

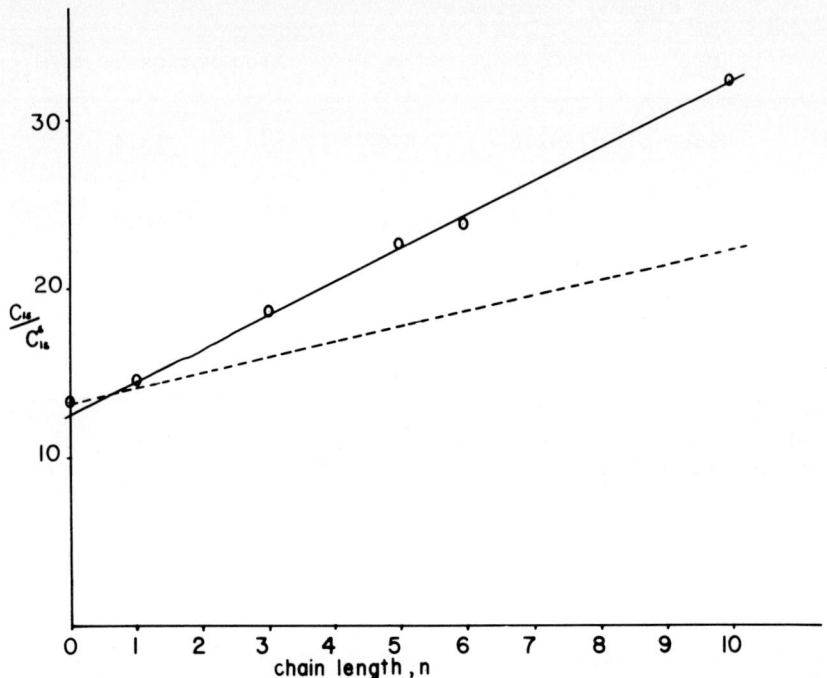

Fig. 27. Correlation of relative intensity of direct photo-ionization to show energy shake up structure with chain length of alkane component in alkane-styrene copolymers (lower correlation theoretically derived on basis of statistical sampling of repeat unit).

the particular case of n = 3. The model with a phenyl group specifically oriented at the surface come remarkably close to the experimental data, and it may fairly be claimed that for this system in addition to providing information on composition; shake up satellites also provide an interesting insight into the possible surface morphology.

Other examples of the exploitation of shake up phenomena includes for example the investigation of the relationship of surface to bulk domain structure in AB block copolymers[36]. These

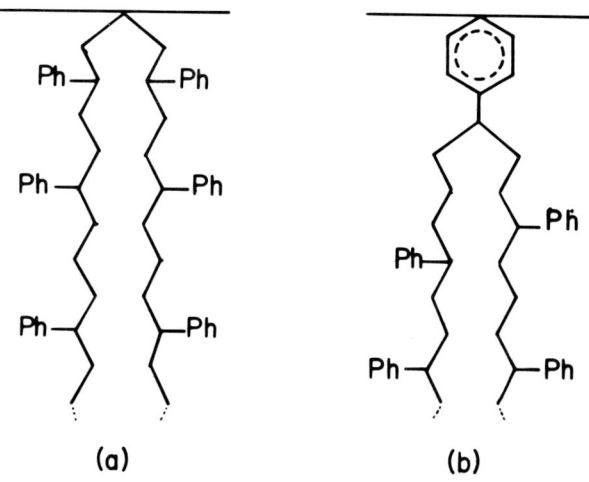

Fig. 28. Possible structures for alkane-styrene copolymers which would lead to non statistical sampling of the repeat unit (for the particular case of n = 3.

studies considerably extend the scope of ESCA as one of the most important shots in the polymer chemist and physicists' locker for studying aspects of structure bonding and reactivity relating to the surface regions of polymers.

ACKNOWLEDGEMENTS

I would like to acknowledge the contributions made to the development of the ESCA Program in Durham by a succession of willing and able research workers. In the most recent installment of the ESCA applied to polymers program, particular mention should be made of Jim Peeling, Ron Thomas and Alan Dilks. I would also like to thank the Science Research Council and Institute of Petroleum for providing equipment.

REFERENCES

1. D. T. Clark, *Advances in Polymer Friction and Wear,* Ed. L. H. Lee, Plenum Press, 5A, 241 (1974).
2. D. T. Clark, Structure and Bonding in Polymers as Revealed by ESCA, in *Proceedings of the NATO Advanced Study Institute,* NAMUR, Belgium (1974), Ed. J. Ladik and J. M. Andre, Plenum Press, New York (1975).
3. D. T. Clark, W. J. Feast, W. K. R. Musgrave and I. Ritchie, J. Poly. Sci. Poly. Chem. Edn., 13, 857 (1975).
4. D. T. Clark, Chapter 9 in *Structural Studies of Macromolecules by Spectroscopic Methods,* Ed. K. Ivin, J. Wiley, London (1976).
5. D. T. Clark, I. W. Scanlan and J. Müller, Theoretica Chim. Acta., 35, 341 (1974).
6. D. T. Clark, Chemical Aspects of ESCA, in *Electron Emission Spectroscopy,* Ed. W. Dekeyser, D. Reidel Publishing Co., Dordrecht, Holland, pp. 373-507 (1973).
7. U. Gelius, E. Basilier, S. Svensson, T. Bergmark and K. Siegbahn, J. Electron Spectroscopy, 2, 405 (1974).
8. (a) A. E. I. Scientific Apparatus Ltd., Manchester, England.
   (b) Hewlett Packard, Palo Alto, California, U.S.A.
9. D. T. Clark, et al., University of Durham, Lab Report.
10. Cf. Ref. & and H. Fellner-Feldegg, U. Gelius, B. Wannberg, A. G. Nilsson, E. Basilier and K. Siegbahn, J. Electron Spectroscopy, 5, 643 (1974).
11. D. T. Clark, et al., Unpublished Data.
12. R. D. Chambers, D. T. Clark, D. Kilcast and S. Partington, J. Poly. Sci. Poly. Chem. Edn., 12, 1647 (1974).
13. D. T. Clark, D. B. Adams, A. Dilks and H. R. Thomas, J. Electron Spectroscopy, 8, 51 (1976).
14. D. T. Clark, A. Dilks, J. Peeling and H. R. Thomas, Disc. Faraday Soc., 60, 183 (1975).
15. Cf. Ref. 3 and D. T. Clark and W. J. Feast, J. Macromol. Sci. Revs. Macromol. Chem., C12 (2), 191 (1975).
16. D. T. Clark, A. Paton and W. Salaneck, J. App. Phys., 47, 144 (1976).
17. Cf. Ref. 6 and
    (a) D. T. Clark and D. Kilcast, J. Chem. Soc. A, 3286 (1971).
    (b) D. T. Clark, D. Kilcast and W. K. R. Musgrave, J. Chem. Soc. D, 516 (1971).
    (c) D. T. Clark, D. Kilcast, D. B. Adams and W. K. R. Musgrave, J. Electron Spectroscopy, 1, 227 (1972).
    (d) D. T. Clark, D. Kilcast, D. B. Adams and W. K. R. Musgrave, J. Electron Spectroscopy, 6, 117 (1975).
18. Cf. D. T. Clark, J. Müller and I. W. Scanlan, Theoretica Chim. Acta, 35, 341 (1974).
19. (a) D. T. Clark, D. B. Adams and D. Kilcast, J. Chem. Soc. Disc. Faraday Soc., 54, 182 (1972).
    (b) D. T. Clark, D. Kilcast, W. J. Feast and W. K. R. Musgrave, J. Poly. Sci. Poly. Chem. Edn., 11, 389 (1973).
    (c) D. T. Clark, W. J. Feast, I. Ritchie, W. K. R. Musgrave,

M. Modena and M. Ragazzini, J. Poly. Sci. Poly. Chem. Edn., 12, 1049 (1974).
20. D. T. Clark and H. R. Thomas (in preparation).
21. D. T. Clark and H. R. Thomas, J. Poly. Sci. Poly. Chem. Edn., 14, 1671 (1976).
22. D. T. Clark, W. J. Feast, H. R. Thomas and P. Tweedale (in preparation).
23. L. E. Sutton, Tables of Interatomic Distances, Chemical Society, London, Special Publication, No. 18 (1965).
24. K. Nakanishi, Infrared Absorption Spectroscopy, Holden-Day Inc., San Francisco (1964).
25. R. Manne and T. Aberg, Chem. Phys. Lett., 7, 282 (1970).
26. D. T. Clark and A. Dilks, J. Poly. Sci. Poly. Chem. Edn., (1976) in press.
27. H. H. Jaffe and M. Orchin, *Theory and Applications of Ultraviolet Spectroscopy*, John Wiley and Sons Inc., New York (1962).
28. J. N. Murrell, *The Theory of the Electronic Spectra of Organic Molecules*, Methuen and Co., Ltd., London (1963).
29. Cf. D. T. Clark and D. B. Adams, Theoretica Chem. Acta., 39, 321 (1975).
30. (a) H. Basch, Chem. Phys. Lett., 37, 447 (1976).
    (b) I. H. Hillier and J. Kendrick, Faraday Trans. II, 7, 1369 (1975).
    (c) D. T. Clark, J. Müller, M. F. Guest and W. Rodwell (in preparation).
31. Cf. (a) K. Siegbahn, C. Nordling, G. Johansson, J. Hedman, P. F. Heden, K. Hamrin, U. Gelius, T. Bergmark, L. O. Werme and Y. Baer, *ESCA Applied to Free Molecules*, North Holland, Amsterdam (1969).
    (b) T. A. Carlson, M. O. Krause and W. E. Moddeman, J. Phys. (Paris), Cr-76, 32 (1971).
32. D. T. Clark, A. Dilks and H. R. Thomas (in preparation).
33. D. T. Clark and A. Dilks, J. Poly. Sci. Poly. Chem. Edn., 14, 533 (1976).
34. D. H. Richards, N. F. Scilly and F. J. Williams, Polymer, 10, 603 (1969).
35. Cf. B. Wunderlich, *Macromolecular Physics*, Volume 1, Academic Press, New York (1973).
36. Cf. 14 and D. T. Clark, J. Peeling and J. M. O'Malley, J. Poly. Sci. Poly. Chem. Edn., 14, 543 (1976).

Sputter-Induced Compositional Change During ESCA/Sputtering of Polymers

Dwight E. Williams

Analytical Services, Dow Corning Corp.
Midland, Michigan 48640
                    and
Lawrence E. Davis

Physical Electronics Industries
Eden Prairie, Minnesota 55343

   The extent of sputter-induced elemental compositional changes (artifacts) in polymers was determined from ESCA band areas. A variety of organic and organosilicone polymers was studied. Pristine surfaces of known, depth-invariant composition were prepared by vacuum fracture. Measured elemental calibration factors agreed well with literature values. Artifacts were evaluated by ESCA after various sputtering durations. The extent of the artifacts depended on the specific polymer and sputtering conditions. The artifacts were generally small or negligible for silicone polymers but were very large for the organic polymers and sputtering conditions used.

INTRODUCTION

   Many of the major techniques presently available that permit highly surface selective chemical analysis of polymers directly indirectly depend upon ion beam sputtering. Such sputtering is inherent in the SIMS and ISS techniques and is extremely useful in ESCA or Auger analysis in order to obtain compositional information as a function of depth. We believe such information is essential for general purpose analysis since thin underlying layers not immediately detected by these techniques are often more relevant than the initial surface. For example, such layers are often uncovered by exposure of the polymer to a different mechanical or chemical environment. On the other hand, adventitious contamination of the relevant surface is likely to have occurred prior to analysis unless rigorous, time consuming and often impractical precautions against such contamination are taken. However, since the nature of the top surface layer which actually exists in a particular environment will control many important surface properties (such as adhesion and release, wettability, boundary layer lubrication, etc.), it is very desirable to obtain a thin layer-by-layer analysis of the surface.

The present study is a preliminary attempt to determine the validity of sputtering techniques and in particular of ESCA/sputtering analysis to obtain depth-compositional profiles of selected polymers. Others have shown that many metal oxides, sulfides and halides are reduced or preferentially sputtered by the ion beam and a variety of other sputtering artifacts are known to occur.[1,2,3] By analogy to inorganics, it is likely that sputtering artifacts also occur for polymers and vary with the particular polymer (e.g., elements, amounts, morphology and chemical structure) and sputtering conditions (voltage, current density, angle and bombarding species).

We feel that the real issue is not whether sputtering artifacts occur, but rather that the following questions should be investigated:

1. What is the magnitude of the artifact for the material of interest?

2. Can the artifact be tolerated for the particular problem or application?

3. Can the sputtering conditions be adjusted to reduce the artifact to tolerable levels?

We have ESCA/sputter analyzed several kinds of homogeneous polymer surfaces as an initial step towards grappling with these issues as they pertain to polymers.

In order to create a polymer surface of known composition and reduce the possibility of initial depth gradients, polymer bars were fabricated and then fractured in vacuum (10^{-8} Torr). Calibration factors relating the area of the photoelectron lines (band area) to elemental composition were then obtained for the initial surface. These factors were used to obtain elemental composition of the surface after it had been subjected to a varying sputter dosage. Unfortunately, the actual sputtering depth was unknown, and for the sake of obtaining some insight into how sputtering artifacts vary with depth, we have assumed that the sputtering rate was equal to that of silicon metal.

EXPERIMENTAL

Apparatus

The samples were analyzed on the PHI ESCA spectrometer using a Mg anode at 400 W. The PHI vacuum fracture accessory was used to fracture scribed polymer bars 3 mm square. This device employs flexural stress in such a way as to avoid contamination of the newly created surface. The surface was somewhat uneven but was usually within 10° of being perpendicular to the bar axis. The sample surface normal was 60° with respect to the electron analyzer axis. The analyzer resolution was set a 1 ev in order to limit the analytical area to that of the sample bar end.[4]

A PHI argon ion gun at 20° from sample surface normal was used for sputtering. Sputter rates at 2 kv were measured by Auger depth profiling of films of Si, SiO_2 and Ta_2O_5 whose thicknesses were determined by interferometry or be anodic current-voltage-thickness relationships. Rates for all three materials were identical to within ± 15% at 2 kv. Sputter rates of Si at other voltages were calculated from the direct proportionality to current density and from tabulated values of ion yield vs. voltage.[5] The ion current density was 55, 48 and 33 $\mu a/cm^2$ at 2, 1 and 0.5 kv as measured with a Faraday cup using a 30 mil aperture. These measurements also showed that the current density was constant across the sample surface (3 mm square bar). At 2, 1 and 0.5 kv, the sputter rate for Si was 55, 32 and 18 Å/min. These rates were used to calculate sputter depths in polymers.

A shielded heated filament at 10V bias was used to supply electrons to reduce sample charging during sputtering of the organic polymers and the silicone resin II. However, we do not believe this neutralization had any real effect. For example, ESCA showed that the composition of cast samples of the silicone rubber after prolonged sputtering with neutralization is identical to the initial composition of a vacuum fractured bar.

Materials

Polymer bars, 3.1 mm wide, were cut from a 3.1 mm thick cast polymer plate with a water-cooled bandsaw. Polymer compositions as determined by wet chemical analysis are shown in Table 1. The Saran sample was a semicrystalline experimental copolymer containing 89.5/10.5 mole % polyvinylidene/vinyl chloride and no other additives. The amorphous polyurethane was prepared from equivalent amounts of toluene diisocyanate and a trifunctional polypropylene glycol, and probably contained various impurities. The amorphous silicone samples were representative of a range of

Table 1 Bulk Polymer Composition (Atom %)*

Material	%O	%N	%C	%Cl	%Si
Silicone Rubber	24.0+.8	–	53.6+.3	–	22.4+.4
Silicone Resin I	18.6+.8	–	65.6+.5	–	15.8+.4
Silicone Resin II	10.4+.7	–	79.1+.5	–	10.5+.3
Polyurethane	18.7+.3	9.62+.12	71.7+.5	–	–
Polymethylmethacrylate	28.1+.5	–	71.9+.5	–	–
Saran	–	–	53.2+.3	46.8+.3	–

*+ values are 90% confidence limits to the analysis.

composition and cross-link density and included a brittle rubber
and two rigid resins. All of the samples were prepared from compatible monomers and should have had no gross inhomogeneities
within the bulk. However, the polyurethane sample contained some
minute, even dispersed air bubbles and the Saran sample contained
some regions which were decomposed by the cutting heat. Fractures
through these regions were distinguished by a large oxygen ESCA
peak and loss of chlorine: Such fractures were discarded. Of
course, preferential fracturing through microscopic impurity
regions, or in the case of the Saran sample, through PVC - rich
or - poor domains is also possible and such effects would invalidate our assumption that the composition of the initial fracture
surface is identical to that of the bulk (however, see DISCUSSION).

The polymethylmethacrylate sample was prepared by irradiation of
reagent grade methyl methacrylate which had been purified by
passage through an alumina column. This highly pure homopolymer
should have contained no impurities through which a preferential
fracture could occur.

Measurements and Calculations

Scans, 20 eV (2 eV/inch), were taken of each element. Line
widths were 1.8 ev or more. The singlet peaks displayed by the
silicone polymers increased in width by different amounts for
each element after sputtering. For example, the O-1s, C-1s and
Si-2p linewidths of silicone rubber increased by 20, 30 and 40%
after 200 Å were sputtered at 2 kv. The organic polymers also
showed chemically shifted multiplets for carbon whose relative
intensities were altered by sputtering. Quite often initially
broad singlets became narrow and vice versa after sputtering.
These effects mandated the use of band areas rather than peak
heights for elemental analysis. Examples of the spectral changes
are shown in Figures 1-3.

Band areas were measured two to three times with a planimeter
and were reproducible to within $\pm$ 1/2%. The baseline was drawn
through the noise envelope between the lowest point on each side
of the peak. The probable major source of systematic error was
due to variable overlap with satellite and inelastic loss peaks
which occurs when the bandshape changes after sputtering.

Atom % values given here consider only the elements detected
by ESCA. Normalization was used in order to eliminate the effect
of changes in emission induced by variations in fracture topography. This procedure also eliminated the effect of the increase
in emission displayed by all elements after sputtering the silicone polymers. Apparently a change in surface density or roughness occurs. A rearrangement of a simplified expression for photo-

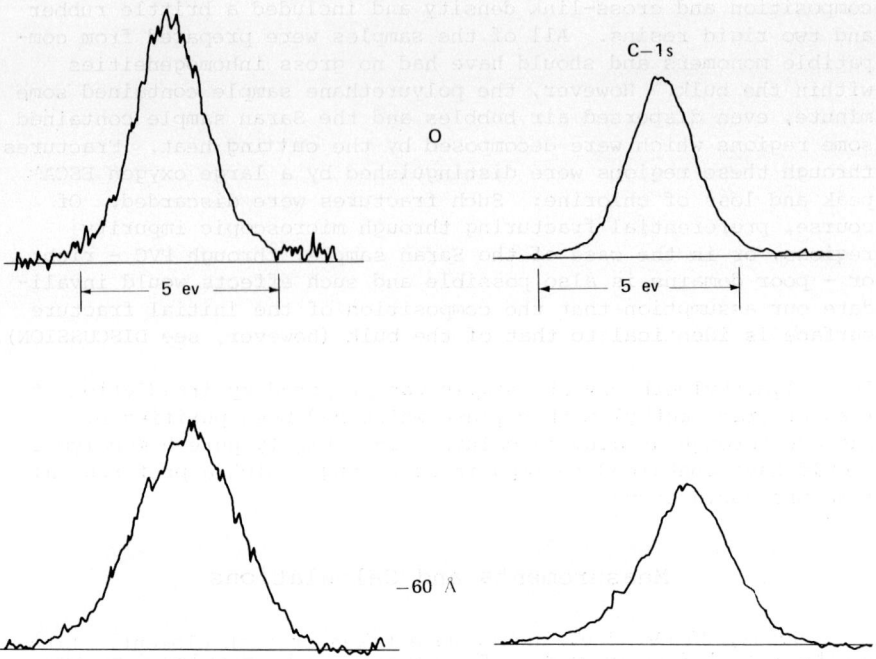

Fig. 1. Silicone rubber bandshapes before and after 2 kv sputter.

emission[6] was employed:

$$P_j = 100 \frac{I_j R_j}{\Sigma_j I_j R_j} \qquad (1)$$

where P_j is the atom % of element j, I_j is the band area and R_j is an empirical calibration factor which is related to the photoelectron emissivity, the electron loss coefficient and the transmission characteristics of the electron analyzer. R is taken to be unity for C-1s since one of the R_j is an arbitrary parameter.

For general use, the calibration factor for a given element is usually assumed to be independent of the matrix although small variations may occur as a result of the matrix dependency of various loss processes. We use an independent set of calibration factors for each polymer matrix in order to verify our assumption that initial depth gradients are minor.

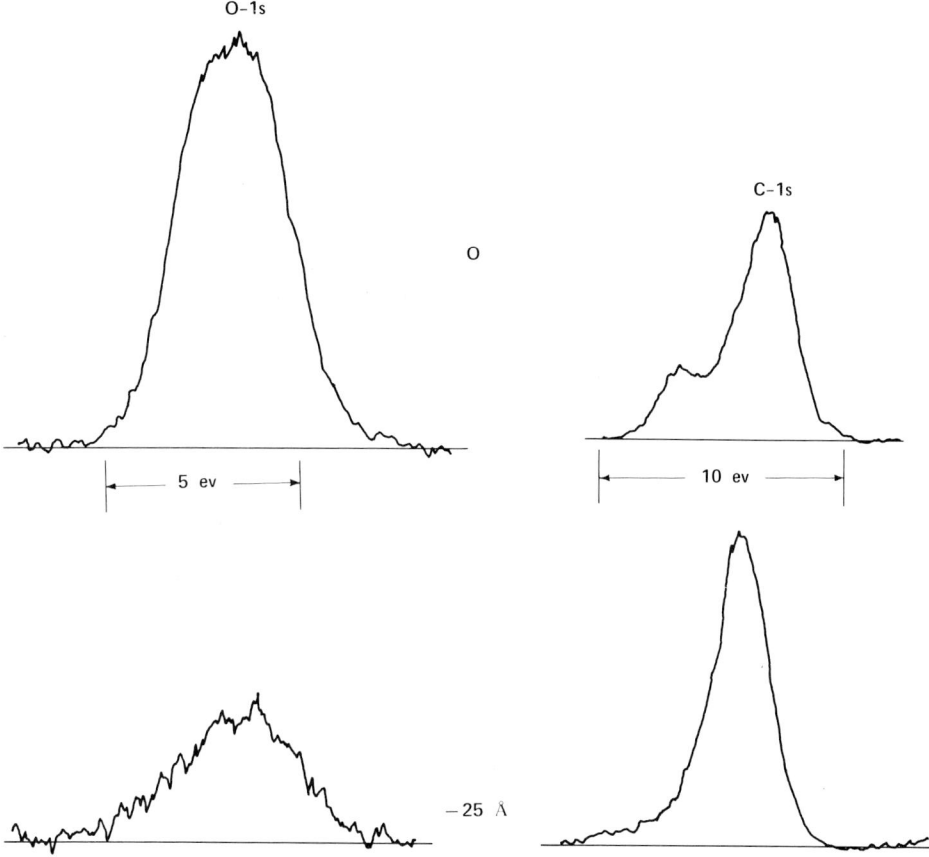

Fig. 2. Polymethylmethacrylate bandshapes before and after 2 kv sputter.

RESULTS AND DISCUSSION

Initial Fracture Composition

Evaluation of sputtering artifacts requires that the initial surface composition be substantially the same as that of the bulk. This issue is of course critical since these artifacts cannot be detected unless initial gradients are negligible. The four major sources of deviations from such a state are handling (or cutting) contamination, selective volatiles desorption, adsorption, and preferential fractures through inhomogeneous regions. The fracture technique removes the first source of gradients. We have occasionally observed unusual surface compositions of air fractured

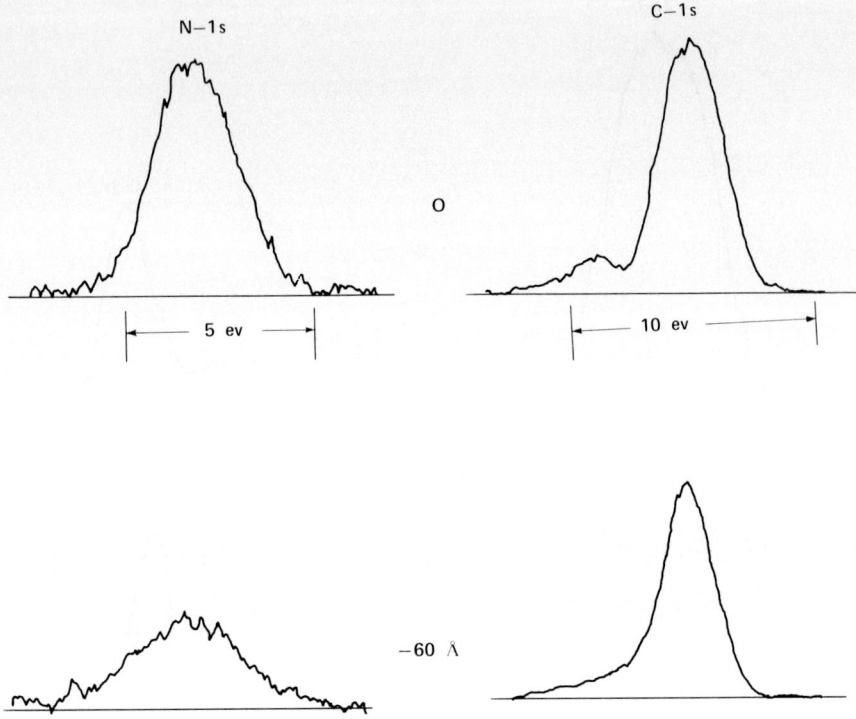

Fig. 3. Polyurethane bandshapes before and after 2 kv sputter.

samples which are probably due to this cause. It is unlikely that the second source is of any consequence in these high molecular weight polymers before sputtering. The third and particularly the fourth source are of most concern for our conditions.

Although polymer surfaces usually are considered to be inert relative to clean metal surfaces, this is probably not entirely true for a freshly fractured surface. The fracture of polymers is known to generate free radicals which may well possess enhanced reactivity to volatiles such as CO_2, etc.[7] These radicals could react with residual gases at 10^{-8} Torr and increase the carbon or the oxygen content (depending on the particular species) at the initial surface. We do feel that these low pressures greatly reduce the amount of adsorption since strong chemisorption would be expected to be no more than one monolayer thick (much less than ESCA escape depth).

We consider the fourth source of initial gradients to be the most likely one to impair the interpretation of our results, particularly for the Saran and the polyurethane samples. The former was a semicrystalline copolymer and the latter, although amorphous, was difficult to mix thoroughly before cure. The remaining polymers were both amorphous and quite homogeneous as indicated by their transparency, and of course, the polymethylmethacrylate sample was a pure homopolymer.

It is of interest to compare the variability in the composition of replicate fractures against that of replicate scans of the same fracture. These data are shown in Tables 2 and 3. The variability between replicate fractures is considerably greater, suggesting that there may be different amounts of adsorbed impurities or of fractured impurity regions on the fracture surfaces. However, only the polyurethane showed a statistically significant (1% level) difference by a t test applied to observed differences between fracture compositions. In addition, the actual variability between fractures is reasonably small.

More insight into whether initial gradients are present was obtained from measured calibration factors shown in Table 4. There were no statistically significant (5% level) differences between calibration factors for the same element measured in different polymers, even though there were large differences in the nature of the polymers. This fact strongly suggests that the initial gradients were negligible in comparison with sputtering artifacts of concern. If adsorption or preferential fracture effects occurred in such a way to make this wide a variety of polymers display virtually identical calibration factors, it would be truly remarkable.

For general analytical work, the weighted average calibration factor shown in Table 4 should be useful. Note that these agree well with theoretical values and reasonably well with experimentally derived literature values. However, our value for Si differs from both "theoretical" and literature values. These differences are probably due to the intrinsic angular dependence of photoemission and to the greater width of the Si-2p band.

Sputtering of Silicone Polymers

Figures 4-8 show the surface composition of the silicone polymers as a function of sputtering depth at various voltages. Sputtering artifacts may be recognized by a systematic change in composition with depth. Such changes occur to a limited extent for the Resin II at 0.5 (not shown) and 2 kv sputtering voltages as well as for the Resin I at 2 kv. The rubber shows no objectionable artifacts at either 0.5, 1, or 2 kv sputtering, nor does Resin I at 0.5 kv, although small changes hidden by the data scatter probably do occur. It is of interest that Resin II

Table 2 Relative Variability in the Measured Composition

Material	± % Variability[*]				
	O	N	C	Cl	Si
Silicone Resin II	4	–	2	–	1
Polyurethane	3	3	1	–	–
Saran	–	–	1	1	1

[*]Relative deviation between replicate scans of the same fracture expressed as the 90% confidence limit.

Table 3 Relative Variability of Composition of Replicate Fractures

Materials	% Variability[*]				
	O	N	C	Cl	Si
Silicone Rubber	10	–	3	–	5
Silicone Resin I	1	–	2	–	9
Silicone Resin II	16	–	1	–	9
Polyurethane	10	6	3	–	–
Saran	–	–	4	4	–

[*]Relative deviation between replicate fractures expressed as the 90% confidence limit.

Table 4

ESCA Calibration Factors[a]

Material	O-1s	N-1s	C-1s	Cl-2p	Si-2p
Silicone Rubber (f=2)	0.321 ±0.025	-	1	-	0.931 ±0.018
Silicone Resin I (f=2)	0.353 ±0.015	-	1	-	0.953 ±0.063
Silicone Resin II (f=1)	0.319 ±0.043	-	1	-	0.984 ±0.060
Polyurethane (f=3)	0.360 ±0.026	0.524 ±0.022	1	-	-
Polymethylmethacrylate	0.361	-	1	-	-
Saran (f=2)	-	-	1	0.414 ±0.021	-
Weighted Average	0.344 ±0.012	0.524 ±0.022	1	0.414 ±0.021	0.937 ±0.017
Theory[b]	0.30	0.53	1	0.44	0.79
Literature[c]	0.4	0.6	1	0.46	0.48

[a] ± values are 90% confidence limits (f = degrees of freedom) based upon the variance between replicate fractures and upon the precision and accuracy of the bulk analyses reported in Table 1.

[b] Inverse ratio of theoretical cross-section to that of carbon[8] corrected for $(K.E.)^{-1}$ transmission of analyzer and compensated for approximate $(K.E.)^{1/2}$ escape depth dependence.[9]

[c] Derived from a recent compilation of empirical peak heights ratios using a Mg anode and a retarding field analyzer scan.[10]

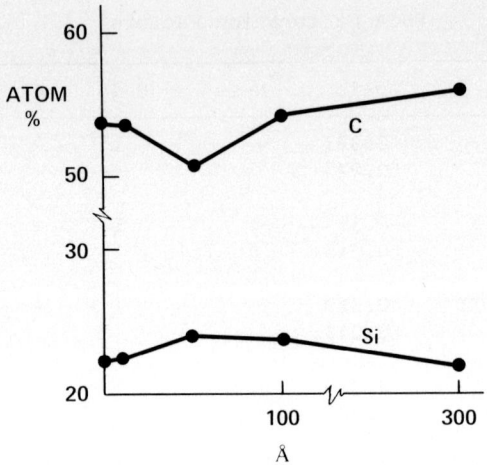

Fig. 4. Silicone Rubber Composition vs. 0.5 kv Sputter Depth.

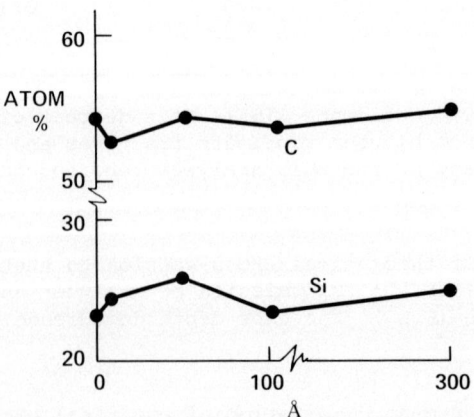

Fig. 5. Silicone Rubber Composition vs. 2 kv Sputter Depth.

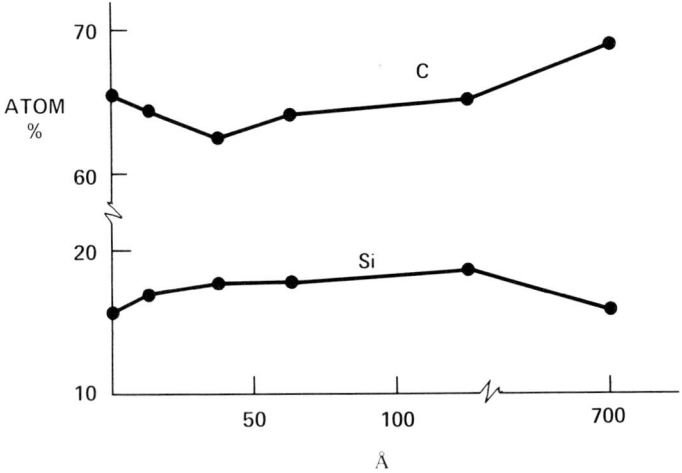

Fig. 6. Silicone Resin I Composition vs. 0.5 kv Sputter Depth.

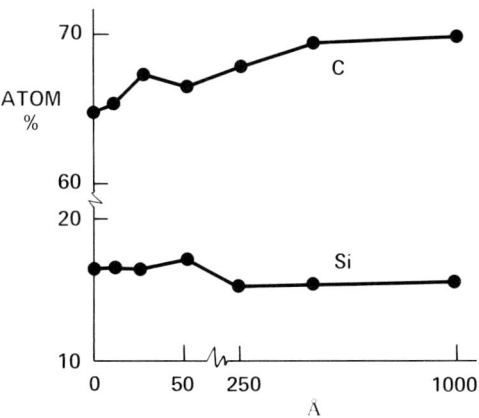

Fig. 7. Silicone Resin I Composition vs. 2 kv Sputter Depth.

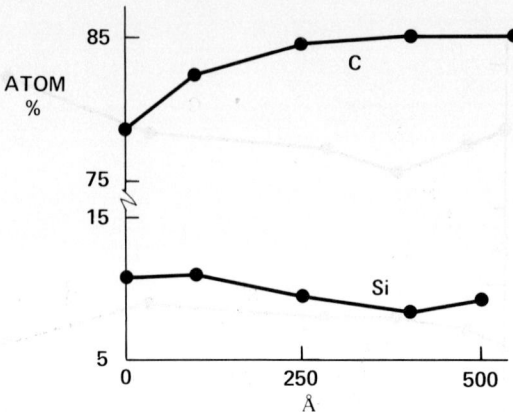

Fig. 8. Silicone Resin II Composition vs. 2 kv Sputter Depth.

displays a C-1s shake-off satellite which is removed by as little as 25 Å of sputtering.

It is evident that ESCA/sputtering is a promising method for the analysis of silicone polymers. However, the variability of the analysis is increased by sputtering as shown in Table 5. The numbers suggest that in favorable cases, such polymers can be analyzed by ESCA/sputtering to within about ±20% relative with very high reliability (90% confidence limits). For example, we have found that the initial surface of cast slabs of some of these polymers display very high compositional gradients but that prolonged sputtering for 1000Å reaches the homogeneous bulk substrate.

Sputtering of Organic Polymers

The organic polymers displayed very large artifacts with the sputtering conditions we used. These artifacts arise both from preferential sputtering and from changes in inelastic scattering which accompany the induction of a compositional gradient. There may also be changes of loss processes in addition to that of inelastic scattering. Eq. (1) is valid only for thick homogeneous samples in which the various loss processes are constant (inelastic scattering, shake-up, shake-off).

As Figures 9-13 illustrate, the atom % values change very rapidly during sputtering but then seem to reach an equilibrium level which depends on the particular element, polymer and sputtering voltage.

Table 5 Compositional Variability During Sputtering of Silicone Polymers

Material	Voltage kv	Max Depth Å	% Variability [a]		
			O	C	Si
Rubber	0.5	300	22	10	13
	1	300	22	17	37
	2	300	33	10	23
Resin I	0.5	700	21	8	25
	1	100	25	5	11
	2	1200	23	9	8

[a] Relative deviation computed from the variance between the initial composition and that of several intermediate sputter depths up to a maximum depth indicated; expressed as the 90% confidence limit.

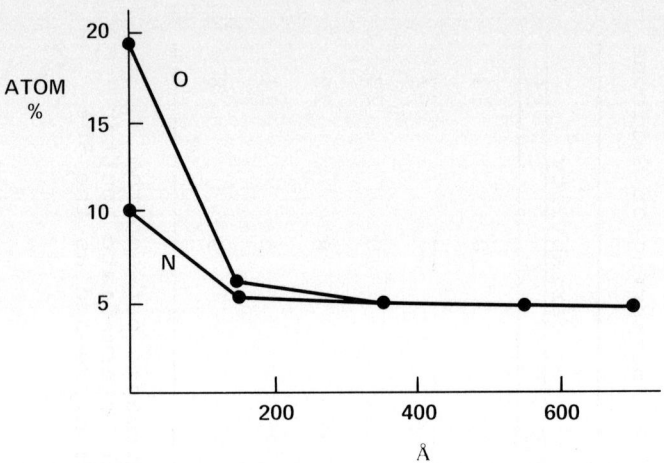

Fig. 9. Polyurethane Composition vs. 0.5 kv Sputter Depth.

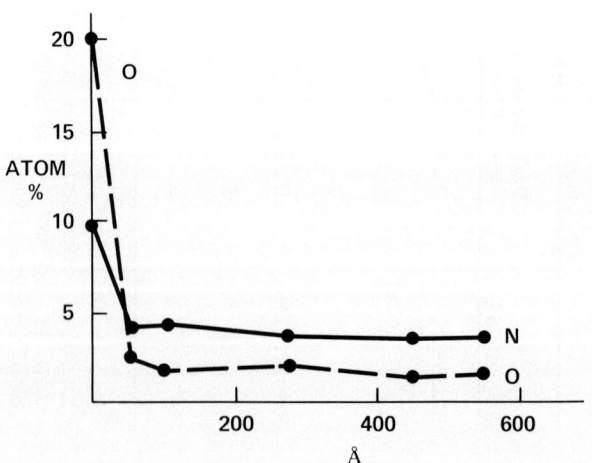

Fig. 10. Polyurethane Composition vs. 2 kv Sputter Depth.

ESCA/Sputtering of Polymers   69

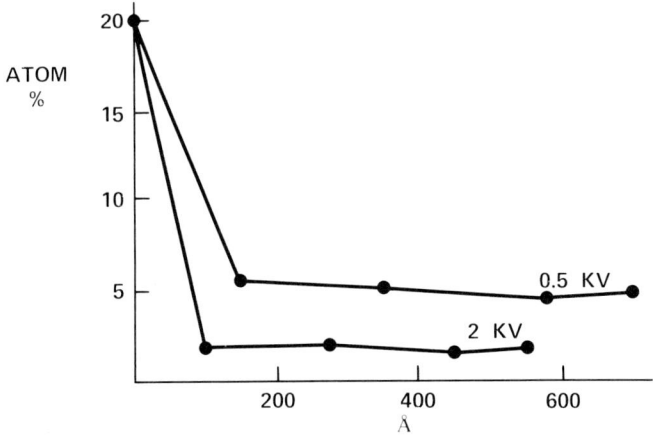

Fig. 11. Oxygen Atom % in Polyurethane vs. Sputter Depth at 0.5 and 2 kv.

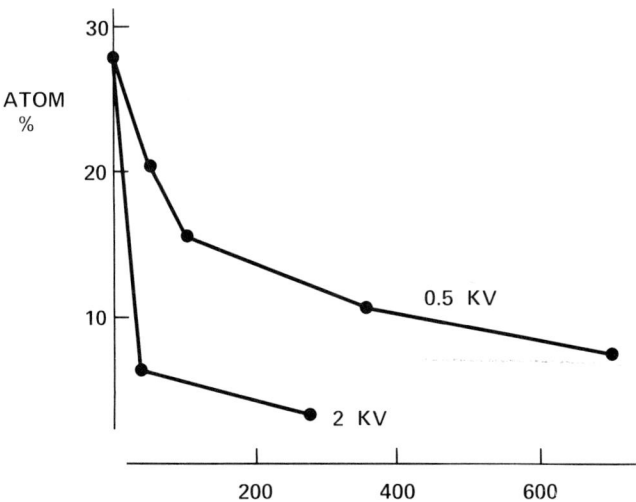

Fig. 12. Oxygen Atom % in Polymethylmethacrylate vs. Sputter Depth at 0.5 and 2 kv.

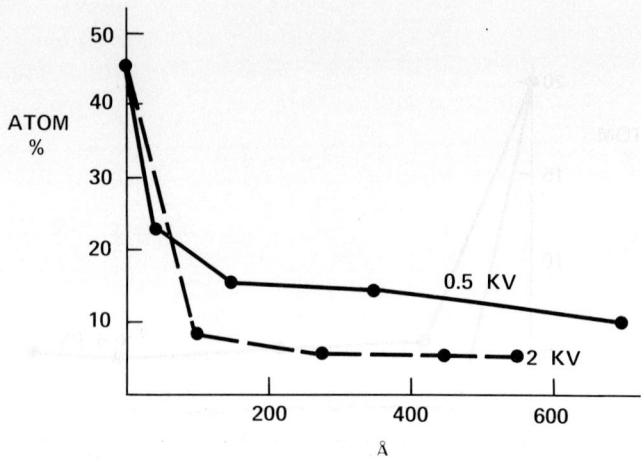

Fig. 13. Chlorine Atom % in Saran vs. Sputter Depth at 0.5 and 2 kv.

This conclusion is independent of the uncertainty in the sputtering rate. Our data suggest that lower sputter voltages tend to reduce artifacts (Figures 11-13). All of these polymers show this effect. Possibly, this is due to a contraction of the zone of sputter damage as the voltage is reduced, as was suggested by Kim et al.[1] for metal oxides. However, the different sensitivities to the sputter voltage of N and O in the same matrix suggest that the situation here is more complex than in metal oxides.

It is clear that the magnitude of these compositional changes cannot be explained by minor initial gradients, and major gradients surely would have been detected by major variability in the initial calibration factors as discussed previously. However, it is comforting that the pure homopolymer-polymethyl methacrylate did not display qualitatively different results.

CONCLUSIONS

ESCA/sputtering has been shown to have considerable promise for the surface analysis of silicone polymers. Such analysis for the organic polymers should still be useful but in a different fashion. Straightforward quantitative analysis of these organic polymers obviously is not feasible when these particular sputtering conditions are used. However, a simple knowledge of the presence or absence of an element as a function of depth may often be useful. Our results suggest that some quantity of these preferentially sputtered elements continues to exist within the ESCA escape depth and hence can be used as a tag to recognize discontinuities under the surface.

We feel it is likely that these artifacts are controlled by kinetic rather than by thermodynamic considerations. Thermodynamically, the silicones should lose volatile products rich in carbon and silicon under the very high local temperatures that exist in the sputter zone. However, depolymerization/volatilization is a much more facile process so an interplay of kinetic factors is at work. Kinetic barriers are also primarily responsible for the much greater thermal stability of silicones relative to organic polymers.

There is certainly an incentive to explore other sputter conditions. The major disadvantage of lowering sputter voltage is a reduction in beam current and sputter rate due to electrostatically induced beam-spreading. Possibily, the use of heavier ions or of a neutral atomic beam would alleviate this problem.

ACKNOWLEDGEMENTS

We would like to thank Drs. Arnold Gatzke, Gary LeGrow, George Vogel, and Mr. Ray Berta and John Uhlmann for supplying the polymer samples.

REFERENCES

1. K.S. Kim, W.F. Baitinger, J.W. Amy and N. Winograd, J. Electron Spectros., $\underline{5}$, 351 (1974).
2. L.I. Yin, S. Ghose, and I. Adler, Applied Spectros., $\underline{26}$, 355 (1972).
3. G.K. Wehner, "Aspects of Sputtering in Surface Analysis Methods" in *Methods and Phenomena*, S.P. Wolsky and A.W. Czanderna, Eds., Elsevier Press, N. Y. (1975).
4. P.W. Palmberg, J. Vac. Sci. and Technol., $\underline{12}$, 379 (1975).
5. L.I. Maissel and R. Glang, *Handbook of Thin Film Technology*, McGraw-Hill, N. Y. (1970), p. 4-40.
6. R.W. Swingle, Anal. Chem., $\underline{47}$, 21 (1975).
7. A. Peterlin, J. Magn. Res., $\underline{19}$, 83 (1975).
8. J.H. Scofield, Lawrence Livermore Laboratory, Report No. UCRL-51326 (Han. 1973).
9. C.J. Powell, Surface Sci., $\underline{44}$, 29 (1974).
10. H. Berthou and C.K. Jorgensen, Anal. Chem., $\underline{47}$, 482 (1975).

Characterization of Chemically Modified Cottons by ESCA

Donald M. Soignet

Southern Regional Research Center
Agricultural Research Service
U. S. Department of Agriculture
New Orleans, Louisiana 70179

    ESCA has been used to study the composition of textiles and the changes produced when these textiles are treated with a variety of reagents capable of altering the surface and/or the bulk properties of the sample. From a measure of the relative intensities of the C_{1s} and O_{1s} signals obtained from blends of cotton and polyester, a determination of the percent composition was made. Conventional flame retardants for cotton fabrics consist of nitrogen and phosphorus resins which form polymers within and on the surface of the fibers. The oxidation state of the phosphorus was monitored as catalyst, reagent, and drying and curing conditions were varied. Additionally, the location of the flame retardant polymer relative to the fiber surface was determined. Some fluorochemical polymers applied to cotton produce both oil repellency and soil release properties. An ESCA examination of these treated fabrics in the dry and water-wet state indicate that some polymer inversion occurs in the wetting process.

INTRODUCTION

    Cotton can be modified by a variety of chemical and physical means to impart desired end-use properties to the textile. To be effective, the chemical modification should take place at a particular location relative to the surface of the fabric. Although electron microscopy in conjunction with x-ray fluorescence can be useful in some cases[1], it cannot be used with many of the chemical finishes employed in the textile industry. ESCA has potential value for studying chemically modified cotton materials because it can probe the surface of material to yield semiquantitative elemental analyses and information about the oxidative and chemical environment of atoms present on that surface.

In this presentation, we will discuss the use of ESCA in determining the oxidative changes which occur when cotton fabric is treated with a variety of flame-retardant finishes. Additionally, we will show that ESCA can be successfully used to locate crosslinking reagents and oil-repellent reagents when they have been applied, separately or simultaneously, to cotton and cotton/polyester blends. The use of ESCA as an analytical tool for determining the ratio of cotton to polyester in blends will also be demonstrated.

EXPERIMENTAL

Textile Finishes

Cotton printcloth was treated with three typical flame retardant systems usually employed with cotton fabric. Two systems, 1) tetrakis(hydroxymethyl)phosphonium hydroxide/amide (THPOH/amide)[2], and 2) tetrakis(hydroxymethyl)phosphonium chloride/urea (Thpc/urea)[3] involve a padding step, a drying step, and a heat curing step. The tetrakis(hydroxymethyl)phosphonium hydroxide/ammonia (THPOH/NH_3)[4] process also involves a padding and a drying step. The curing, however, occurs in a chamber of ammonia gas. In each process, a polymer containing nitrogen and phosphorus is formed and it is the polymer which is responsible for the flame retardant properties of the treated fabrics. The durability of the flame retardancy to laundering can usually be enhanced by subjecting the cured fabric to a post-treatment with dilute H_2O_2 solution.

Soil/oil resistant properties can be imparted to cotton by treatment of the textile with a fluorochemical polymer. FC-218, a commercially available fluorochemical possessing these properties, is available from 3M Co.* This reagent is applied from an emulsion followed by a cure step. Reagents which produce resiliency in cotton can be incorporated into the emulsion, which results in a fabric with both resilient and flame retardant properties[5]. Soil/oil resistant properties have also been imparted to cotton by treatment with fluorochemical and other reagents which when cured form copolymers. An example of this is the use of 1,1-dihydroperfluorooctylamine (POA) in conjunction with tetrakis (hydroxylmethyl)phosphonium chloride (Thpc) as described by Connick and Ellzey[6].

---

*Mention of companies or commercial products does not imply recommendation or endorsement by the U. S. Department of Agriculture over others not mentioned.

ESCA Analysis

ESCA spectra were obtained using a Varian IEE 15 spectrometer equipped with a magnesium anode. All-cotton samples were mounted onto the sample holder by means of "Scotch Magic Transparent Tape". Textile samples were examined in the form of fabric anc fabric ground to a homogeneous state. The latter were prepared by grinding the fabric sample to pass a 20-mesh screen in a Wiley mill, followed by ball-milling. Ball-milling was performed in both an air and CO_2 atmosphere. Particle size of the ball-milled samples was on the order of one micron.

DISCUSSION

Flame Retardant Finishes

The P_{2p} spectra of fabric treated by the THPOH/amide (A), Thpc/urea (B), and THPOH/NH_3 (C) methods are shown in Figure 1. The upper spectra are of samples prior to H_2O_2 treatment and the lower spectra after H_2O_2 treatment. In each case, samples treated with H_2O_2 exhibited a narrow peak with maxima at 133 eV that is characteristic of pentavalent phosphorus.

The upper spectra are broad, and in two instances (B and C) a maximum and a shoulder at lower binding energy can be discerned. The broad spectrum of A also suggests the presence of two contributors to the peak. Each spectrum was separated into two contributors representative of P^{+3} and P^{+5} species. Treatments A and B deposited approximately 10% of the phosphorus on the surface as P^{+3}, whereas treatment C deposited 40% as P^{+3}. These experiments were the first that clearly demonstrated that ESCA could be used to study oxidative changes in elements of flame retardant polymers.

ESCA not only provides elemental analyses and the oxidative states information of elements on the surface of fabrics, but it can be used to investigate the elemental distribution throughout a sample. The intensity of an ESCA spectrum of an element which is uniformly distributed throughout a sample is independent of particle size of the sample under observation. The intensity from an element localized on the fabric sample surface is greater for a fabric sample than for one that is ball-milled. Thus a comparison of signal strengths with decrease in particle size of finished fabric leads to an understanding of reagent location within the fabric.

Cotton flannelette fabric was treated with two conventionally used flame retardants, namely, Thpc/urea and THPOH/NH_3. In each method, a polymer is formed from the reaction of an N-methylol phosphine and a nitrogen compound. In the Thpc/urea process, after a fabric is padded with a solution of the phosphorus-

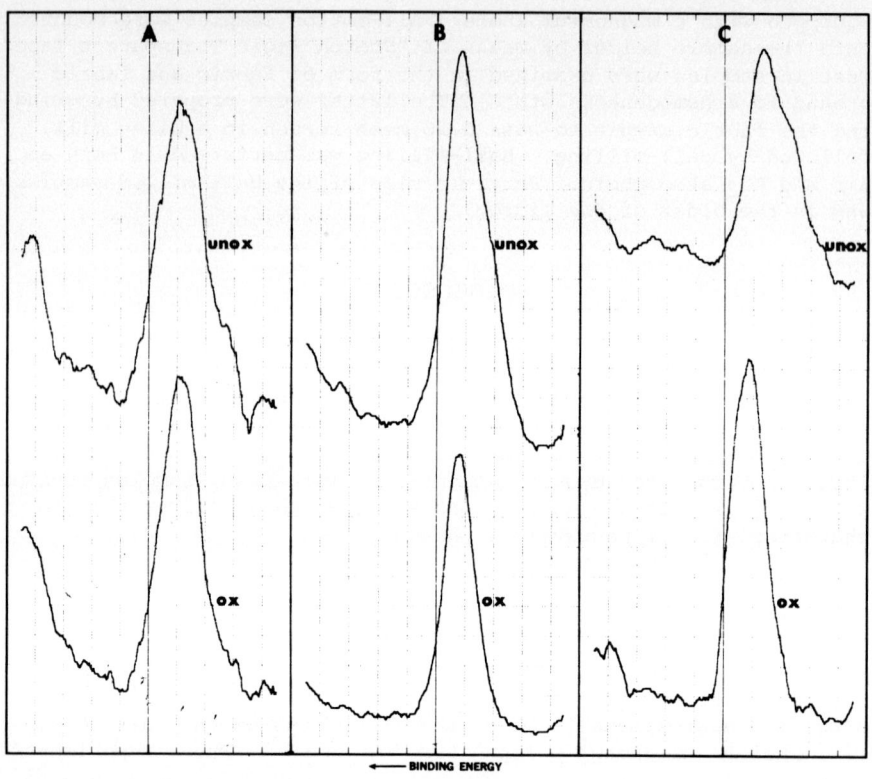

Fig. 1. P_{2p} ESCA spectra of cotton flannelette samples before (UNOX) and after (OX) H_2O_2 treatment. A - THPOH/Amide; B - Thpc/Urea; C - THPOH/NH_3. Initial binding energy of each P_{2p} spectra was 145 eV with a sweep width of 20 eV.

containing and the nitrogen-containing reagents, it is then subjected to heating, during which period the polymer forms. In the THPOH/NH_3 process, the fabric is padded with the solution of THPOH only, then is passed through a chamber containing ammonia gas. The flame-retardant polymer is formed on the fabric while in the ammoniating chamber. When the phosphorus and nitrogen reagents are applied simultaneously from the pad bath, they are distributed uniformly throughout the sample and reaction does not occur until a higher temperature is reached. Unless the reagent migrates during the heating process, the resulting polymer from such a reaction should also be uniformly distributed. In the THPOH/NH_3

system, the phosphorus reagent is evenly distributed by the padding process, but the NH_3 vapor contacts the phosphorus first at the surface. Instantaneous polymer formation on the surface may restrict penetration of the ammonia, and uniform distribution of the polymer may not be achieved. Spectra in Figure 2 clearly show the flame-retardant polymer from the Thpc/urea system is uniformly distributed throughout the fabric, and that from the THPOH/NH_3 is deposited more on the surface than throughout the fabric. These conclusions are based on the observed intensities of the N_{1s} and P_{2p} decrease significantly as particle size of the sample decreased.

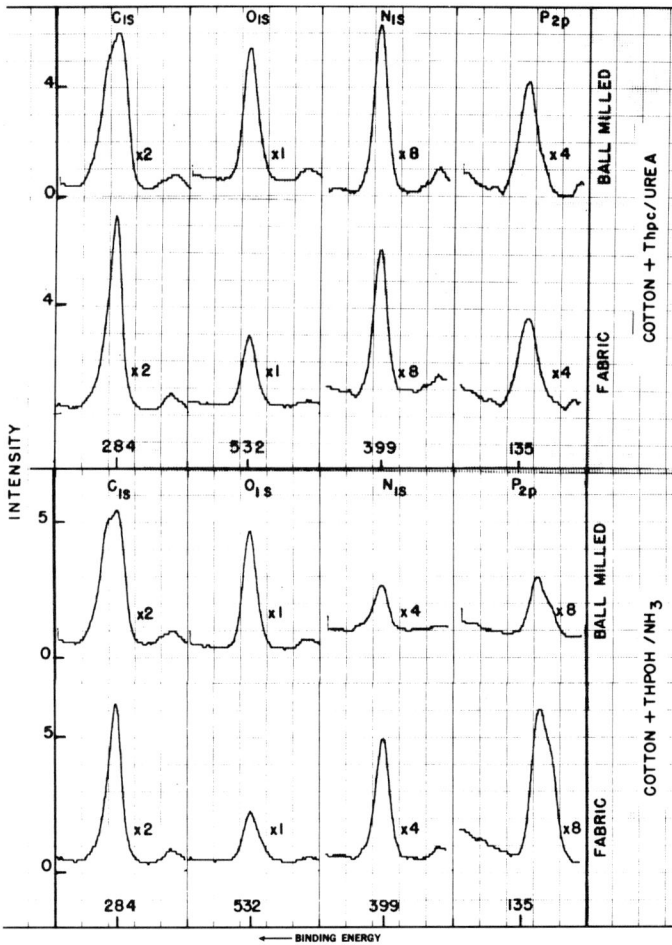

Fig. 2. Comparisons of ESCA spectra of treated cotton flannelette before and after ball-milling. Thpc/urea in upper and THPOH/NH_3 in lower spectra. Sweep width 20 eV.

The intensity of the O_{1s} spectra of the ball-milled samples of each treated fabric is greater than that of the corresponding O_{1s} spectra of the treated fabric. In addition, the C_{1s} spectra of the ball-milled samples contain a component at higher binding energy. This higher binding energy component is not observed in the spectra of the treated fabrics. These observations suggest that either the samples undergo oxidation during the ball-milling or that the surface of the treated fabric differs chemically from the bulk. An examination of the P_{2p} spectra of the THPOH/NH$_3$ samples shows the presences of both a high binding energy component (P^{+5}) and a low binding energy component (P^{+3}) for both the fabric and the ball-milled sample. The ratio of P^{+5}/P^{+3} is slightly higher in the ball-milled sample than in the fabric sample. This does indicate some oxidation but not the extent indicated by the C_{1s} and O_{1s} spectra. To minimize the amount of oxidation that could occur during ball-milling, dry ice was added together with the sample to the milling apparatus. The C_{1s} and O_{1s} spectra from the CO_2-ball-milled sample was the same as that obtained when CO_2 was not present. The ratio of P^{+5}/P^{+3} was the same for the CO_2-ball-milled samples as for the fabric sample. This latter experiment clearly shows that the differences observed in the C_{1s} and O_{1s} spectra of the fabric and ball-milled samples is principally due to differences in composition of the surface and the bulk and not due to oxidation which occurs during the ball-milling process.

The nonuniform distribution of another reagent on cotton is illustrated by ESCA spectra of cotton treated with an oil-repellent/soil release reagent, FC-218. This reagent is a fluoropolymer that also contains nitrogen and sulfur. The reagent is applied to the fabric from an emulsion and the padded fabric is given a heat-cure step. In Figure 3 are the spectra of such a treated sample examined as the fabric (3A) and as the ball-milled sample (3B). The spectrum of the fabrics shown strong N_{1s}, P_{2p}, and F_{1s} signals and a small C_{1s} signal at the high binding energy (292 eV), attributed to carbon atoms attached to fluorine atoms. All signals were reduced dramatically when samples were ball-milled.

When N-methylol ureas are used to impart resiliency to fabrics, they are applied from a pad bath, and crosslinking of cellulose occurs during a heat-curing step. An ESCA study of fabrics crosslinked with a variety of the N-methylol ureas showed that the N_{1s} signal strength did not vary with sample particle size, therefore these reagents were uniformly distributed. When these water-soluble N-methylol ureas are applied in an oil-in-water emulsion containing an oil-repellent/soil release reagent in the oil phases, the cured fabrics possess the properties attributed to both reagents. ESCA spectra of such a treated sample (Figure 4) showed that the N_{1s} signal strength remained constant even though particle size of the sample changed. However, the F_{1s}, P_{2p}, and C_{1s} signals associated with perfluoro groups decreased as the particle size decreased. Such data indicate that each reagent type is distributed

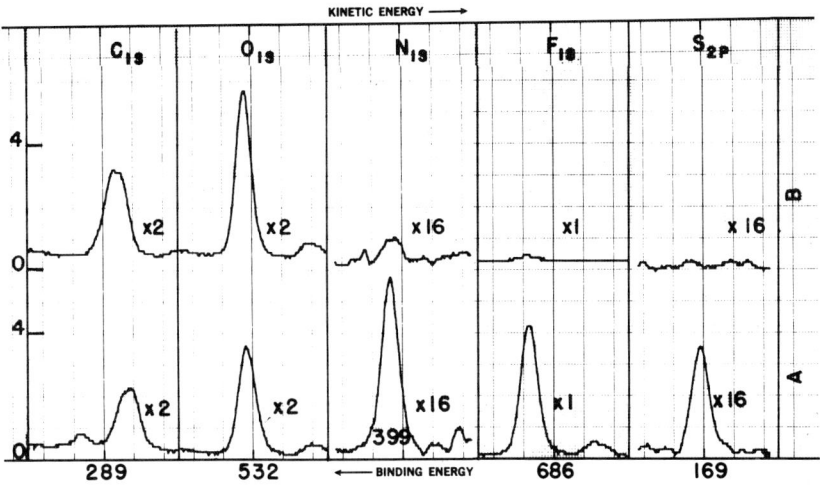

Fig. 3. ESCA spectra of FC-218-treated cotton fabric (A) and ball-milled treated cotton (B). Sweep width 20 eV.

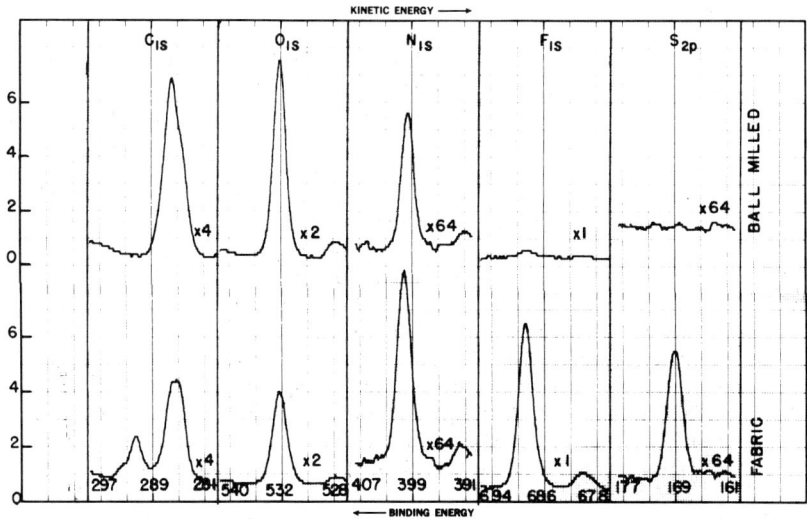

Fig. 4. ESCA spectra of fabric and ball-milled samples of cotton printcloth treated with an emulsion of a perfluorocarbon (FC-218) and dimethyloldihydroxyethylene urea (Permafresh PR-183). Sweep width 20 eV.

from the emulsion as if the other had not been present.

In earlier work with a fluorochemical polymer similar to FC-218, both oil repellency and soil release properties were imparted to cotton fabric. Sherman et al.[5] postulated that the polymer on the surface of the fabric underwent an inversion when wet. In the dry state, the surface was thought to be populated with hydrophobic fluoro groups that fold under when wetted to yield a hydrophilic surface in the wet state. If these two orientations do exist, and if they can be maintained during ESCA examinations, data in support of the inversion theory might be obtained. In an attempt to secure such data, a fabric treated with FC-218 was soaked in water and then immersed in liquid nitrogen. While maintaining the sample at -60°C, a temperature at which the frozen water sublimes in the high vacuum of the spectrometer, the intensities of the F_{1s}, N_{1s}, and S_{2p} signals were recorded. It was assumed, that under these conditions, the polymer configuration on the fabric surface would be similar to that on the surface of a wet sample. The intensities of the F_{1s}, N_{1s}, and S_{2p} signals were also recorded at -60°C for the treated fabric that had not been soaked in water. The F/N and F/S peak intensity ratios were greater for the dry sample than for the wet one. These differences might be anticipated if the hydrophobic segment, which contains the fluoro groups, is on the surface of the dry sample and the sulfur-containing groups, which are part of the hydrophilic reagent, are on the surface of the wet sample.

To satisfy consumer demand for particular physical properties, the textile industry produces fabrics blended of cotton and polyester. The weight percent of the polyester and the chemical nature of the polyester vary among cotton/polyester blends. In cotton, the carbon atoms are attached to at least one other carbon atom and one oxygen atom. The oxygen atoms are either hydroxyl or ether oxygens. In all-polyester, at least one carbon and one oxygen are part of a carbonyl group. The electron density about the carbon or oxygen atoms of the carbonyl group differs from that of the carbon or oxygen atom of cellulose, therefore the core binding energies of each element should differ in cotton and in polyester. If the differences are significant, a comparison of C_{1s} and O_{1s} spectra might be used to determine the composition of a blend. In addition, if the elemental composition of carbon or oxygen in cotton differs from that in the polyester, a comparison of C_{1s} and O_{1s} signal intensities might also define the blend composition.

Blends of cotton (45% C, 49% O, and 6% H) and a polyethylene terphthalate (PET-type) polyester (62% C, 33% O, and 5% H) were examined. Weight of the polyester in the blends ranged from 15% to 65%. Figure 5 shows the C_{1s} and O_{1s} signals of the cotton, of the polyester, and of the blends. All spectra were of ball-milled samples. The C_{1s} spectrum of the polyester is characterized by peaks at 288.4 and 285.0 eV and a shoulder at 286.9 eV. The C_{1s}

spectrum of cotton, however, contains a maximum at 286.6 and a shoulder at 284.9 eV. As the blend composition changes, the spectral characteristics of the C_{1s} signal changes also. Differences in O_{1s} spectra are also observed for the blends. The polyester, which contains two types of oxygen atoms, contains two maxima at 533.4 and 531.0 eV, whereas the cotton samples produced one maximum at 533.4 eV. The width of the O_{1s} peak of the cotton sample is narrow (2.0 eV). This width increases to a maximum of 3.1 eV at 100% polyester. We have found that the C_{1s}/O_{1s} peak area ratios obtained from the ball-milled samples of the blended fabrics vary linearly with the percentage of polyester in the blend. A correlation such as that shown in Figure 6 for the polyester-cotton blend can be used to determine the percentage composition of an unknown blend.

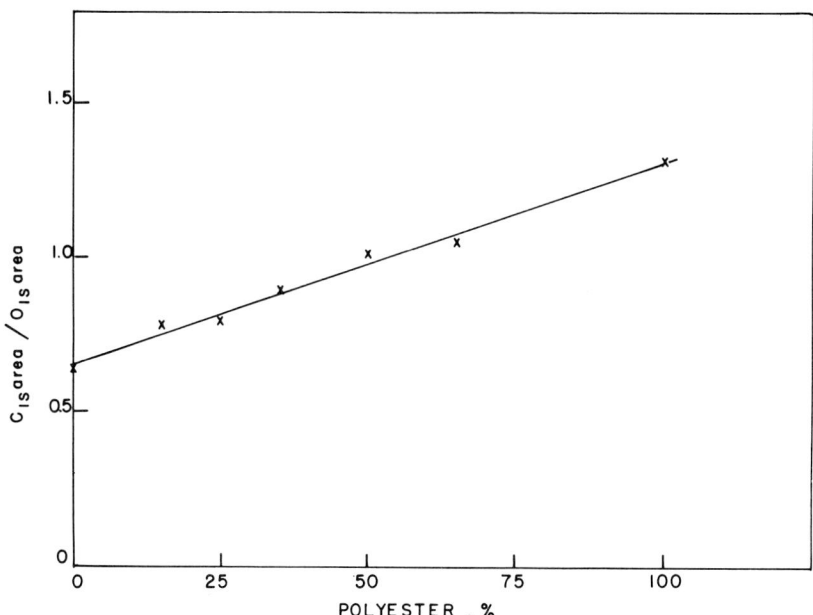

Fig. 5. Variation of C_{1s} and O_{1s} spectra of ball-milled samples of cotton/PET-type polyester blends with percentage composition of polyester. Sweep width 20 eV.

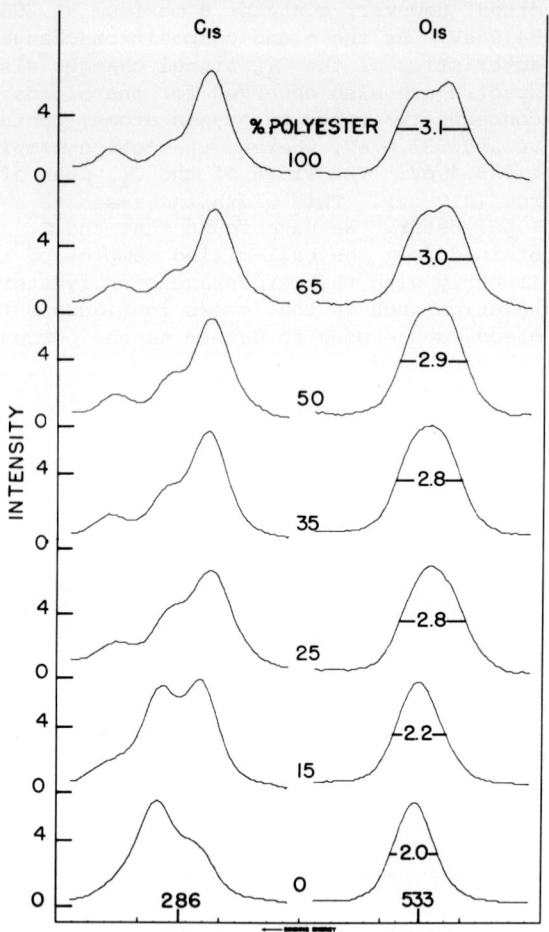

Fig. 6. Variation of C_{1s}/O_{1s} peak areas vs percentage of polyester in a cotton/PET-type polyester blend. ESCA spectra of ball-milled samples

CONCLUSION

These studies clearly demonstrate that ESCA can be used to detect differences in oxidative state of phosphorus in flame retardants and to monitor changes that occur during the application of flame-retardant reagents. Additionally, by measuring the signal strengths of elements associated with a particular finish, the location of the finish relative to the fabric surface can be ascertained. The ESCA technique has also been shown to be a tool for determining composition of cotton/polyester blends.

REFERENCES

1. W. R. Goynes and J. H. Carra, "Application of Energy Dispersive X-Ray Analysis to Textile Fibers", Proceedings, 33rd Meeting, Electron Microscope Society of America, 78 (1975).
2. J. V. Beninate, E. K. Boylston, G. L. Drake, Jr., and W. A. Reeves, "Conventional Pad-Dry-Cure Process for Durable Flame and Wrinkle Resistance with Tetrakis(Hydroxymethyl)Phosphonium Hydroxide (THPOH)", Text. Res. J., $\underline{38}$, 267, (1968).
3. D. J. Donaldson, F. L. Normand, G. L. Drake, Jr., and W. A. Reeves, "A Durable Flame Retardant Finish for Cotton Based on Thpc and Urea", J. Coated Fabr. $\underline{3}$, 250, (1974).
4. J. V. Beninate, E. K. Boylston, G. L. Drake, Jr., and W. A. Reeves, "Application of a New Phosphonium Flame Retardant", Am. Dyest. Rept. $\underline{57}$, 74, (1968).
5. P. O. Sherman, S. Smith and B. Johannessen, "Textile Characteristics Affecting the Release of Soil During Laundering. Part II: Flurochemical Soil-Release Textile Finishes", Text. Res. J., $\underline{39}$, 449 (1969).

Surface Analysis of Plasma Treated Wool Fibers by X-Ray Photoelectron Spectroscopy

Merle M. Millard

Western Regional Research Laboratory
Agricultural Research Service
U.S. Department of Agriculture
Berkeley, California   94710

    Chemical changes on the surface of wool fiber as a result of low temperature plasma treatment were analyzed using X-ray photoelectron spectroscopy. Extensive changes in the electron spectra of carbon, nitrogen, and oxygen were observed. The carbon 1s electron spectra were deconvoluted into three electron lines. Two lines at higher binding energies resulted from surface oxidation. Two nitrogen and two oxygen 1s lines were present before and after plasma treatment. However, the intensity of these two lines changed considerably with treatment. The surface atom concentration before and after treatment was calculated from electron line intensities. The principal result of plasma treatment was found to be extensive surface oxidation of the wool fibers.

INTRODUCTION

    Low temperature plasma treatment has been found to be an effective method for improving various properties of wool yarn[1,2]. Felting shrinkage is lowered to an acceptable level while yarn strength and abrasion resistance are increased.

    Plasma treatment has been used extensively as a technique for modifying the surface properties of materials[3]. Changes in surface properties resulting from plasma treatment include wettability, molecular weight of the surface layer, and chemical composition of the surface.

    An extensive literature exists on chemical treatments to effect shrinkproofing of wool yarn.[4] Most recent studies have focused on chemical treatments that affected the outer surface of the wool fiber. The effectiveness of shrinkproofing chemical treatments correlates with the ability of the reagent to modify the surface as opposed to the interior of the fiber[5-8]. However, the analysis and detection of surface chemical changes on fibers proved to be difficult.[9] For instance, scanning electron micrographs of plasma treated wool fibers have failed to detect significant surface changes.[2]

Previous work in this laboratory determined that X-ray photoelectron spectroscopy is a powerful technique for the surface analysis of textiles.[10-12] In a preliminary note, the sulfur 2p electron spectra of plasma treated wool fiber was reported[11]. The principal change in the sulfur 2p electron spectra as a result of plasma treatment was the appearance of an intense line due to oxidized sulfur.

This paper reports changes in the carbon 1s, nitrogen 1s, and oxygen 1s electron line spectra from wool fibers after plasma treatment. Changes in the surface atom concentration were estimated from electron line intensities. Donnet et al.[13] observed the appearance of carbon 1s lines at higher binding energy when graphite fibers were oxidized. The spectra were not convoluted in this preliminary report.

EXPERIMENTAL

Materials and Reagents

Technical grade oxygen and nitrogen were used without further purification. A 6.3 oz/yd^2 plane weave, undyed, wool fabric was used for plasma treatment.

Plasma Treatment

The apparatus used for plasma treatment of wool fabric has been described previously[14]. The plasma treatment conditions were similar for oxygen, nitrogen and air. Fabric samples were placed in the region of plasma generation between the external electrodes. Each plasma was maintained at a pressure of 1 torr for five minutes with 50 watts of power coupled into the gas. For afterglow treatment, air was used to generate the plasma. A luminous yellow green afterglow extended beyond the area of plasma generation. In this case, wool fabric samples were placed 40 cm downstream from the electrode region.

Surface Analysis

Photoelectron spectra were measured using a Varian IEE 15 Spectrometer with a Mg K_α X-ray source. The application of this technique for surface analysis of textiles has been described in considerable detail[10,12]. Samples of woven fabric were attached to double back tape and fastened to the sample holder, which consisted of a rod that could be lowered into the X-ray target area.

Carbon and nitrogen spectra were obtained using a 10 second sweep over a region 20 eV wide. Data from 35 scans was accumulated and used for deconvolution. Oxygen spectra were obtained using a 10 second sweep over a 20 eV wide region, accumulating 30 scans of data.

The data was fitted with Lorentzian or Gaussian functions by a non-linear least squares fitting program. This program and its application for the deconvolution of electron line spectra has been described in detail[15]. The program works by varying the parameters describing each line shape, such as the position of the line, the area under the line, and the width of the line until a "best" fit is obtained by the criterion of a minimum in the value of $(E_i-L_i)^2$, where E_i is the ith experimental point and L_i the ith Lorentzian or Gaussian point. Inelastic scattering will lead to a tail on the high binding energy side of the spectrum. The tail may be constant in magnitude or behave as an exponential function. In addition to the option of a leading constant or exponential tail, a parameter giving the tail height to peak height ratio is introduced. To accomodate the background under the lines, a linear background may be added to the above line shape functions. The background is specified by its slope and intercept. The fundamental line shape cannot be predicted and the usual approach is to try combinations of line shapes, tail functions and backgrounds until a reasonable approximation to the given experimental curve is obtained.

Full width at half maximum (FWHM) of lines from solid samples are dependent on the resolution capability of the instrument. Using the Varian IEE instrument, line widths obtained on powder samples of pure amino acids were about 2.5 eV. Satisfactory fits of the experimental electron spectra were obtained by fixing the line widths and comparing the results allowing the line widths to converge in the program. Using this trial and error procedure, the best fits were obtained using lines with FWHM values of 2.6 eV for carbon and oxygen and 2.4 eV for nitrogen. The combination of line parameters was selected that gave the best fits for a series of spectra. The untreated wool fiber control spectrum was the most difficult to fit and the electron line parameters that gave a reasonable fit for this spectrum usually gave a reasonable fit for the spectra of treated samples. Carbon electron line spectra were fitted with Lorentzian functions with a leading constant tail. Oxygen and nitrogen electron lines were fitted with Gaussian functions with an exponential tail. Binding energies (BE) were referenced to the carbon 1s electron line at lowest energy, assumed to occur at 285.0 eV. Sample charging shifts never amounted to more than a few tenths of an eV.

RESULTS AND DISCUSSION

Carbon 1s Electron Spectra

The carbon 1s electron line spectra obtained from oxygen, nitrogen, and the afterglow plasma treatment are shown in Figure 1. The parameters for the carbon electron lines are given in Table 1.

Table 1

Carbon 1s Electron Line Parameters.
Spectra From Plasma Treated Wool Fiber.

Sample	Binding energy (eV)	Relative line area	Shift in (eV)
Untreated wool fiber	285.0	1.0	0.0
	286.07	0.092	1.07
	287.77	0.043	2.77
Nitrogen plasma treated wool fiber	285.0	1.0	0.0
	286.43	0.28	1.43
	288.46	0.31	3.46
Afterglow plasma treated wool fiber	285.0	1.0	0.0
	286.56	0.24	1.56
	288.42	0.33	3.4
Oxygen plasma treated wool fiber	285.0	1.0	0.0
	286.55	0.23	1.55
	289.04	0.45	4.04

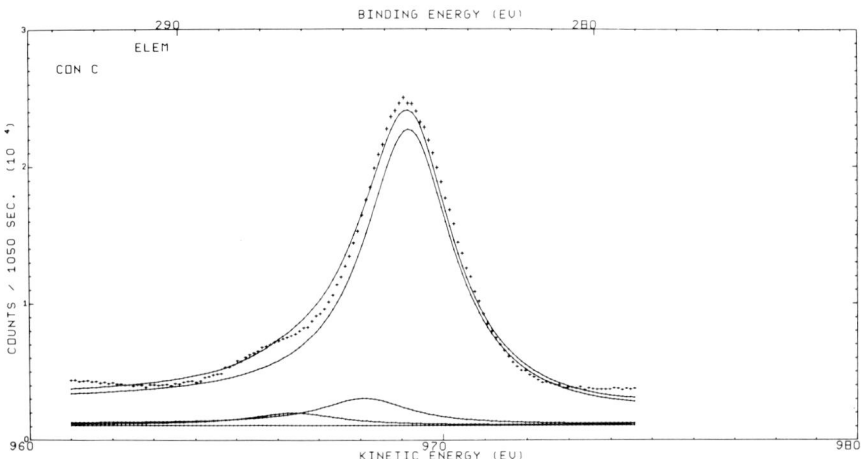

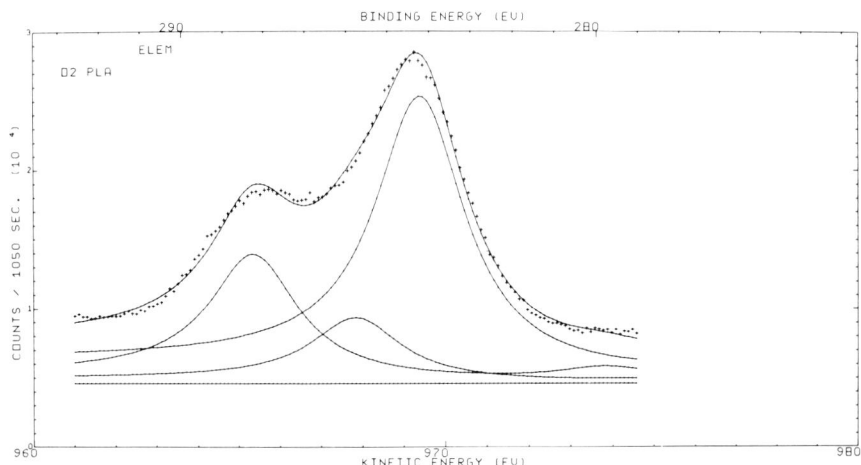

Fig. 1. Carbon 1s electron line spectra. A. Untreated wool fiber. B. Oxygen plasma treated wool fiber.

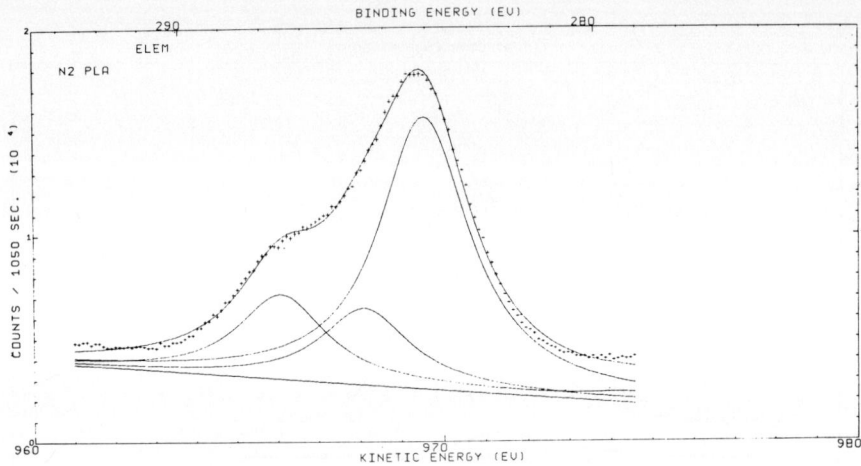
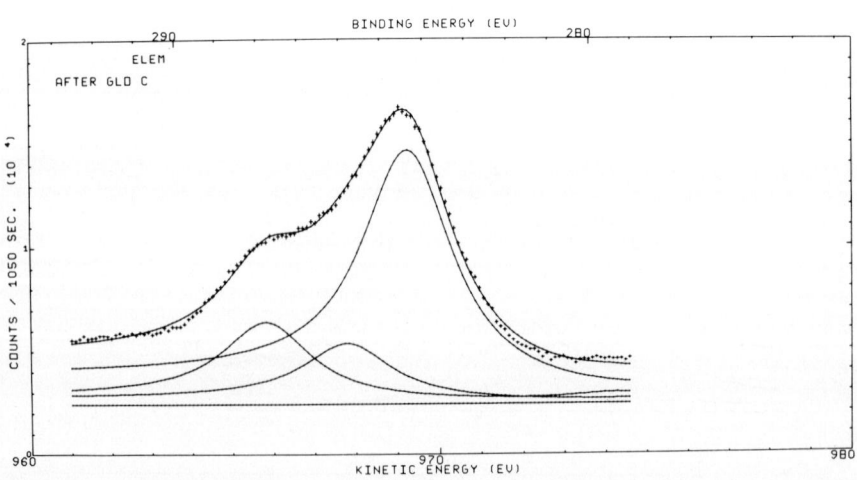

Fig. 1. C. Nitrogen plasma treated wool fiber. D. Wool fiber exposed to afterglow region from air plasma.

The carbon 1s electron line spectra were fitted with three lines. The position of the two high binding energy lines shifted several tenths of an eV going from the control spectra to the spectra from the oxygen plasma treated sample. The intermediate electron line shift of 1.07 eV from the 285 eV line in the control and 1.5 eV in the plasma treated samples is in the region reported by others for carbon containing a single bond to oxygen.[16] The high binding energy line shifts from 2.77 eV in the control to 4.04 eV in the spectra from the oxygen plasma treated sample. This electron line is in the region reported for carbonyl carbon[13,16]. Individual functional groups falling into each of the above broader structural classifications such as alcohols and ethers or carboxylic acids and ketones cannot be differentiated. As the surface population of carbon atoms with higher binding energies increases, the composite line at higher binding energy shifts to higher values. The obvious feature to note concerning these spectra is the increase in intensity of electron lines at higher binding energies. These lines represent oxidized carbon species and as expected, the line in the carbonyl carbon region was most intense for the oxygen plasma treated samples.

Oxygen 1s Electron Spectra

Oxygen 1s electron line spectra for untreated wool fiber and plasma treated wool fiber are shown in Figure 2. Line positions and intensities are given in Table 2. The oxygen electron spectra could be resolved into two lines. The relative intensity of the higher oxygen electron line decreased by a factor of about 3 after plasma treatment. The assignment of these two oxygen lines is difficult. Very little data in the literature exists on the binding energies of structurally different oxygen atoms bound to carbon. Lindberg et al.[17] reported oxygen 1s lines for a number of carbon compounds containing sulfur. The majority of these compounds contain oxygen bound to sulfur, carbon, and nitrogen. The oxygen 1s line in compounds containing oxygen bound to sulfur and in carbonyl and carboxylate groups is around 532 eV. The high binding energy line around 532.2 eV is probably due to lines arising from oxygen in carbonyl, carboxylate, sulfone, sulfoxide, and sulfonic acid. This interpretation is supported by the presence of electron lines due to oxidized sulfur. The sulfur 2p electron spectra from plasma treated samples contains a strong line at 168 eV due to sulfonic acid or sulfate indicating the presence of oxygen on sulfur. The sulfur 2p electron line spectra from wool fiber exposed to the afterglow region of an air plasma is shown in Figure 3.

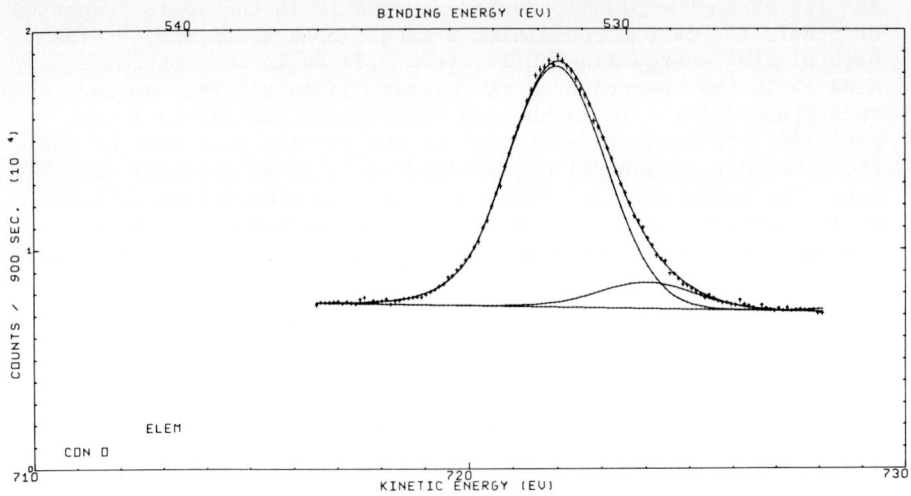

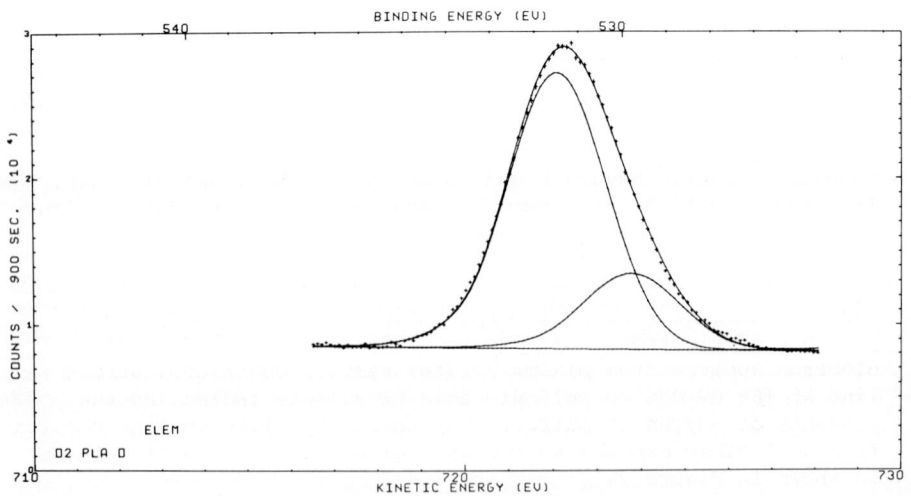

Fig. 2. Oxygen 1s electron line spectra. A. Untreated wool fiber. B. Oxygen plasma treated wool fiber.

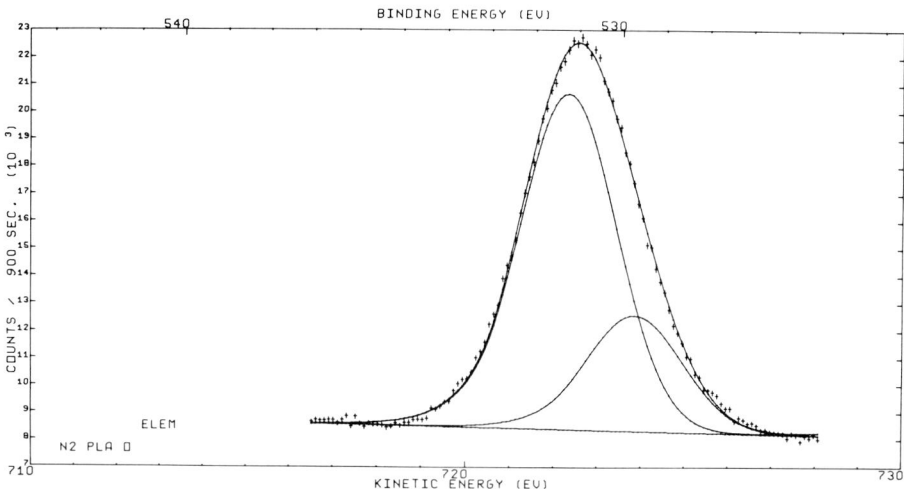

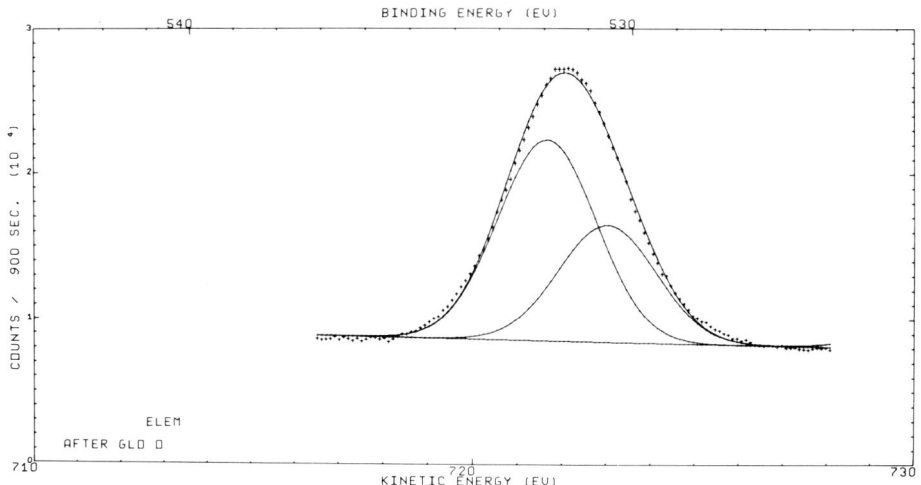

Fig. 2. C. Nitrogen plasma treated wool fiber. D. Wool fiber exposed to afterglow region from air plasma.

Table 2

Oxygen 1s Electron Line Parameters.
Spectra From Plasma Treated Wool Fiber.

Sample	Binding energy (eV)	Relative line area	Shift in (eV)
Untreated wool fiber	530.04	1.0	0.0
	532.06	9.19	2.02
Nitrogen plasma treated wool fiber	530.73	1.0	0.0
	532.25	2.89	1.52
Afterglow plasma treated wool fiber	530.87	1.0	0.0
	532.41	3.1	1.53
Oxygen plasma treated wool fiber	530.61	1.0	0.0
	532.29	3.62	1.68

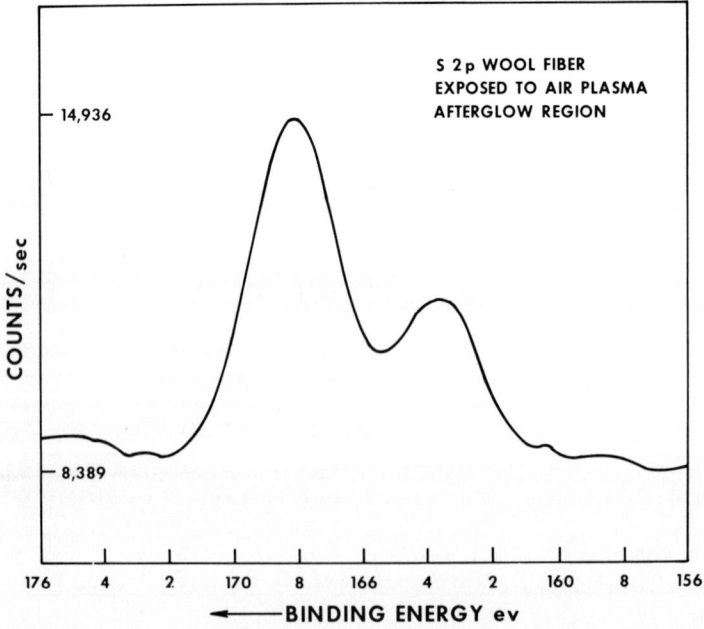

Fig. 3. Sulfur 2p electron line spectra from wool fiber exposed to the afterglow region of an air plasma.

The lower binding energy line around 530.5 eV in Figure 2 is the region reported for metallic and refractory oxides. The oxygen 1s electron line has been reported at 530.5, 530.7 eV in metal aluminates[18], 530.7 eV for UO_2[19], and 530.9 eV for chemisorbed PbO[20]. Grunthaner[21] observed oxygen 1s electron lines at 532.2 and 530.9 eV from oxidized iron dithiolene complexes. The latter line was attributed to oxide oxygen and the former to sulfoxide oxygen. Barber et al. reported oxygen 1s spectra from a graphite that had been exposed to atomic oxygen at room temperature and 673°K. Two oxygen 1s lines were present. The higher binding energy line was attributed to carbonyl oxygen and the lower binding energy line to oxygen with a negative charge[22]. The low binding energy 1s line increases in intensity by a factor of 3 upon plasma treatment. Apparently, a surface oxide species is formed on the surface of the wool fiber as a result of plasma treatment.

Nitrogen 1s Electron Spectra

The nitrogen 1s electron line spectra are shown in Figure 4 and the parameters obtained from the deconvoluted spectra are given in Table 3. Two lines are present in these spectra, and the ratio of the two lines changes considerably upon plasma treatment. The high binding energy line decreases by a factor of three in intensity upon plasma treatment. This behavior is very similar to that observed for the oxygen 1s line spectra.

Table 3

Nitrogen 1s Electron Line Parameters. Spectra From Plasma Treated Wool Fiber.

Sample	Binding energy (eV)	Relative line area	Shift in (eV)
Untreated wool fiber	398.03	1.0	0.0
	399.94	8.9	1.9
Nitrogen plasma treated wool fiber	398.6	1.0	0.0
	400.3	3	1.67
Oxygen plasma treated wool fiber	398.95	1.0	0.0
	400.54	2.74	1.59

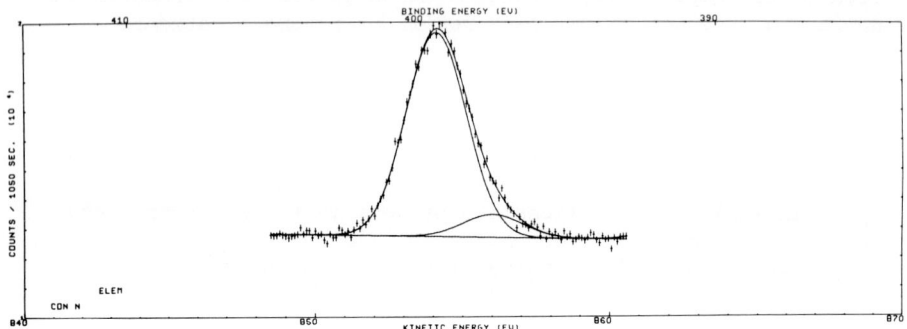

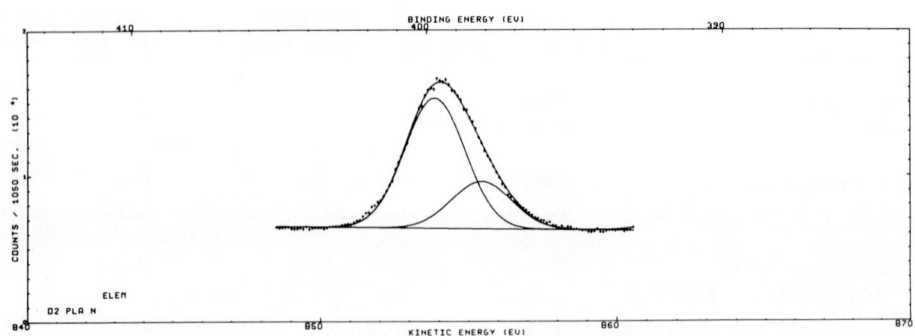

Fig. 4. Nitrogen 1s electron line spectra. A. Untreated wool fiber. B. Oxygen plasma treated wool fiber.

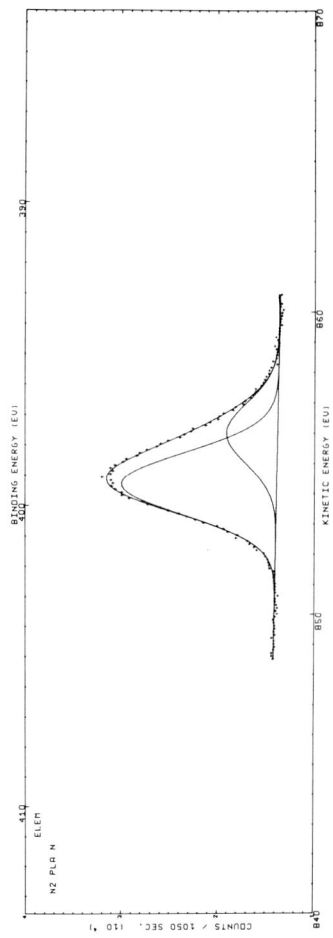

Fig. 4. Nitrogen 1s electron line spectra. C. Nitrogen plasma treated wool fiber.

A correlation of these lines with some structural feature on the surface of the wool fiber is somewhat difficult. The binding energy shifts for nitrogen are small and a considerable number of structurally different nitrogen atoms could be present on the side chains of the amino acids on the surface of the fiber. In addition to the nitrogen atoms on the side chains, nitrogen in the polypeptide chain could also give rise to a nitrogen electron line. Few reports exist in the literature concerning the nature of the nitrogen 1s electron lines observed from proteins. Klein and Kramer[23] reported the nitrogen 1s spectra of the endoplasm from light red kidney bean and L-isoleucyl-L-alanine together with cystine. The nitrogen line for the dipeptide could be deconvoluted into two lines. A higher binding energy line coinciding in position with the line from cystine was assigned to the amino nitrogen. The line at lower binding energy was assigned to amide nitrogen formed during the peptide bonding. The nitrogen line from the kidney bean endoplasm was deconvoluted into two lines. A low intensity line at lower binding energy was attributed to amide nitrogen and the intense line at higher binding energy was attributed to amino nitrogen. Brolin et al.[24] measured the nitrogen 1s electron line spectra of insulin and observed a single line.

One possible interpretation of the nitrogen 1s line spectra observed from wool fiber would be to attribute the lower line to the amide nitrogen of the polypeptide and assign the higher line to the other nitrogen atoms in the side atoms. Upon plasma treatment some of the side chain nitrogen containing groups could be broken away by ion etching or oxidation leaving more exposed polypeptide nitrogen. This would explain the decrease in intensity of the high binding energy line upon plasma treatment.

Nordberg et al.[25] have reported nitrogen 1s electron shifts for a large number of compounds. The nitrogen 1s line for organic amines is in the region of 398.0 eV. Protonation of the amine shifts the nitrogen 1s line to 400.4 eV. The low line observed from the wool fiber could be due to amine type nitrogen and the higher line due to protonated amine. Plasma treatment could reduce the number of protonated amines thus leading to a decrease in the intensity at the higher binding line. These interpretations are tentative and much more work will have to be done in order to understand the nature of the nitrogen electron spectra from proteins.

Surface Composition from Electron Line Intensities

The electron line intensities can be related to atomic composition using Wagner's sensitivity factors[26]. The carbon, oxygen, and nitrogen atomic ratio on the surface can be calculated from the relative elemental line intensities and this data is given in Table 4. The elemental composition of wool fiber obtained by chemical analysis is $C_1O_{.33}N_{.28}$. This ratio reflects the composition of the bulk. The elemental ratio for the control calculated from electron line intensities is $C_1O_{.33}N_{.12}$. This ratio for oxygen plasma treated fiber is $C_1O_{.96}N_{.53}$ and for nitrogen plasma treated fiber the ratio is $C_1O_{.55}N_{.36}$. The elemental ratio calculated from XPS surface analysis is in fair agreement with that calculated from bulk analysis. The nitrogen content is low. The surface elemental ratio changes appreciably upon plasma treatment. Approximately three times as much oxygen and four times as much nitrogen are on the surface after oxygen plasma treatment. The increase in the ratio of oxygen to carbon correlates with the increase in the carbon electron lines at higher binding energy.

Table 4

Elemental Electron Line Intensities

Sample	Maximum line intensity (counts/sec)	Total line intensity	Corrected line intensity
Control			
Carbon	48,600		180,000[a]
Oxygen	31,100		59,700[b]
Nitrogen	8,780		21,000[c]
Oxygen-plasma			
Carbon	23,300	39,100	145,000[a]
Oxygen	56,700	72,300	139,000[b]
Nitrogen	24,100	32,400	77,000[c]
Nitrogen plasma			
Carbon	31,500	50,000	185,000[a]
Oxygen	39,700	52,500	100,000[b]
Nitrogen	21,400	28,200	67,000[c]
Afterglow			
Carbon	28,200	44,300	164,000[a]
Oxygen	51,250	68,000	131,000[b]
Nitrogen	14,250		

[a] Intensity divided by 0.27.
[b] Intensity divided by 0.52.
[c] Intensity divided by 0.42.

REFERENCES

1. A.E. Pavlath and R.F. Slater, Appl. Polymer Symposia, 18(II), 1317 (1971).
2. A.E. Pavlath, in "*Techniques and Applications of Plasma Chemistry*", Chapter 4, A.T. Bell, J.R. Hollahan, Editors, John Wiley and Sons, Inc., New York, 1974.
3. M. Hudis, in "*Techniques and Applications of Plasma Chemistry*", Chapter 3, A.T. Bell, J.R. Hollahan, Editors, John Wiley and Sons, Inc., New York, 1974.
4. J.D. Leeder and J.H. Bradbury, Text. Res. J., $\underline{215}$ (1971).
5. M.W. Andrews, A.S. Inglis, F.E. Rothery, and V.A. Williams, Text. Res. J., $\underline{33}$, 705 (1963).
6. J.H. Bradbury, G.E. Rogers, and B.K. Filshie, Text. Res. J., $\underline{33}$, 617 (1963).
7. J.H. Bradbury, Text. Res. J., $\underline{31}$, 735 (1961).
8. J.H. Bradbury, J. Text. Inst., $\underline{51}$, 1226 (1960).
9. M.W. Andrews, A.S. Inglis, and V.A. Williams, Text. Res. J., $\underline{36}$, 407 (1966).
10. M.M. Millard, K.S. Lee, and A.E. Pavlath, Text. Res. J., $\underline{42}$, 460 (1972).
11. M.M. Millard, Anal. Chem., $\underline{44}$, 828 (1972).
12. M.M. Millard and A.E. Pavlath, Text. Res. J., $\underline{82}$, 460 (1972).
13. J.B. Donnet, H. Dauksch, J. Escard, and C. Winter, C. R. Acad. Sci., $\underline{275(C)}$, 1219 (1972).
14. M.M. Millard, J.J. Windle, and A.E. Pavlath, J. Appl. Polym. Sci., $\underline{17}$, 2501 (1973).
15. C.S. Fadley, Ph.D. Thesis, University of California, Berkeley, California, 1970 LRL Report, UCRL 19535.
16. U. Gelius, P.F. Heden, J. Hedman, B.J. Lindberg, R. Maune, R. Nordberg, C. Nordling, and K. Siegbahn, Physica Scripta, $\underline{2}$, 70-80 (1970).
17. B.J. Lindberg, K. Hamrin, G. Johansson, U. Gelius, A. Fahlman, C. Nordling, and K. Siegbahn, Phys. Sci., $\underline{1}$, 286 (1970).
18. J.L. Ogilvie and A. Wolberg, Appl. Spectrosc., $\underline{26}$, 401 (1972).
19. G.C. Allen and P.M. Tucker, J. C. S. Dalton, $\underline{5}$, 470 (1973).
20. K.S. Kim and N. Winograd, Chem. Phys. Letters, $\underline{19}$, 209 (1973).
21. F.J. Grunthaner, Ph.D. Thesis, California Institute of Technology, Pasadena, California, 1974.
22. M. Barber, E.L. Evans, and J.M. Thomas, Chem. Phys. Letters, $\underline{18}$, 423 (1973).
23. M.P. Klein and L.N. Kramer, in "*Symposium: Seed Proteins*", The Avi Publishing Co., Westport, Conn., p. 265, 276, 1976.
24. S.E. Brolin, B.J. Lindberg, A. Fahlman, K. Hamrin, G. Johnsson, C. Nordling, and K. Siegbahn, International Diabetes Federation, Proceedings of the 6th Congress, Stockholm, p. 410 (1967, 1969).
25. R. Nordberg, R.G. Albridge, T. Bergmark, U. Erickson, J. Hedman, C. Nordling, K. Seigbahn, and B.J. Lindberg, Ark. Kemi, $\underline{28}$, 257 (1968).
26. C.D. Wagner, Anal. Chem., $\underline{44}$, 1050 (1972).

Plasma Modification of Polymers Studied by Means of ESCA

D. T. Clark and A. Dilks

Department of Chemistry
University of Durham
South Road, Durham City, England

The surface modification and synthesis of polymers by glow discharge techniques forms an area of intense research activity both in industrial and university laboratories. The complexities of the processes involved are gradually being unravelled and in this ESCA as a spectroscopic tool has a key role to play. Our interests have centered around the interaction of polymer surfaces with plasmas excited (inductively coupled RF) in a variety of gases and monitoring changes in surface and subsurface compositions by means of ESCA. This has involved characterizing the plasmas' and investigating the relative importance of photoemitted and direct energy transfer processes at surfaces. The work carried out thus far demonstrates the great potential for electron spectroscopy as a whole in this important area.

INTRODUCTION

The plasma treatment of polymers and solids in general has been the subject of considerable research interest over the past decade[1]. The distinguishing feature of surface modification effected by means of plasmas is that the process may be selectively controlled such that bulk properties are unaffected. An important area of application is in the glow discharge treatment of polymer surfaces to improve adhesive bonding. The surface nature of such modifications make it difficult to employ conventional techniques for understanding the processes which occur during surface modification by means of plasmas excited in a variety of gases. Symptomatic of the difficulties of employing conventional techniques which do not have specific surface sensitivity is the unresolved question of the relative roles of direct and radiative energy transfer in effecting surface modifications[1]. Although no hard scientific

evidence has been produced, the general consensus seems to be that
reaction at the very surface of a sample may be associated with
either or both of direct energy transfer from species in the plasma
and the U.V. component of electromagnetic radiation emitted from
the plasma. For the bulk, however, the evidence would appear to
be in favor of a mechanism dominated by radiative energy transfer.

In an extensive series of publications[2], we have shown how
ESCA as a spectroscopic tool may be used to investigate various
aspects of the structure, bonding and reactivity of polymeric
systems. A particular feature arising from the strong dependence
on kinetic energy of the mean free path for photoemitted electrons
is the possibility of employing the technique for analytical depth
profiling in which surface, subsurface and bulk may be differentiated. ESCA, therefore, provides a very powerful tool for studying surface modifications of polymers, and indeed in previous papers
we have detailed investigations of the surface fluorination of
polyethylene and the surface modification of ethylene-tetrafluoroethylene copolymers initiated by Argon ion bombardment in a simu-
lation of the CASING procedure[3]. These preliminary investigations
of the Argon ion treatment showed that the reactions were essentially
confined to the surface region. Since the light output from the
DC discharge employed was essentially collimated, and hence irradiation confined to a small fraction of the total surface area the
results obtained represent prima facie evidence that ions and
metastables in an inert gas plasma can initiate surface
modifications.

In this paper, we describe a detailed investigation of the
interaction of polymer films with RF plasmas excited in inert
gases. For a variety of reasons, the system subjected to the most
detailed scrutiny involves an ethylene/tetrafluoroethylene copolymer
system and inductively coupled RF plasmas excited in Argon. The
rationale for studying the particular copolymer system (52% TFE)
may be summarized as follows. Firstly, the polymer has been the
subject of an intensive ESCA investigation previously, and the
structure is known to be largely alternating[4]. The shift in binding energy for the C_{1s} levels is sufficiently large that the signals arising from photoemission from the tetrafluoroethylene components are well resolved with respect to the ethylene components
(shift $\sim$ 4.7 eV). It is, therefore, relatively easy to monitor
changes in structure arising from plasma treatment by monitoring
the components of the C_{1s} levels and also the F_{1s} levels. The F_{1s}
and F_{2s} levels span a substantial range in kinetic energy for the
photoemitted electrons, and the monitoring of these levels therefore provides a convenient means of establishing the homogeneity
or otherwise of the surface regions of the sample[5]. The dominant
features in the polymer structure of alternation provides a convenient mechanism for crosslinking arising from the effective elimination of HF, and the system therefore provides a simple prototype
for more complicated systems for which the number of information

levels available from the ESCA experiment is considerably reduced.

The choice of an electrodeless inductively coupled RF plasma excited in inert gases for the study of surface modification of polymers enables close consideration of all of the variables which are likely to be of importance. The nature of the experiment allows, for example, a wide range of pressures and flow rates to be investigated and provides a convenient means since the power loading can be continuously varied of performing kinetic studies as a function of power. The facility for pulsing the glow discharge extends the power range which may be spanned by providing a stable discharge at relatively low power loadings. The effective power range may, therefore, be varied over three orders of magnitude. In addition, an inductively coupled RF plasma allows considerable flexibility in terms of reactor design and configurations for introducing and removing samples.

In this work, we have addressed ourselves to the following primary points:

  (i) Since plasmas are a copious source of electromagnetic radiation extending from the vacuum ultraviolet to the visible and since for the former cross sections can be very large even for saturated systems, it is clear that the surface reaction could also contain a significant contribution arising from radiative energy transfer from the plasma[7]. A technique such as ESCA which allows one to differentiate the surface from the subsurface and bulk should, in principle, be capable of shedding light on the relative importance of direct and radiative energy transfer as far as the surface is concerned.

  (ii) For a given polymer system, how does the surface reaction vary with the parameters involved: power, pressure, flow rate, inert gas, reactor configuration and for a given power loading the effect of continuous or pulsed mode of operation.

Although the interaction of RF plasmas with solids in general has been an active area of research in both industrial and academic laboratories, there have been few attempts to characterize the plasmas involved in terms of the energy distributions of electrons, ions and metastables. Indeed, such information is only semiquantitatively available for very simple systems although the broad theoretical framework is reasonably well understood[6].

The salient features with regard to RF plasmas excited at relatively low powers < 100 watts are as follows. The energy distributions for the neutral species and for positive ions corresponds approximately to the ambient temperature, whilst the electron temperature is thought to be considerably higher with a Maxwellian distribution which in the pressure range 1 - 100 torr typically peaks in the range 0 - 10 eV. The average electron energy is a function of both the power loading and the pressure. The mean free paths for electrons in the energy range 0 - 10 eV in polymer samples is almost certainly > 25Å in which case very little of the surface and subsurface reaction can arise from energy transfer involving the electrons. For plasmas excited in Argon, the important neutral species capable of undergoing energy transfer to a surface are almost certainly the relatively long-lived metastable $^3P_{2,0}$ states (energies 11.55 and 11.72 eV), for which the dominant energy transfer process may well be via Penning ionization processes[8]. The first ionization potentials of Argon $^2P_{3/2}$ 15.759 eV, $^2P_{\frac{1}{2}}$ 15.937 eV is considerably higher than that for a typical polymer sample so electron transfer at the surface is likely to be of some importance. Indeed, the interaction of Argon ions with surfaces form the basis for a surface sensitive spectroscopic technique developed by Hagstrum (Ion Neutralization Spectroscopy)[9]. The mean free paths for both Argon ions and metastables would, therefore, be expected to be such that surface energy transfer processes should dominate with these species.

The electromagnetic radiation associated with RF plasmas excited in Argon is predominantly in the vacuum ultraviolet with the ArI (∼ 1048 Å, 1067 Å) and ArII (∼ 920 Å, 932 Å) resonance lines superimposed on a continuum extending to longer wavelength (most intense part of continuum in the range 1000 - 1400 Å)[7]. For photon energies in the range spanned by the continuum and the discrete line spectra (8.85 eV - 13.47 eV), the total attenuation cross sections are undoubtedly dominated by the photoionization component. We have previously studied the valence bands of ethylene-tetrafluoroethylene copolymers with a soft x-ray source, and the available photon energy range spans the initial portion encompassing orbitals dominated by contributions from F_{2p} lone pairs and orbitals with large contributions from C_{2p} and H_{1s} orbitals (C-H, C-C and C-F bonding orbitals)[2]. (Note that this is with respect to the Fermi level as energy reference.) It is interesting to note that for the experimentally determined attenuation cross section for polyethylene in the corresponding photon energy region (k = 2 x 10^5 cm^{-1} at 1300 Å) 10% of the light at that wavelength is absorbed in the top 50Å[10]. The production of ions in the polymer lattice provides a variety of mechanisms for crosslinking which we consider in a later section. It is clear, however, that in addition to fragmentation and isomerization the neutralization of ions by low energy electrons yields highly energetic species capable of further transformations.

It is clear from this brief discussion that the energy distribution of electrons in plasmas is of some importance. Since the prime interest in this work is the investigation by means of ESCA (which involves accurately measuring the energy distribution of photoemitted electrons) of the surface modifications effected by means of plasmas it seems logical to attempt to employ the double focusing electrostatic analyzer to sample the electron distributions in plasmas. It should be noted that the few systems for which experimental data is available concerning inductively coupled RF plasmas largely refer to pressures in the range 1 - 100 torr, whereas the work reported in this paper refer to pressures in the range $10^{-2} - 10^{-1}$ torr. The increased mean free paths would suggest that the electron temperature should be somewhat higher for the lower pressure range.

Before considering the ESCA investigations of the plasma modification of polymers, we therefore briefly describe in Section 3 some preliminary measurements we have made on the energy distributions of electrons sampled from RF plasmas excited in inert gases.

EXPERIMENTAL

The ethylene-tetrafluoroethylene copolymer samples used in this investigation were in the form of films ∿ 60 micron thick. The method of presenting samples to the plasma varied depending on the reactor design involved.

Three reactor designs were used. Reactors A and B (Fig. 1) were mounted in a greaseless vacuum system and pumped by a two stage rotary pump with a speed rating of 50 L min.$^{-1}$.

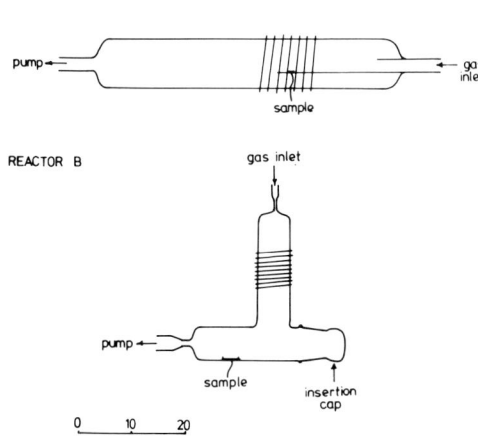

Fig. 1. Reactor designs A and B used in RF glow discharge modifications.

Pressures were recorded using Pirani type vacuum gauges and the inert gas introduced via a leak valve. Reactor A consisted of a pyrex tube 52 cm. long and 9 cm. diameter, with 1" inlet and outlet tubes at its ends. The discharge was excited by a 2.5 µH copper coil centered 10 cm. from the inlet tube. Samples were mounted on a glass platform positioned at the center of the RF coil. Reactor B consisted of 6 cm. diameter pyrex tubing in an inverted 'T' shaped configuration. The overall dimensions were 28 cms. long and 27 cm. high with inlet and outlet tubes ¼" and ½" diameter respectively, the inlet being at the top. The discharge was excited in the vertical limb by a 3.5 µH copper coil centered 13 cm. from the inlet tube. Samples were mounted in a stainless steel frame capable of holding two samples side by side with one covered by quartz or pyrex glass (1mm. thick) for U.V. studies. The frame was inserted into the reactor via a removable cap opposite the outlet tube, and positioned 6 cm. from the outlet.

Samples were removed from reactor A or B and mounted on a probe tip by means of double sided Scotch tape for analysis.

Reactor C (Fig. 2) was mounted directly onto the insertion port of the ESCA spectrometer and pumped by the rough vacuum system of the spectrometer via a ¼" tube. Pressures were recorded using a thermocouple vacuum gauge. Reactor C consisted of a pyrex tube 16 cm. long and 5 cm. diameter sandwiched between stainless steel flanges by 'O' ring seals and enclosed in a copper mesh screen, to prevent RF interference with the electronics of the spectrometer. The discharge was excited by a 4 µH copper coil wound centrally on the pyrex tube. Samples were mounted on a ½" stainless steel probe, 60 cms. long, (by means of double sided Scotch tape) which was capable of passing through the reactor, on 'O' ring seals and into the spectrometer for analysis of the samples without being exposed to the atmosphere.

Plasmas were excited, in all cases, using a Tegal Corporation RF Generator capable of delivering a power output from 0.05 w. to 100 w., continuously variable. A pulsing facility was used at lower power levels, giving greater stability to the plasma. Tuning of the RF power was achieved by an L-C matching network and monitored by the standing wave ratio using a Heathkit HM102 RF power meter.

Research grade Argon was used and purification achieved by a sorption train which removed hydrocarbons, water, carbon dioxide and oxygen. Flow rates were measured by either of two methods. The first, simply a bubble flow meter at the exhaust of the pump. The second, by measuring the initial rise in pressure as a function of time when the pump was shut off.

Spectra of the samples were recorded on an AEI ES200 B spectrometer using $Mg_{K\alpha_{1,2}}$ radiation and under the conditions

REACTOR C

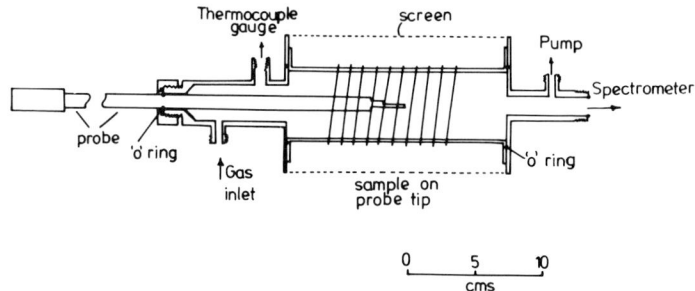

Fig. 2. Reactor configuration (employed in kinetic studies).

employed in this investigation the $Au_{4f_{7/2}}$ level at 84.0 eV binding energy, used for calibration purposes, had a FWHM of 1.15 eV. Area ratio measurements were carried out with a DuPont 310 curve resolver. In all cases, the measured binding energies are quoted with a precision of $\pm$ 0.1 eV and area ratios $\pm$ 5 %.

Some Observations on the Energy Distributions of Electrons Sampled From RF Plasmas Excited in Inert Gases

In studying a polymer by means of ESCA, the sample is conventionally mounted onto a probe tip attached to a metal probe inserted into the source of the spectrometer by means of an insertion lock arrangement (typical pressures in the source $\sim$ 5 x 10^{-8} torr). A horizontally mounted Henke type x-ray gun located at the rear of the sample chamber irradiates an area of the sample (normally inclined at $\sim$ 45° to the horizontal) $\sim$ 1 cm. x 0.8 cm. (length x breadth) and photoemitted electrons enter a vertically mounted retarding lens system and thence the double focusing electrostatic analyzer (10" mean diameter) by means of an adjustable slit mechanism (dimensions $\sim$ 1.2 cm. x 0.060, 0.120, 0.200 cms. The relatively long path length between the sample and the analyzer and detector requires pressures below $\sim 10^{-5}$ torr in the analyzer region to obviate any effects due to scattering involving the extraneous residual atmosphere (typically the pressure in the analyzer region is $\sim$ 10^{-9} torr). The possibility of irreversible damage to the electron multiplier arising from operation at relatively high pressures must also be kept in mind. The source region of the spectrometer in which the sample is irradiated is separated from the lens system by means of a teflon sleeve valve which when fully opened in normal operation provides (via restriction of pumping speed) a pressure differential between source and analyzer of approximately two orders of magnitude. With this constraint, it is not possible to excite a discharge in the source region of the spectrometer in the pressure range of interest in this study and directly sample the electrons. It is clear, therefore, that any such experiment requires a differential pumping arrangement.

The series of experiments which we have performed are different in character to those which have previously been reported in that the sampling of the plasma has been indirect but direct measurements have been made of the energy distribution. By contrast, most of the previous work has involved direct sampling of the plasma (by means of electrical probes) but an indirect estimate of energy distributions[6,11].

The experimental arrangement is indicated schematically in Figure 3 and involved the generation of plasmas in a ½" pyrex tube directly pumped with a 2 stage 120 litre/min. rotary pump connected to a ballast volume maintained at $\sim 10^{-3}$ torr. The tube traversed the spectrometer source vertically below the entrance slit to the analyzer via insertion locks. A slit $\sim$ 1 cm. long x 0.1 cm. wide machined along the length of the tube was aligned with the entrance

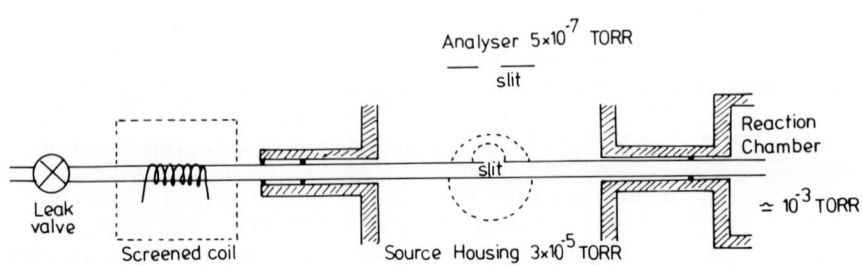

Fig. 3. Experimental arrangement for the investigation of electron energy distributions in RF plasmas.

slit to the analyzer and a 30 turn coil of 1/8" copper tubing extending over 6" was located external to the source housing, the center of the coil being approximately 12 ins. from the center point of the slit. With the separately pumped insertion lock seals based on viton O rings, it was possible to rotate the tube to investigate the effect of alignment of the slits.

The differential pumping across the slit in the tube (the source region of the spectrometer was directly pumped by a 4" diffusion pump) was such that for pressures in the range 10^{-3} - 10^{-2} torr in the discharge region the pressure in the source was maintained below 5×10^{-5} torr and that in the analyzer below 5×10^{-7} torr. Inert gas was metered via a purification train into the tube and the region encompassing the coil was screened by means of a grounded wire mesh cage which contained the electronics. Experiments in which secondary electrons were generated in the source (by means of U.V. irradiation of a metal sample through a quartz viewing port) and in which a plasma was struck in a closed end tube (pumping via the rotary pump of the insertion lock system) demonstrated the lack of any radiative RF interference in the source region.

The results described below pertain to plasmas excited in Argon, as a function of pressure and input power. The basic philosophy behind the experiments was to investigate the possibility of sampling the plasma in terms of electron energy distribution by means of an analyzer located external to the tube in which the plasma was excited. There are, of course, distinct difficulties in such an endeavor, and the results reported here are such that only a few tentative conclusions may be drawn since many more experiments would be required to evaluate the relative importance of all of the likely contributing factors.

Initial experiments indicated that a plasma excited in the tube produced an electron flux readily detected with standard counting equipment and showing a structured distribution in energy. Rotating the slit axially out of alignment with the entrance to the analyzer decreased the overall count rates confirming that the electrons being sampled emanated from the slit in the tube. No attempt has been made to measure the average potential of the plasma with respect to earth, and since the pyrex or quartz tubes were in any case insulated from the spectrometer there is a distinct possibility that a significant potential exists between the entrance slit to the analyzer, the plasma overall and the tube in which the glow discharge was sustained. Although there was considerable emission in the visible in the region of the coil, such emission was less evident for the section of tube aligned with the spectrometer slit.

The results from the first series of experiments are shown in Fig. 4 where the power ratings refer to output from the generator. In each case, the matching network was adjusted to minimize the

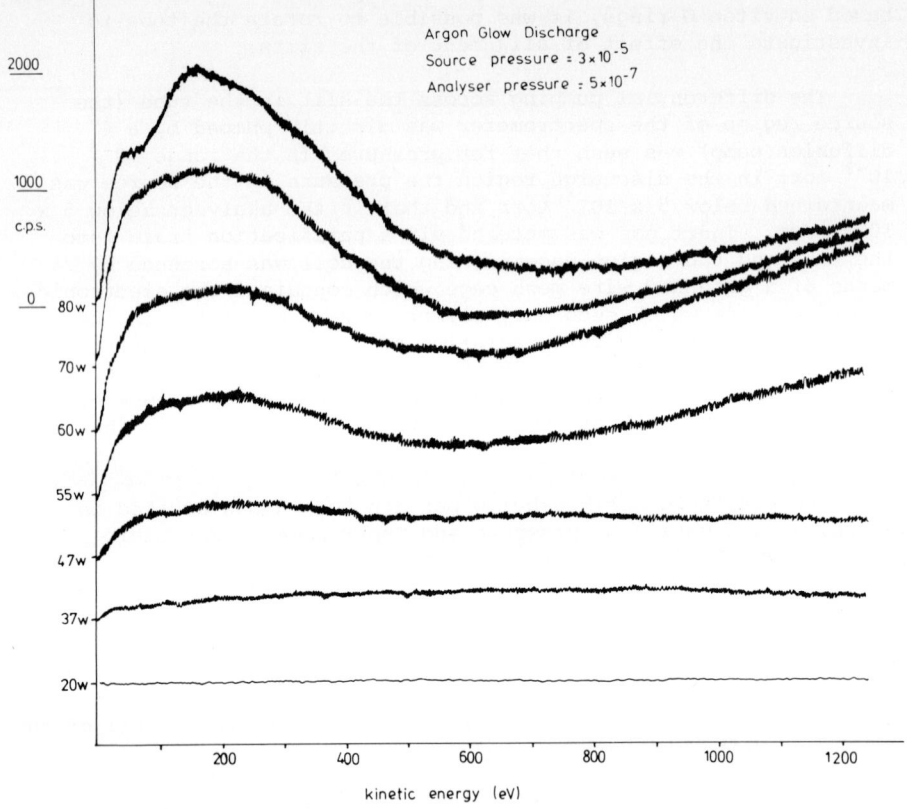

Fig. 4. Kinetic energy distributions for electrons sampled from RF plasmas excited in Argon.

standing wave ratio, the latter becoming progressively larger at higher power loadings (1.5 at 20 watts - 2.3 at 80 watts). At 20 watts, the count rate across the whole energy range (1250 eV) is negligible; however, as the power is increased a broad energy distribution extending over ~ 500 eV develops with a tail to higher kinetic energy. At 80 watts, the maximum for the detected electrons centered ~ 150 eV has a count rate of ~ 2×10^3 counts/sec., and further structure develops in the low energy region with a shoulder centered ~ 30 eV. The tail extending above ~ 500 eV appears to be variable in nature as is clearly evidenced from the data shown in Fig. 4. Having established that an electron flux emanating from the plasma (or rather, downstream of it) can be readily detected and exhibits a structured energy distribution, the effect of retuning the matching network on the low energy structure was investigated and the results are shown in Fig. 5.

The structure at low energy (centered $\sim$ 30 eV) acquires considerable intensity on retuning and separates into a distinctive peak with the semblance of a Maxwellian type distribution. The pressure dependence of this structure was investigated by adjusting the metered flow of Argon into the discharge tube. The pressures in the discharge tube itself were not monitored during the experiments since the RF plasma interferes with both the thermocouple

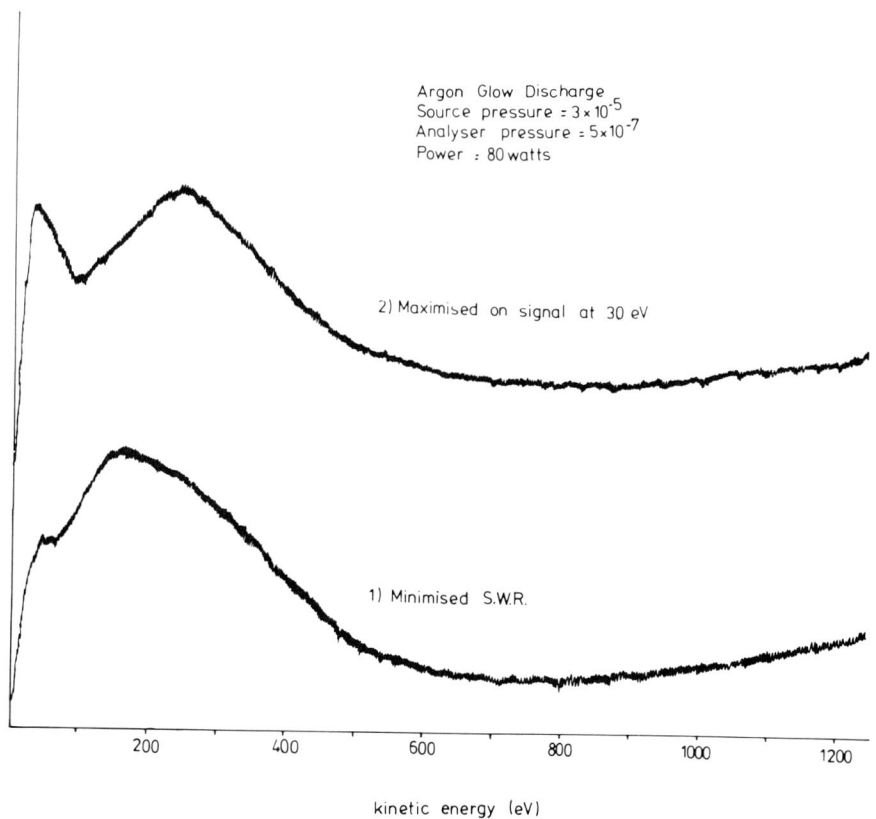

Fig. 5. Effect of retuning the matching network on the KE distribution of electrons sampled from an RF plasma.

and pirani gauge readouts in the inlet system, however, the pressures are believed to be somewhere in the range $5 \times 10^{-3} - 5 \times 10^{-2}$ torr. The pressures measured directly in the spectrometer source by means of an ion gauge were, however, recorded during each experiment and reflect the pressure trends in the discharge tube. It is clear from the data in Figure 5 that the electron count <u>increases</u> considerably on going to lower pressures consistent with an increase in mean free path in the discharge tube.

The salient features from these experiments are therefore as follows: An electron flux emanating from the plasma may be detected and depends markedly on the power and pressure in the expected manner. The energy distribution seems to comprise three components. A low energy component with energy ∿ 30 eV at the peak maximum whose intensity can be enhanced by retuning the matching network. A broad distribution centered around ∿ 150 eV (minimized SWR) or ∿ 230 eV (maximized signal on 30 eV peak) depending on the tuning circuit, and a long tail to higher kinetic energy which tended to be less reproducible and in any case did not vary in a consistent manner either with pressure or power input to the plasma. The interpretation we place on these data must of necessity be somewhat speculative; however, it is tempting to assign the Maxwellian type distribution at low energy to the general distribution of electrons in the plasma. It should be pointed out, however, that there are problems associated with the energy reference. The kinetic energy of the electrons are measured with respect to the fermi level of the spectrometer, however, there is no guarantee that the average potential of the plasma approximates to earth or that the exit slit to the pyrex tube is at earth potential. Indeed, the higher energy structures would suggest that space charge effects at the edges of the slit might be of some importance. Further factors which might be considered are the strong fields arising inside the spectrometer source from the plasma which could well account for the high energy components of the distributions shown in Figs. 4 - 6. Such questions require a very complete and extensive series of experiments to be carried out and these will form the basis of a future communication. At this stage, therefore, we may merely note that the possibility exists of combining ESCA studies with investigations of energy distributions, and the preliminary results outlined here form an interesting and very useful basis for such studies.

Preliminary Experiments on the RF Glow Discharge Modification of Polymer Surfaces

The initial experiments previously reported on the Argon ion bombardment of the ethylene tetrafluoroethylene copolymer at relatively low beam currents indicated that on the ESCA depth sampling scale the reaction was extremely rapid < 1 minute. The preliminary experiments involving the interaction of a similar polymer sample with an RF plasma in Argon were therefore aimed at

Plasma Modification of Polymers 113

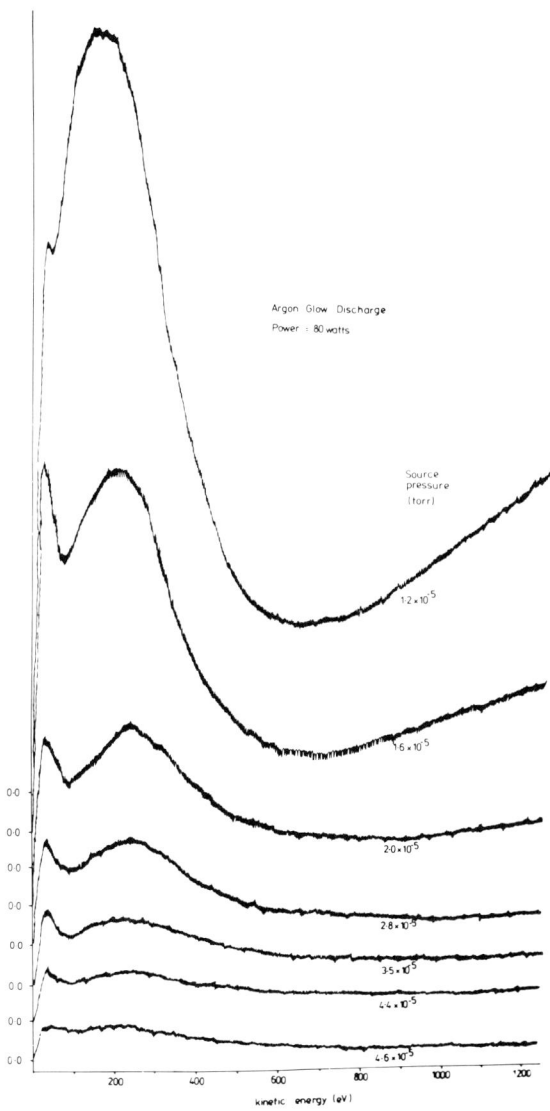

Fig. 6. KE distribution of electrons sampled in RF plasmas excited in Argon as a function of pressure.

establishing a suitable time scale for the reaction as a function of pressure, power and flow rate.

As a starting point using reactor configuration A, samples were treated for 30, 45 and 60 secs. at a pressure of $\sim 2.5 \times 10^{-2}$ torr. with a continuous, inductively coupled RF glow discharge with a power loading of 25 watts using unpurified Argon. The C_{1s} and F_{1s} core level spectra revealed extensive reaction indicating that shorter time scales and lower power loadings were required for the initial experiments. Whilst our previous studies indicated a small extent of surface oxidation for the polymer system[4], the glow discharge treated samples showed evidence of increased oxygen content which could arise from a variety of sources. Later work, in fact, provided strong evidence that the major part of the oxygen contamination did not arise from either impurities in the Argon or from exposure of the samples to atmosphere subsequent to glow discharge treatment. At this stage, the most likely source of such contamination is from desorption of oxygen species from the walls of the reactor under the influence of the plasma. Further experiments established that within wide limits the effect of pulsing the discharge for the same average power loading was identical to that for a continuous discharge. This is of some importance since as experiments progressed it became evident that in order to follow the reaction on a reasonable time scale, low average power loadings were required < 1 watt. Such low power loadings are most readily reproduced in a pulsed discharge with appropriate adjustment of the pulsing sequence. For an average power loading of 1 watt, for example, it was possible to operate a 25 watt discharge with a ratio of time on to time off of 1/25th, the time scales being in the micro-second range. The results for a series of polymer samples exposed to the discharge for varying periods of time are shown in Fig. 7.

The changes in the core level spectra are qualitatively and indeed semi-quantitatively similar to those previously reported for Argon ion treatment[12]. Namely, a decrease in the intensities of the core levels corresponding to $\underline{C}F_2$ and C-$\underline{F}$ environments and the buildup of intensity at a binding energy appropriate to $\underline{C}F$ linkages as would arise from cross linking in the surface and subsurface of the polymer sample.

The changes in intensities as a function of time for the various core levels are shown in Fig. 8 and Fig. 9.

It is clear that the extraneous O_{1s} signal rapidly reaches a constant (reproducible) value as might be expected if such contamination arises from desorption from the walls. The smooth nature of the curves demonstrates the reproducibility of the plasma over the region in which the samples were located although separate experiments indicated that the overall rate of reaction could be varied by locating the samples in differing positions with respect

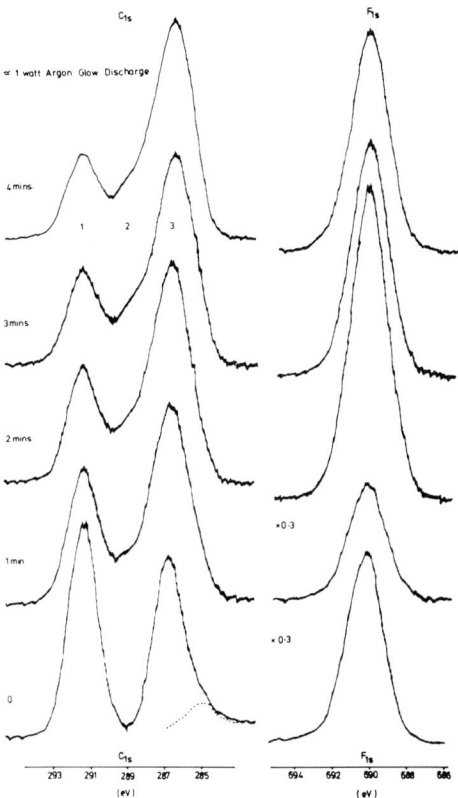

Fig. 7. Core level spectra for a series of samples of an ethylene-tetrafluoroethylene copolymer subjected to glow discharge modification.

to the RF coil. This was particularly evident if samples were located not coaxially with the discharge tube but close to the reactor walls. It is clear from the data in Fig. 9 that the loss in intensity of the high binding energy $\underline{CF_2}$ component is accompanied by an increase in intensity of a lower binding energy component consistent with the production of $\underline{CF}$ structural features, and that overall the integrated intensity of the C_{1s} signals (as is evidenced from Fig. 7) increases significantly with time. We will show in a later section that these observations are consistent with a cross-linking mechanism.

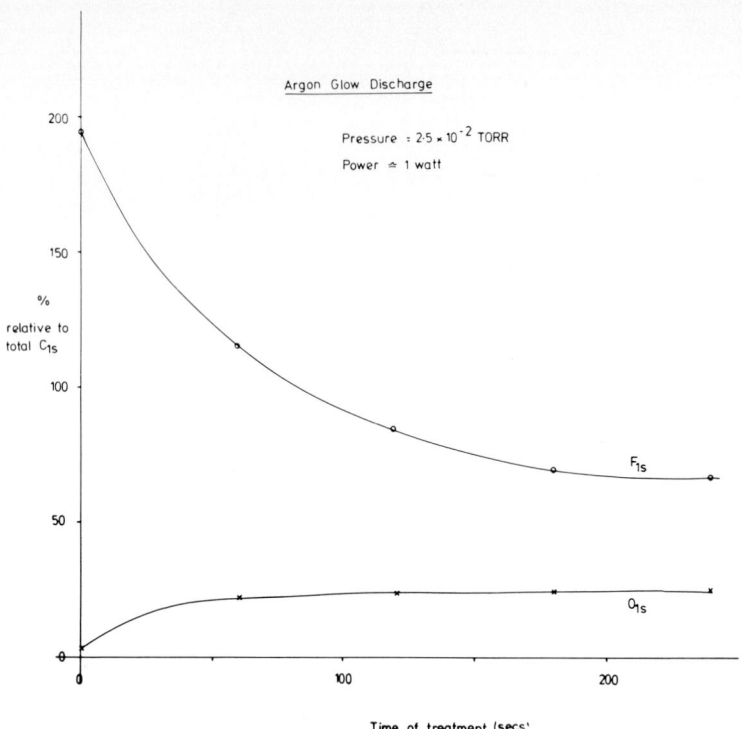

Fig. 8. Variation in relative intensities of F_{1s} and O_{1s} signals with reaction time.

Over a very limited pressure range, it was found that the extent of reaction for a given power loading showed a linear dependence such that reaction was much faster at lower pressure consistent with the longer mean free paths of the reactive species.

The surface nature of the reaction is readily shown by following the intensity ratios of the F_{1s} and F_{2s} levels as a function of time, since it is known that these levels span an appreciable difference in escape depth dependence[5]. If the surface differs, therefore, from subsurface and bulk in terms of its chemical composition with the surface region being relatively lower in fluorine content, the larger mean free path for photoemitted electrons from

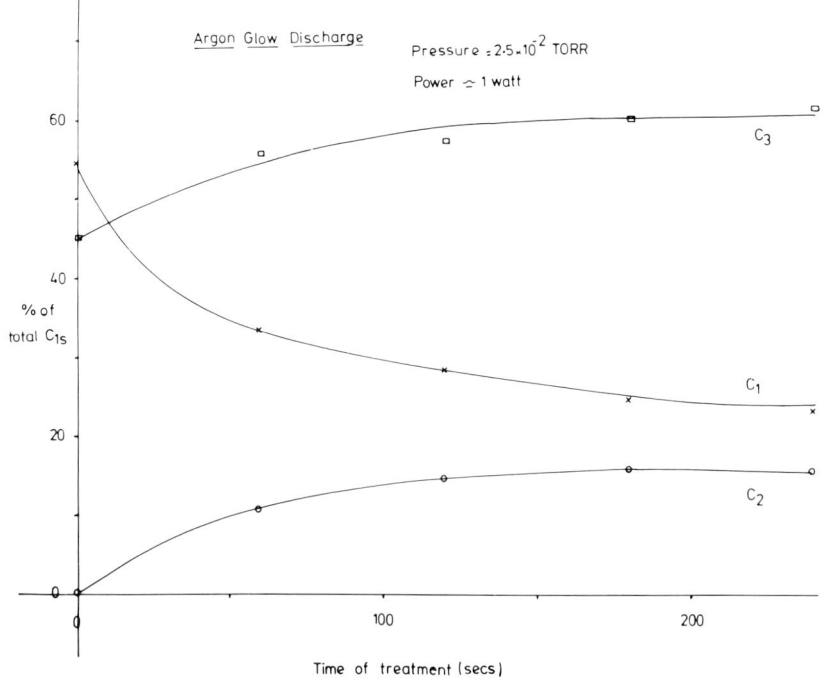

Fig. 9. Variation in relative intensities of components of C_{1s} levels as a function of reaction time.

the essentially core-like F_{2s} levels should lead to a decrease in the overall intensity ratio as is clearly seen from the results in Fig. 10.

A further manifestation of the changes in structure and bonding in the surface regions is the variation in shift in kinetic energy scale for the treated samples arising from the equilibrium buildup of charge (surface charging) on the sample under irradiation[13]. This is shown in Fig. 11.

It is interesting to note that for a given photon flux from the x-ray gun the equilibrium static charge for the untreated copolymer is strikingly similar to that for polyvinylidene fluoride of similar overall stoichiometry, whereas for the treated samples the plateau value approaches quite closely that found for polyethylene which would strongly suggest that the fluorine content of the first monolayer is quite low.

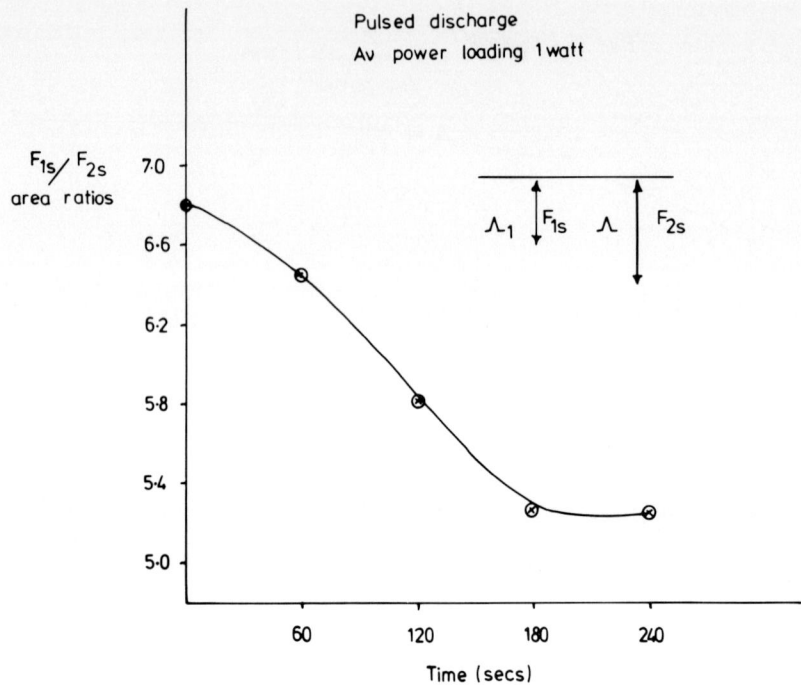

Fig. 10. F_{1s}/F_{2s} integrated intensity ratios as a function of glow discharge treatment.

Further indications of the inhomogeneous nature of the surface regions of the polymer system come from estimates of the apparent carbon to fluorine stoichiometries (based on the known sensitivity factors for the C_{1s} and F_{1s} levels for unit stoichiometry) from the integrated intensities of the component peaks of the C_{1s} levels and a comparison of the relative intensities of C_{1s} and F_{1s} levels. Since the escape depths for photoemitted electrons are in the order $\Lambda\ C_{1s} > \Lambda\ F_{1s}$, the apparent stoichiometry from the relative intensity ratio of the two core levels is always lower in fluorine than that derived from an analysis of the component peaks of the C_{1s} levels. Thus for a homogeneous sample (on the ESCA depth sampling scale > 100 Å, the sensitivity corrected area ratios for the F_{1s} levels (arising from $\underline{C}F_2$ and $\underline{C}F$ environments) relative to the $\underline{C}F_2$ component of the C_{1s} levels must always be > 2. For the discharge treated samples, this ratio, in fact, is always < 2 indicating the inhomogeneous nature of the samples.

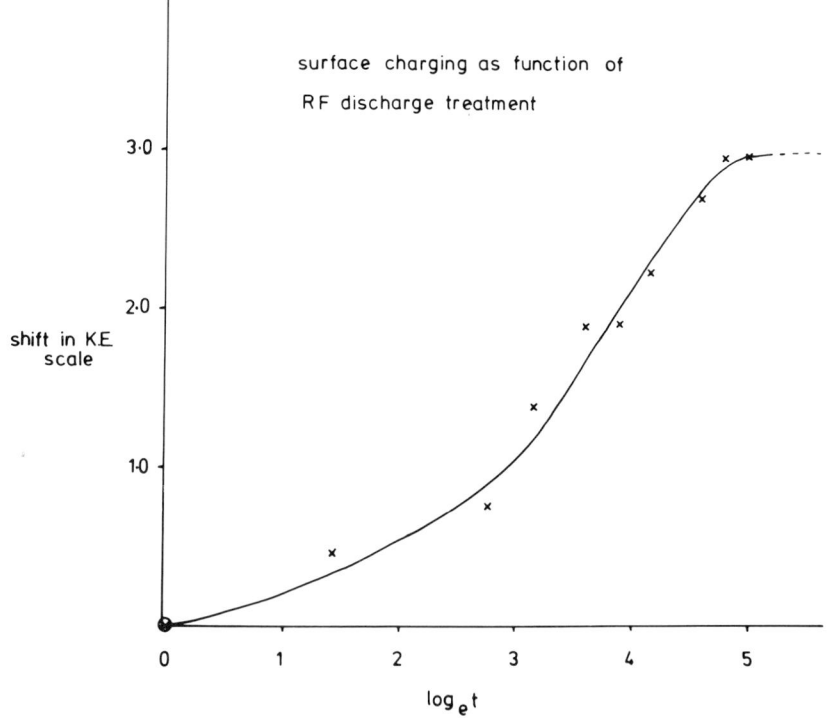

Fig. 11. Surface charging shift in KE scale as a function of RF discharge treatment.

Since the experiments suggested that even at relatively low power, side reactions, (most probably arising from species desorbed from the reactor walls) contributed significantly to the overall reaction, the possibility arises that hydrocarbon contamination from desorbed species could contribute to the increase in intensity of the lower binding energy component of the C_{1s} levels. This may readily be discounted, however, from further experiments which will now be described. A further point, worthy of some consideration, is the relative importance of surface ablation processes. To investigate the possibility of hydrocarbon contamination, a sample of gold whose ESCA spectrum revealed slight hydrocarbon contamination of the surface was subjected to an Argon plasma under similar conditions to those employed in the previously described work. Re-investigation of the core level spectra showed that after treatment for 60 seconds at a power loading of 1 watt the hydrocarbon contamination was in fact significantly reduced which provides some evidence that even at low powers low molecular weight material can be removed from surfaces. This provides further evidence for the likely involvement of surface desorbed oxygen species, since the Argon used in these experiments was passed through a sorption train which should have removed any oxygen, carbon dioxide, water, or hydrocarbon impurities.

A series of experiments were performed in which the reactor was coated by plasma polymerization of a pentafluoroethane monomer to a depth of ∿ 3 - 4000 Å. Gold, silver and glass substrates were placed in the reactor during the polymerization so that the composition of the polymer could be determined by ESCA. The C_{1s} spectra were highly structured showing $\underline{C}$, $\underline{C}F$, $\underline{C}F_2$ and $\underline{C}F_3$ groups. A sample of polystyrene film was then treated in the coated reactor by an Argon plasma at a power loading of 100 watts for 30 mins. (Polystyrene was chosen because of its relatively high reactivity towards oxygen containing species in the plasma, as determined from previous work.) Subsequent analysis of the ESCA spectrum of the treated film revealed a highly fluorinated system. The C_{1s} spectrum was again highly structured showing $\underline{C}$, $\underline{C}F$, $\underline{C}F_2$ and $\underline{C}F_3$ groups. The only reasonable mechanism for this is via ablation of low molecular weight fragments from the surface of the polymer coating and grafting of the fragments onto the polystyrene film.

This experiment demonstrates that ablation of the surface of a polymer can be achieved at high power loadings, although at lower power levels the process would be expected to be relatively unimportant. Under the conditions of the previously described work (∿ 1 watt), desorption of hydrocarbon and low molecular weight oxygen containing species might be expected to be the dominant processes as far as the internal surface of the reactor is concerned. Furthermore, once hydrocarbon has been removed from the walls of the reactor it is less likely that it would be replaced before subsequent treatment of polymer films, compared with the oxygen containing species (i.e. O_2, CO_2 and H_2O) which are at a higher partial pressure in the laboratory atmosphere to which the reactor was exposed in loading and unloading samples.

The close similarity of the surface modified samples by means of either Argon ion bombardment or low powered RF discharge in Argon strongly suggests that at the outermost surface reaction is dominated by direct energy transfer from charged and/or metastable species in the plasma.

We have previously emphasized, however, that plasmas are copious sources of electromagnetic radiation and there is an extensive body of literature on the modification of polymers initiated by U.V. radiation[14]. Indeed, there has been considerable discussion in the literature on the relative importance of direct and radiative energy transfer processes, however the techniques which have previously been brought to bear on this problem have not enabled the initial stages of reactions involving the outermost few tens of Angstroms to be delineated[1].

In the pressure range pertinent to this work and indeed most previous investigations, the photon flux is highest in the vacuum ultraviolet, however there are significant outputs in the U.V. and visible regions[7]. For an ideal ethylene-tetrafluoroethylene copolymer, the lowest excited states, corresponding to $n(F_{2p}) \rightarrow \sigma^*$ and $\sigma \rightarrow \sigma^*$ transitions, correspond to absorption in the vacuum U.V., and only end absorption would be available for the U.V. and visible regions. In a largely alternating ethylene-tetrafluoroethylene copolymer, it is inevitable that elimination of HF occurs to a minute extent (not detectable by most conventional techniques) and the unsaturated centers so produced could give rise to absorption in the U.V. A similar possibility arises from surface oxidation features (> C=O) known to be present in submonolayer coverage in the polymer films studied[4]. Such features could conceivably be important, if the oscillator strengths for the transitions were orders of magnitude larger than for absorption and photoionization processes occurring at much shorter wavelengths in the vac. U.V. with the considerably more intense radiation from the plasmas in that region. The available data strongly mitigates against this possibility, since the total attenuation coefficients increase quite sharply in going from the U.V. to vac. U.V. and at shorter wavelengths the total cross sections are dominated by Rydberg transitions converging on ionization limits and by direct photo-ionization processes[10,15].

This situation may well be modified for polymer systems with chromophores with a large number of available excited states in the U.V./visible region.

The evidence is fairly conclusive, therefore, that short wavelength radiation can lead to modification of polymers and undoubtedly such processes compete with direct energy transfer from species in the plasma which because of their diffusion controlled nature would be expected to be of most importance for reactions at the immediate surface. The fact that short wavelength radiation can effect crosslinking at the polymer surface may be demonstrated by prolonged x-ray irradiation of the polymer in the spectrometer source, the modification being monitored by studying the C_{1s} levels. With an x-ray gun power loading of 180 watts, only a fraction of which will appear in the characteristic line spectra and bremmstrahlung, surface modification proceeds at a finite rate some 10^3 times slower than for glow discharge treatment at 1.0 watt. (This rate is extremely slow compared with the typical time scale for measurement of the core level spectra, and surface modification arising during the exposure to the x-ray source is therefore negligible.) The photon energy distribution from a conventional x-ray source is of course completely different from that emitted from an RF plasma and only a minute fraction of the radiation will occur in the vacuum U.V. region spanned by the latter. Since cross sections for absorption, photoionization, etc. are strongly dependent on photon energy, it is clear that the most that can be inferred from this experiment is that short wavelength electromagnetic radiation can cause crosslinking at the surface of ethylene-tetrafluoroethylene copolymers.

In order to elucidate the relative importance of surface reactions arising from direct and radiative energy transfer, a special sample mount was developed which enabled polymer samples to be located in the plasma either directly in contact with the plasma or exposed only to the electromagnetic radiation from the plasma through a quartz window.

Using reactor configuration B, samples of the copolymer were located symmetrically with relation to the RF coil with one of the samples being covered by 1 mm. thick quartz slide. The C_{1s} levels for the polymer samples treated for 60 seconds at 1 watt are shown in Fig. 12. It is clear that the sample exposed to the electromagnetic radiation has not undergone any appreciable reaction. The cut off for quartz is $\sim$ 1600 Å, and for a slide $\sim$ 1 mm. thick there will be some attentuation of longer wavelength radiation[7]. The experiment, therefore, suggests that the radiative energy transfer component of the reaction is dominated by wavelengths < 1600 Å as might have been anticipated. To confirm this, experiments were carried out in which samples covered with a quartz slide were exposed to the electromagnetic radiation from RF plasmas run at full power (100 watts) for a period of 30 minutes. The C_{1s} spectra

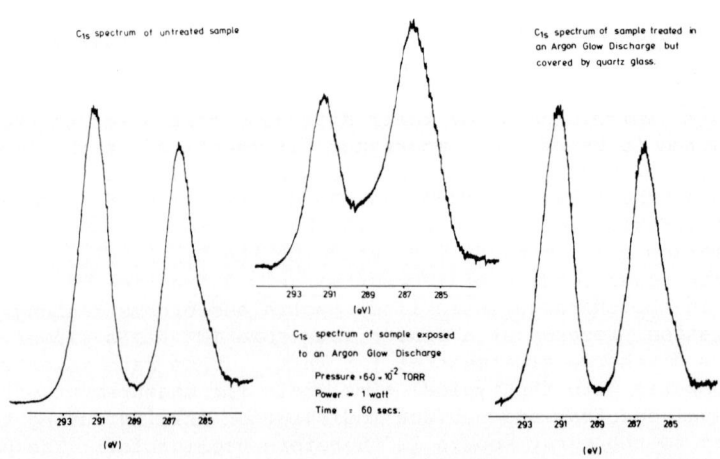

Fig. 12. C_{1s} levels for samples of TFE-Ethylene copolymer exposed to an RF plasma excited in Argon.

for these samples showed them to be completely unreacted. For
comparison purposes, experiments were also carried out with samples
of polystyrene and polyethylene terephthalate both of which absorb
strongly in the U.V. At high power and after substantial periods
of irradiation, both systems showed some evidence of reaction, and
more detailed investigations of these systems will be reported in
due course.

Kinetic Studies of the RF Glow Discharge Modification
of Ethylene-Tetrafluoroethylene Copolymers
by Plasmas Excited in Argon

The general conclusion from the experiments outlined above is
that any radiative energy transfer component is likely to be dominated by radiation with wavelength < 1600 Å. The experiments
described thus far, suggested the design of a reactor configuration
(C) directly mounted on the ESCA spectrometer source which would
enable detailed kinetic studies to be made on the same sample without exposure to the atmosphere between successive interactions with
a given plasma. The pumping speed for reactor configuration C was
considerably lower than for the previously described reactors (A and
B). At similar total pressures, therefore, the much lower flow
rate gave for an identical power loading a much faster reaction
with configuration C than for either A or B. In order to increase
the time scale for a convenient kinetic study, the total pressure
of Argon at which the discharge was excited was increased to $\sim 10^{-1}$
torr, and the plasma was operated at a greatly reduced average
power loading (~ 0.1 watt).

The core level spectra for a given sample successively treated
in such a plasma are shown in Fig. 13. These results are quite
striking and the decrease in intensity of the F_{1s} and $\underline{C}F_2$ levels
can readily be followed as a function of time.

With such data, it seems worthwhile to attempt a crude semi-quantitative analysis in terms of a kinetic scheme involving both
radiative and direct energy transfer components.

The theoretical model is outlined in the schematic in Fig. 14
where processes occurring in the first monolayer are separated from
the subsurface and bulk reaction. As a starting hypothesis, we
consider that because of the diffusion controlled nature of processes
involving M* (for which the dominant reaction should be associated
with that from Argon ions and metastables) reactions at the surface
should have components from both direct and radiative energy transfer processes. For the subsurface and bulk, however, the dominant
process should be that associated with radiative energy transfer.

If the surface concentration of structural features (designated
X) is X_o^m for the first monolayer of the unmodified polymer surface,

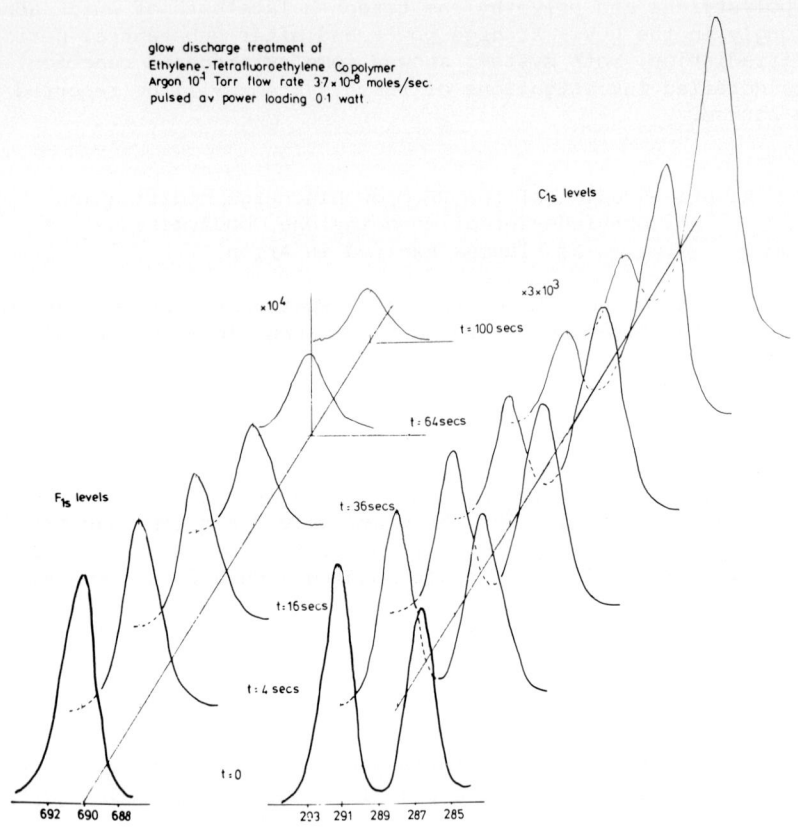

Fig. 13. Core level spectra for a sample of ethylene-tetrafluoroethylene copolymer successively treated with an RF plasma (0.1 watt) in Argon.

then the rate of modification of such sites may be written as (with obvious notation)

$$\frac{dx^m}{dt} = K_M(X_o^m - X^m)M^* + K_L(X_o^m - X^m)I_o(1-e^{-kd}) \quad (1)$$

where K_M is a composite rate constant for processes involving M^*, $I_o(1-e^{-kd})$ represents the fraction of electromagnetic radiation absorbed in the first monolayer of thickness d, with k representing a composite of attenuation coefficients for all the frequencies

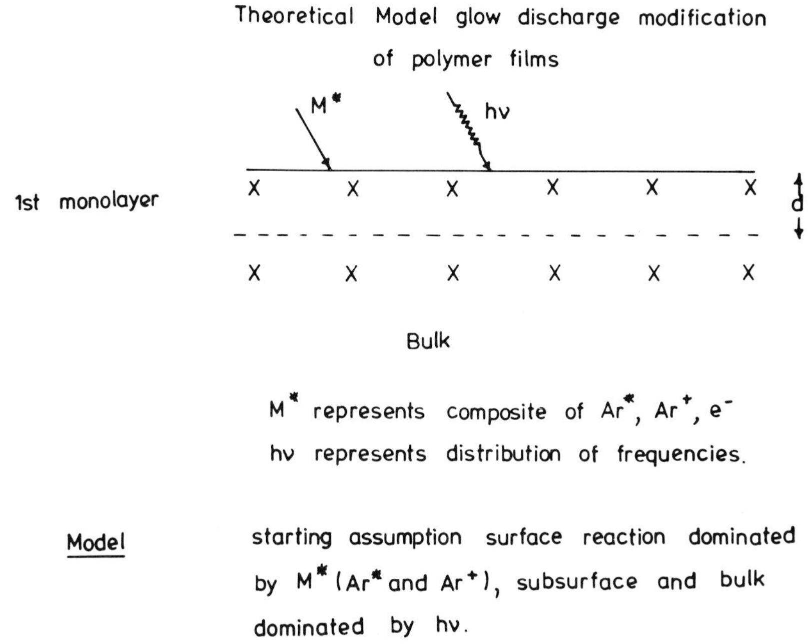

Theoretical Model glow discharge modification of polymer films

1st monolayer

M* represents composite of Ar*, Ar⁺, e⁻
hν represents distribution of frequencies.

Model   starting assumption surface reaction dominated by M* (Ar* and Ar⁺), subsurface and bulk dominated by hν.

Fig. 14. Schematic of theoretical model for interpreting data in Fig. 13.

involved. K_L is a composite rate constant for the radiative energy transfer processes and may well be a function of both the wavelength and intensity distribution of the electromagnetic radiation. Since the flux of M* and electromagnetic radiation bombarding the sample surface is constant, for a given set of operating parameters for the plasma, we may recast (1) in a simplified integrated form in which rate processes involving the first monolayer are encoded in the pseudo rate constant K_m as in equation (2).

$$X_o^m - X^m = X_o^m e^{-K_m t} \tag{2}$$

where $K_m = K_M M^* + K_L I_o (1-e^{-kd})$

For the subsurface and bulk, a parallel treatment gives (3) and (4)

$$\frac{dx^b}{dt} = K_L(x_o^b - x^b)I_o e^{-kd} \tag{3}$$

$$\therefore \quad x_o^b - x^b = x_o^b e^{-K_b t} \tag{4}$$

where $K_b = K_L I_o e^{-kd}$

Now, the proportion of core level signal arising from the first monolayer for a given structural feature is given by

$$(1-e^{-d/\Lambda})$$

where Λ is the escape depth for photoemitted electron of a given kinetic energy. Whilst for the subsurface and bulk the proportion of signal arising is modulated by a factor of $e^{-d/\Lambda}$. The total integrated intensity for a core level monitoring a given structural feature as a function of time is therefore given by equation (5).

$$\frac{x^T}{x_o^T} = (1-e^{-d/\Lambda})e^{-K_m t} + e^{-d/\Lambda}e^{-K_b t} \tag{5}$$

where x_o^T is the total integrated intensity for a given structural feature for the initial system.

It is clear from the data in Fig. 13 that convenient means of monitoring the rate processes involved are provided by the F_{1s} levels and $\underline{CF_2}$ component of the C_{1s} levels. Figure 15 shows a logarithmic plot of the fluorine intensity ratio versus time, and exhibits distinct curvature as might be expected from a two component system. As t becomes large, the dominant contribution is from the component of smaller exponent, and indeed a good linear extrapolation is evident; the slope of which gives an exponent of 0.0117 with intercept 0.56. By replotting the differences for the initial portion of the curve, the exponent and intercept for the process of higher exponent may readily be evaluated as 0.0605 and 0.375 respectively. The surface components of the reactions must obviously be faster than for the subsurface and bulk (at least in the initial stages of reaction which we are considering here) since both direct and radiative energy transfer components are likely to be of importance. and we may therefore assign the larger exponent to this surface component. It is interesting to note that the rate processes for the surface are half an order of magnitude larger than for the subsurface reaction, and since the attenuation coefficients for the electromagnetic radiation are likely to be such that the radiation suffers relatively minor attenuation in passing through the surface, the only consistent conclusion which may be drawn is that the surface reaction is dominated by reactions involving ions and metastables with only a minor contribution from radiative energy transfer whilst the latter dominates processes for the subsurface

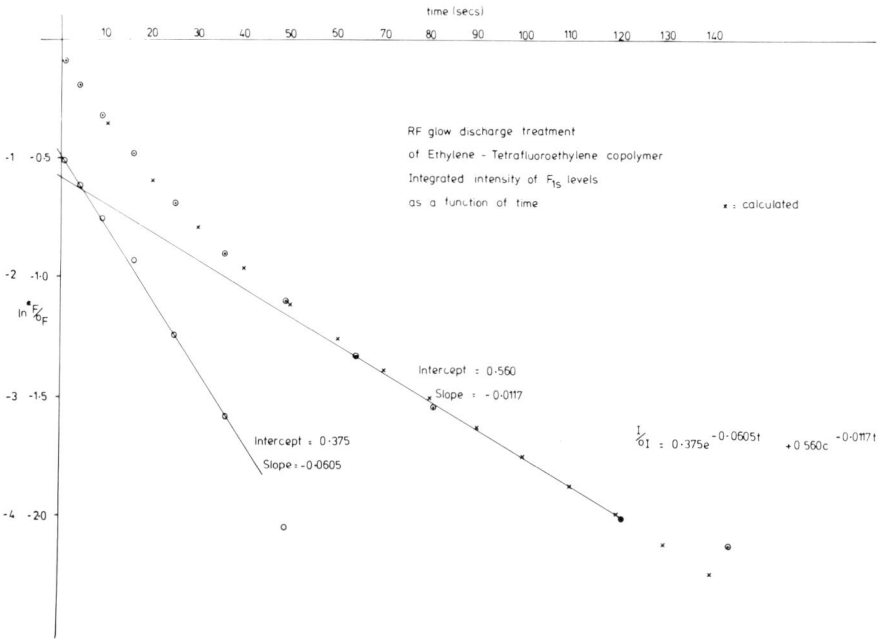

Fig. 15. Integrated relative intensity of F_{1s} core levels as a function of time for glow discharge treatment of an ethylene-tetrafluoroethylene copolymer.

and bulk. As a check on the analysis of the data in terms of a two component system, we have recalculated the theoretical curve (crosses) and comparison with the original data (open circles) shows excellent agreement over the whole time scale.

A similar analysis of the C_{1s} levels associated with the $\underline{CF_2}$ components is shown in Fig. 16 and this also exhibits an excellent correlation with the experimental data. The exponents and pre-exponential factors are somewhat similar to those for the F_{1s} levels, and it is gratifying to note that the sum of the pre-exponential factors is close to unity in both cases.

Having demonstrated that the experimental data may be interpreted in terms of a composite of two processes, we may now inquire as to the significance of the constants evaluated from such a treatment. Previous work in these laboratories have suggested that electron mean free paths for kinetic energies appropriate to

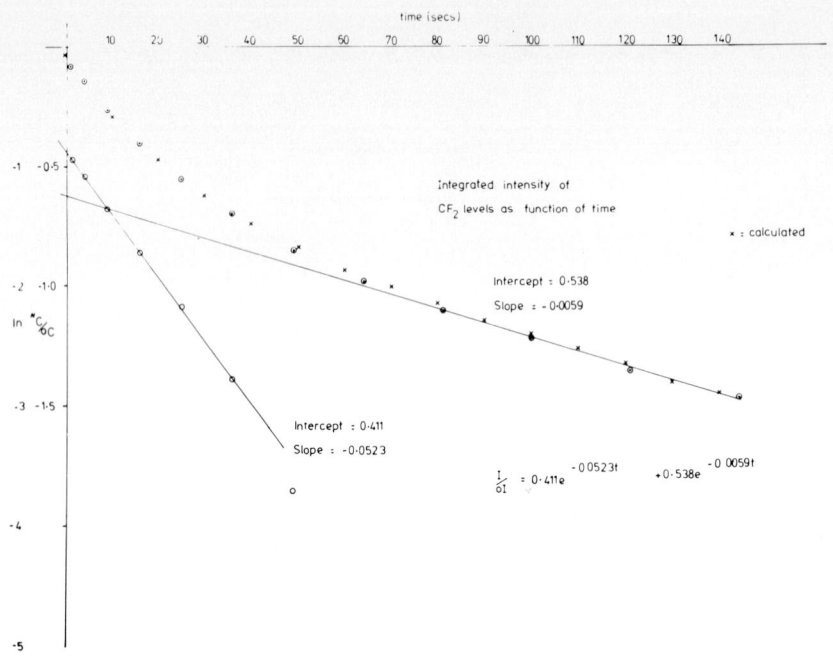

Fig. 16. Integrated relative intensity for CF$_2$ levels as a function of reaction time.

photoemitted electrons from F_{1s} and C_{1s} core levels using $Mg_{K\alpha_{1,2}}$ radiation are in the order $\Lambda_{C_{1s}} > \Lambda_{F_{1s}}$ [5]. This is also apparent from the data for the pre-exponential factors. The analysis[5] of data on the surface fluorination of polyethylene suggests typical escape depths (mean free paths) of 7 Å and 10 Å for the F_{1s} and C_{1s} levels respectively. From these values and the pre-exponential factors, it is possible to compute a value for d. From the F_{1s} level data, this analysis gives a value of ∼ 4 Å whilst for the CF$_2$ levels a value of ∼ 6 Å is obtained. These are in excellent agreement considering the approximations inherent in the analysis, and furthermore are close to the depth appropriate to the first monolayer which is eminently reasonable in terms of the chemistry and physics of the processes involved.

Mechanistic Aspects of the Plasma Modification of Polymer Films

In this section, we briefly consider possible mechanisms which would provide a basis for rationalizing the results. Clearly, the most that can be accomplished at this stage is a likely scenario for the processes involved and more detailed discussions must await the results of further experiments. A key observation in addition to those relating to the relative changes in intensities of the F_{1s} levels and individual components of the C_{1s} levels, is that the total integrated signal intensity for the C_{1s} levels increases for the surface modified species (by a factor of ~ 1.3 for the sample treated for 144 seconds relative to the starting material). All of these observations are consistent with a crosslinking mechanism for surface modification in which $\underline{C}F_2$ sites are converted to $\underline{C}$-F and $-\underline{C}-$ environments. For a given core level, the integrated intensity follows a general relationship of the form:

$$F \alpha\ NK\Lambda(1-e^{-d/\Lambda})$$

where: F is the x-ray flux,
α is a cross section for photoionization,
N is the number of atoms (on which the core level is localized) per unit area,
K is a spectrometer dependent factor and Λ and d are the electron mean free path and film thickness respectively.

For a crosslinking process, the average interchain distance should considerably decrease leading to an increase in N and thus a greater integrated intensity for the C_{1s} levels[16].

We may, therefore, briefly consider possible mechanisms for such a process. The energetic species emanating from the plasma are electrons, Argon ions and metastables and vacuum ultraviolet radiation. Considering firstly the electrons, it is clear that since mean free paths for electrons in the energy range 0 - 30 eV are likely to be at least an order of magnitude larger than for either Argon ions or metastables, their role is likely to be a secondary rather than a primary one as will become apparent. Even so, it is clear that the most likely primary role of the electrons, however small, should involve the production of the parent ion of a localized portion of the polymer chain. Indeed, it will become apparent that this is the most likely primary process in general. We have previously pointed out that valence ionization of the polymer is energetically feasible by direct interaction with either Argon ions or metastables, whilst for the short wavelength vacuum U.V. radiation the largest contributions to the attenuation coefficient is undoubtedly from photoionization and from transitions to diffuse Rydberg states which in the solid closely approximate in properties the ionized states.

A valence ionized polymer chain produced in the condensed media can undergo a number of transformations which we may now consider. The available valence ionized states correspond to removal of F_{2p} lone pairs or electrons dominantly of C-C or C-H bonding characteristics[4]. Such ions could undergo unimolecular reactions such as carbon carbon or carbon hydrogen bond cleavage, however in such a rigid matrix the cage effect would be expected to dominate more particularly for the former reaction, and the more likely process would seem to be ion neutralization by means of the electron flux through the sample. This would lead to electronically excited systems with energies considerably higher than that required for bond cleavage and/or unimolecular elimination of a small molecule (e.g. HF or H_2). In the particular case of carbon hydrogen bond cleavage in the parent ion, the relative ionization potentials and electron affinities would ensure the production of a hydrogen atom and a carbo-cation. The electronically excited neutral systems produced by a quenching process have sufficient localized internal energy for homolytic C-F, C-C and C-H bond cleavage. The greater mobility of hydrogen and fluorine atoms almost certainly dominate the following reaction sequences since the carbon radicals produced form part of the relatively inflexible polymer chain. Considering therefore a typical polymer repeat unit, a reasonable outline of the initial stages of reaction may be elaborated as in Scheme 1. A variety of pathways are therefore available for the production of potential sites for crosslinking, and a consideration of the likely rates of abstraction and elimination would tend to suggest that the effective elimination

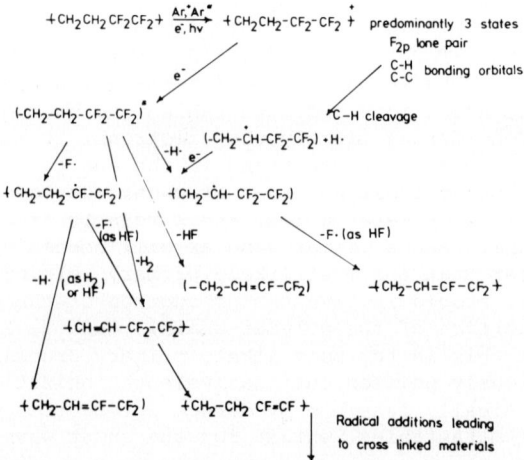

Fig. 17. <u>Scheme 1</u>. Possible reaction sequence for initial stages of glow discharge modification of polymer films.

of HF is an important precursor to crosslinking. The production
of radicals or carbocations in close proximity to any unsaturated
centers provides a ready means of crosslinking by either radical
or electrophilic addition mechanisms. Such a scheme leads to a
straightforward interpretation of the decrease in intensity of the
C_{1s} levels associated with $\underline{CF}_2$ structural features and the increase
in intensity of those at intermediate binding energy associated
with $\underline{CF}$ components. It is interesting to note that for the faster
component of the rate process the composite pseudo rate constant
for disappearance of the $\underline{F}_{1s}$ and $\underline{CF}_2$ signals are approximately the
same whilst for the slower component corresponding to radiative
energy transfer that for the F_{1s} levels is approximately twice
that for the CF_2 levels strongly suggesting that the radiative
energy transfer process is dominated by an overall reaction scheme
in which the signals arising from $\underline{CF}_2$ structural features are
replaced by components with binding energies appropriate to carbon
not directly attached to fluorine in addition to $\underline{CF}$ structural
features. This would not seem unreasonable if the dominant interaction of the short wavelength electromagnetic radiation with the
polymer involved photoionization of F_{2p} lone pairs for which the
cross section should be particularly large, since any $\underline{CF}$ structural
features initially produced could then undergo further transformation.

The work described in this paper, therefore, forms a sound basis
for more extensive studies which will be reported in due course.

ACKNOWLEDGEMENTS

Thanks are due to the Science Research Council for provision
of ESCA instrumentation and a research studentship to one of us
(A.D.). Thanks are due to the Institute of Petroleum for financial
support without which this work would not have been possible. We
wish to acknowledge many helpful discussions and assistance with
instrumentation for the experiments described in Section 3 provided
by H. Ronald Thomas.

REFERENCES

1. M. Hudis, *Techniques and Applications of Plasma Chemistry*,
   Ed. J. R. Hollahan and A. T. Bell, J. Wiley and Sons, New York
   (1974).
2. (a) D. T. Clark and A. Dilks, Part XIII of ESCA Applied to
       Polymers, J. Polymer Sci. Chem. Ed., 1976, in press.
   (b) For a recent review see D. T. Clark in *Structural Studies
       of Macromolecules by Spectroscopic Methods*, Ed. K. J. Ivin,
       J. Wiley and Sons, New York (1976).
3. (a) R. H. Hansen and H. Schonhorn, Polymer Letters, **4**, 203
       (1966).

(b) D. T. Clark in *Advances in Polymer Friction and Wear*, <u>5A</u>, Ed. L. H. Lee, Plenum Press, New York, 1974.
4. D. T. Clark, W. J. Feast, I. Ritchie, W. K. R. Musgrave, M. Modena and M. Ragazzini, J. Pol. Sci. Polymer Chem. Ed., 1974.
5. D. T. Clark, W. J. Feast, W. K. R. Musgrave and I. Ritchie, J. Polymer Sci. Polymer Chem. Ed., <u>13</u>, 857 (1975).
6. Cf. ref. 1 and G. Francis, *Ionization Phenomena in Gases*, Butterworth Publications Ltd., London (1960).
7. J. A. R. Samson, *Technique of Vacuum U. V. Spectroscopy*, John Wiley & Son Inc. (1967).
8. Cf. E. E. Muschlitz, Jr., Science, <u>159</u>, 599 (1968).
9. H. D. Hagstrum in *Techniques of Metals Research*, Ed. R. F. Bunshak, John Wiley & Sons, New York (1972).
10. R. H. Partridge, J. Chem. Phys., <u>45</u>, 1685 (1966).
11. Cf. J. E. Osher in *Plasma Diagnostic Techniques*, Ed. R. H. Huddlestone and S. L. Leonard, Academic Press, New York (1965).
12. Cf. Ref. 3 and D. T. Clark and W. J. Feast, J. Macromol. Sci. Revs. Macromol. Chem., C12 (2), 191 (1975).
13. Cf. D. T. Clark, *Chemical Applications of ESCA*, pp. 373-507 in *Electron Emission Spectroscopy*, Ed. W. Dekeyser, D. Reidel Publishing Co., Dordrecht, Holland (1973).
14. B. Ranby and J. F. Rabek, *Photodegradation, Photo-oxidation and Photostabilization of Polymers*, J. Wiley and Sons, New York (1975).
15. Handbook of Spectroscopy, Ed. J. W. Robinson, Vol. 1, CRC Press, Cleveland, (1974).
16. B. Wunderlich, *Macromolecular Physics*, Vol. 1, Academic Press, New York (1973).

XPS Studies of Polymer Surfaces for Biomedical Applications

Joseph D. Andrade, Gary K. Iwamoto, and Bonnie McNeill

Department of Materials Science and Engineering
University of Utah
Salt Lake City, Utah 84112

Several examples are presented on the use of XPS to study polymer surfaces intended for biomedical applications. These studies include the apparent segregation of nitrogen-rich (polyurethane) and silicon-rich (polydimethylsiloxane) components in a block copolymer due to fabricating and curing conditions; the study of "treated" and untreated polystyrene cell culture dishes; and the examination of the surface of PVC films and catheters. This work is preliminary and must not be considered definitive at this stage. It does serve, however, to indicate the applications of XPS to biomaterials surface science.

INTRODUCTION

The surface and interfacial properties of polymers are important in a variety of biomedical applications, particularly in blood-contact applications. Polymer surface-induced mechanisms appear to be important in coagulation and thrombosis, the two interrelated processes in blood "clotting". Assuming a pure polymer, with no leachable toxic or biochemically active products, the major interfacial interaction appears to be adsorption of plasma proteins. The plasma protein adsorption properties seem to correlate with the materials long term blood "compatibility".[1]

A number of surface properties have been proposed to correlate with blood compatibility. These are tabulated in Table 1.[2-11]

As is evident from Table 1, most investigators have considered only one surface property. Very little work has been done in attempting to measure a spectrum of surface properties and look for

TABLE 1. SURFACE CHARACTERIZATION OF MATERIALS FOR BLOOD COMPATIBILITY CORRELATIONS*

General Class	Surface Property	Measurement	Investigator	Year	Reference
	Wettability	Visual Observation	Lampert	1931	2
	Surface Free Energy, γ_S	Contact Angles	Lyman	1965	3
Surface Energetics	Critical Surface Tension, γ_C	Contact Angles	Baier	1972	4
	Work of Adhesion	Contact Angles	Bischoff	1968	5
	Interfacial Free Energy	Contact Angle and Intermolecular Force Calculations	Andrade	1973	6
	Wettability Spectrum	Contact Angle and Intermolecular Force Calculations	Nyilas	1975	7
Charge or Potential	Zeta Potential	Streaming Potential	Ross, Mirkovitch, et. al. Sawyer, et.al.	1953	8
	Charge	Ion Interactions	Hubbard & Lucas	1960	9
	Conductivity	Bulk Conductivity	Bruck	1973	10
Surface Chemistry	Functional Groups	Direct Synthesis	Falb	1970	2
	Functional Groups	MIR-Infrared Spectroscopy	Baier	1970	11
Cleanliness	Impurities	"Teflon Test"	Baier	1970	11

*This is a very brief, summary table. The references given are representative and not intended to be complete.

their correlation, if any, with protein adsorption and/or blood compatibility, though R.E. Baier and his collaborators[11,12] have pioneered in the use of a variety of surface characterization tools for biomedical material surface.

Although the surface chemistry of polymers is of great interest for biomedical applications, the only technique which has been routinely applied is multiple internal reflection (MIR) infrared (IR) spectroscopy. The MIR method, however, samples of the order of a wavelength, or in the micron range for the conventional infrared spectrum[13]. Thus, MIR-IR is in reality a bulk method or a method for monitoring the subsurface zone[14], though with spectral subtraction techniques true surface sensitivity can be attained.

X-ray photoelectron spectroscopy (XPS) has been used for polymer surface characterization; a number of recent comprehensive reviews are available[15,16] (and this symposium). The application of XPS to the study of freeze-etched hydrogels has been suggested[14]. A recent book is available.[17]

The work reported here is preliminary in nature; it is basically a feasibility study. Although we are using XPS as only one of a battery of surface characterization tools, only XPS studies are reported here. No attempt is made in this paper to correlate XPS results with biological properties, although a correlation matrix between surface properties and biological behavior is being sought.[18]

MATERIALS AND METHODS

Polyvinyl chloride (PVC) catheters were obtained from local commercial sources; "treated" polystyrene tissue culture dishes were from Falcon Plastics, Inc.; Avcothane-51 polyether urethane-polydimethylsiloxane block copolymer[19] was obtained from a commercial balloon sold for cardiac-assist applications.

The data reported were obtained on the following instruments: 1) duPont Instruments Model 650B photoelectron spectrometer using MgKα radiation; 2) the GCA/McPherson ESCA 36 (MgKα); 3) the Physical Electronics Ind. (PEI) ESCA/Auger electron spectrometer; and 4) the Hewlett-Packard 5950B instrument, which utilizes monochromatic Al Kα 1,2 radiation. Samples were generally mounted on double-stick tape, run at 10^{-7} to 10^{-9} torr vacuum at ambient temperatures, and charging shift referenced to the C-1s line. Some samples were mechanically mounted without tape or Indium foil (PEI; GCA/McPherson, Hewlett-Packard).

RESULTS AND DISCUSSION

Avothane-51 is a synthetic elastomer that is widely used for blood-contact applications. Developed by E. Nyilas of the Avco-Everett Corporation, it is a polyether urethane/polydimethylsiloxane copolymer[19,20]. Its blood compatibility is reported to be due to a very low H-bonding capacity at the surface[21], which is highly dependent on the fabrication/dipping/drying conditions[20]. Nyilas has concluded, based on MIR-IR data[20] that "... the distribution as well as the orientation of the silicone component of Avcothane-51 in the solid films of this elastomer is anisotropic. It is this anisotropic distribution that induces the differences between surface molecular structures and the different biological effects ..." (Reference 20, page 80,83).

The variable surface properties are probably due to different ratios of siloxane and urethane components at the surface. This is a good problem for XPS as the N-1s line can be attributed only to the urethane component while the Si-2p line can be attributed only to the silicone component. An Avcothane-51 cardiac-assist balloon was examined. The outer surface of the balloon is designed and fabricated for optimal blood compatibility, while the inner surface never sees blood. Inner and outer (blood-contact) surfaces were examined in the duPont, Physical Electronics, Ind. and GCA/McPherson XPS instruments. The "outer" (blood) surface had a N-1s signal which was clearly evident in fast survey scans on all three instruments. No N-1s signal was evident in survey scans for the "inner" surface. Figure 1 gives the N-1s and Si-2p data taken on the duPont 650B electron spectrometer, corrected to a C-1s line of 285.5 eV. Clearly the "blood compatible" surface contains both urethane and siloxane while the "blood incompatible" surface is largely siloxane.

The use of a higher resolution instrument, such as the Hewlett-Packard 5950B, permits one to distinguish the higher binding energy polyether urethane carbons from those in the polydimethyl siloxane component.

A commercial polystyrene petri dish was examined (Figure 2) and showed the presence of surface oxygen, probably due to oxidation, and silicon, perhaps due to mold release agents. The commercial "treated" dish (probably corona discharged) showed the absence of silicon and a much more intense O-1s signal, as expected (Figure 2). Such surfaces are commonly used for *in vitro* cell culture. The surface properties of the substrate material are known to play an important role in the morphology and the properties of the cell culture.[22] A residue of organic biochemical matter remains on a surface after cells have been cultured on it and then removed. This material has been called cell exudate or substrate-attached material[23]. XPS analysis of the substrate-attached material may

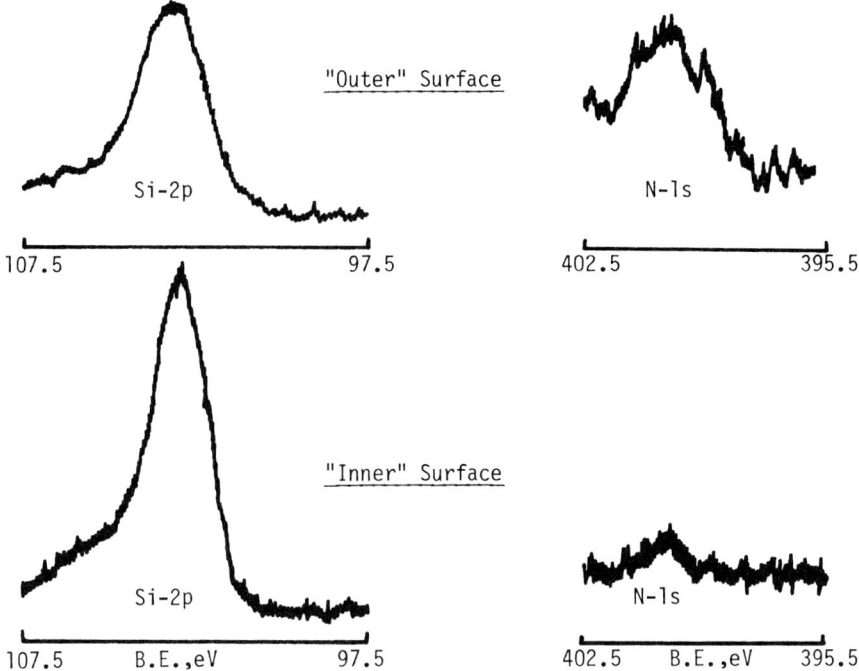

Fig. 1. Avothane-51 cardiac-assist balloon examined in duPont 650B using MgKα. Spectra were charge-corrected to the C-1s line at 285.5 eV. Note that the outer or "blood compatible" surface shows a higher polyurethane content (N-1s) than the inner surface. Note also the change in the Si-2p intensity for the two surfaces. Silicon spectra contained at sensitivity of 500 counts/sec/vertical distance at 0.2 eV/sec scan rate. Nitrogen spectra obtained at sensitivity of 100 counts/sec/vertical distance and 0.05 eV/sec scan rate.

be very important in characterizing cell exudates and cell-substrate interactions[24].

Catheters are tubular devices used to provide access to an animal or patient, generally in the vascular system, to sample blood, administer nutrients or drugs, etc. Little is known about the surface nature of commercial catheters.

Polyvinyl chloride (PVC) is commonly used as a tubing and catheter material. It is highly plasticized, generally with phthalates or adipates. As PVC is susceptible to dehydrochlorina-

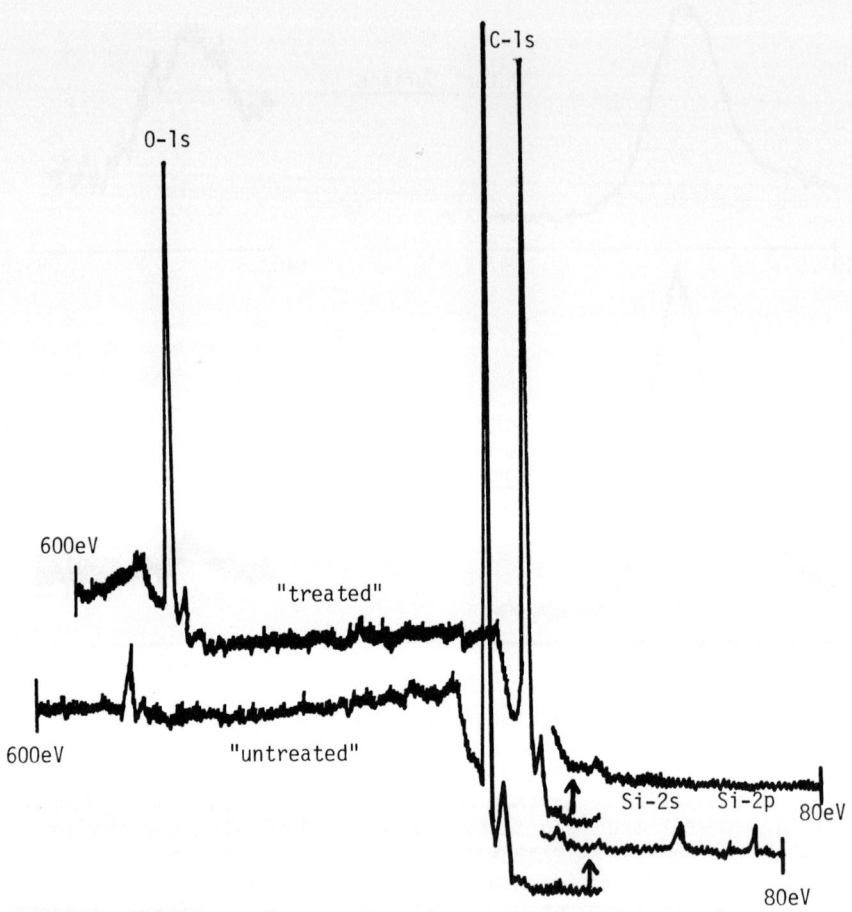

Fig. 2. XPS spectra of a commercial polystyrene cell culture dish, as received. The lower spectrum is for the untreated dish, showing the presence of surface oxidation (O-1s) and silicon (Si-2s, 2p), perhaps due to a mold release agent. The upper spectrum is for the treated, highly wettable dish, showing the absence of silicon and much more extensive surface oxidation. These spectra were obtained on the PEI XPS with MgKα - no charge corrections have been applied.

tion at processing temperatures, it is generally heat stabilized with various metal salts of fatty acids and epoxidized oils. Further, it is known that the presence of phthalate at the surface produces high platelet adhesion, which may be responsible for the relatively poor blood compatibility of common PVC[25].

XPS examination of a pure PVC film (cast from N,N-dimethylformamide solution) produced the spectrum in Figure 3. The relatively pure film, prepared under ambient conditions, shows fairly substantial surface oxygen probably due to oxidation. The spectrum of a commercial PVC catheter is given in Figure 3. Comparison of the two spectra clearly shows the much higher level of surface oxygen in the fabricated material. One must be cautious in such analyses due to the probable presence of a surface layer of plasticizer in the commercial material. A problem with PVC is the apparent radiation "damage" under x-irradiation, leading to a progressive loss in intensity of the Cℓ lines. The decrease in Cℓ-2p intensity appears to be linear with x-ray exposure for at least the first 30 minutes in the duPont 650B XPS. We detected no such effects in the Hewlett-Packard instrument even after 2 hours of x-ray exposure.

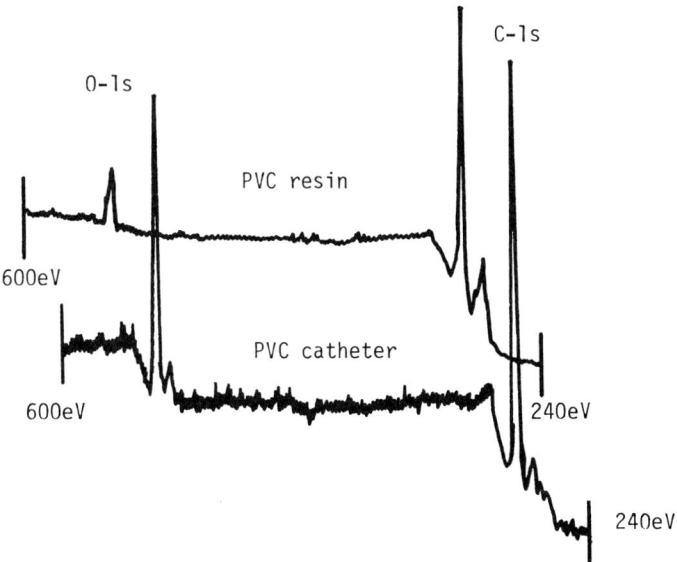

Fig. 3. XPS spectra (MgKα) of a PVC film cast from DMF (upper spectrum) and a commercial PVC catheter (lower spectrum) clearly show evidence of oxygen atoms at the surface. Note the greater degree of oxidation in the catheter material. Upper spectrum obtained on duPont 650B; lower spectrum is from the PEI instrument. The Cℓ-2p lines are intentionally omitted.

ACKNOWLEDGEMENTS

We thank the following persons for generous donations of instrument time and assistance for the evaluation of their XPS instruments:

W. Riggs and C. Ginnard, duPont Instruments, Monrovia, California
J. Rendina and P. Larsen, GCA/McPherson, Inc., Acton, Massachusetts
L. Davis and J. Koenig, Physical Electronics Industries, Inc., Eden Prairie, Minnesota
A. Wolstenholme, AEI, Inc., Elmsford, New York
M. Kelly and H. Harrington, Hewlett-Packard, Inc., Palo Alto, California
P. Williams, VG Scientific, Inc., East Grinsted, England

REFERENCES

1. S.W. Kim and R.G. Lee, in *Applied Chemistry at Protein Interfaces*, R.E. Baier, ed., Amer. Chem. Soc. Adv. in Chem. Series No. 145, 1975.
2. R.I. Leininger, CRC Critical Reviews in Bioengineering, $\underline{1}$, 333 (1972).
3. D.J. Lyman, W.M. Muir and I.J. Lee, Trans. Amer. Soc. Artificial Internal Organs, $\underline{11}$, 301 (1965).
4. R.E. Baier, Bull. N.Y. Acad. Med., $\underline{48}$, 257 (1972).
5. K.B. Bischoff, J. Biomed. Materials Res., $\underline{2}$, 89 (1968).
6. J.D. Andrade, Med. Inst., $\underline{7}$, 110 (1973).
7. E. Nyilas, W.A. Morton, D.M. Lederman, Th. H. Chiu, and R.D. Cumming, Trans. Amer. Soc. Artif. Int. Organs, $\underline{21}$, 55 (1975).
8. P.N. Sawyer, ed., *Biophysical Mechanisms in Vascular Homeostasis and Intravascular Thrombosis*, Appleton-Century-Crofts, 1965.
9. D. Hubbard and G.L. Lucas, J. Appl. Physiology, $\underline{15}$, 265 (1960).
10. S.D. Bruck, Nature, $\underline{243}$, 416 (1973).
11. R.E. Baier, V.L. Gott, and A. Feruse, Trans. Amer. Soc. Artificial Internal Organs, $\underline{16}$, 50 (1970).
12. V.A. de Palma, R.E. Baier, J.W. Ford, V.L. Gott, and A. Feruse, J. Biomed. Materials Res., Biomed. Materials Symp. No. 3, 37 (1972).
13. N.J. Harrick, *Internal Reflection Spectroscopy*, Interscience, N.Y. (1967).
14. J.D. Andrade, R.N. King, and D.E. Gregonis, in *Hydrogels for Medical and Related Applications*, J.D. Andrade, ed., Amer. Chem. Soc. Symp. Series, 1976, in press.
15. D.T. Clark and W.J. Feast, J. Macromol. Sci., Revs. Macromol. Chem., C12 (2), 191 (1975).
16. D.T. Clark, in *Advances in Polymer Friction and Wear*, Lieng-Huang Lee, ed., Part A, Plenum Press, page 241, 1974.

17. T.A. Carlson, *Photoelectron and Auger Spectroscopy*, Plenum Press, 1975.
18. R.N. King, Ph.D. Dissertation, University of Utah, in progress.
19. E. Nyilas, U.S. Patent 3,562,352, February 9, 1971.
20. E. Nyilas, W.A. Morton, R.S. Ward, and P.N. Madras, Annual Report, U.S. Government Report on Contract, AK-1-12506, August, 1973.
21. E. Nyilas, Proc. 23rd Ann. Conf. Engrg. Med. and Biol., November, 1970, page 12-2.
22. L. Smith, D. Hill, J. Hibbs, S.W. Kim, J.D. Andrade, and D.J. Lyman, Polymer Preprints, 16, 186 (1975).
23. L.A. Culp, J. Cell Biology, 63, 71 (1974).
24. R. Van Wagenen, Ph.D. Dissertation, University of Utah, June, 1976.
25. S.W. Kim, R.V. Peterson and E.S. Lee, J. Pharm. Sci., 65, 670 (1976).

Discussion

On the Paper by D.T. Clark

J.D. Andrade (*University of Utah*): How do the density, structure, and orientation of the para-xylylene films influence the mean free path results?

D.T. Clark (*Durham University*): As you probably know, the pyrolylic route to Parylene polymers from paracyclophane precursors has been exhaustively studied particularly by people at Union Carbide. The deposition of p-xylylene species on a metal substrate and the subsequent polymerization reaction undoubtedly leads to uniform films formed in planes parallel to the surface of in our case gold substrates. Each thickness at which the attenuations were measured corresponds to successive deposition and since there is little evidence for crosslinking, this probably leads to parallel planes of polymer. The density of Parylene film is extremely well documented and since our thicknesses are measured directly with a quartz deposition monitor, and it is a trial exercise to show that islanding is not important. The measurement of signal intensities from both substrate and overlayer leaves little room for doubt that the escape depths in organics are relatively short.

L.H. Lee (*Xerox Corp.*): Dr. Clark's excellent presentation has once again brought us to the frontier of recent research with ESCA for polymer surfaces. Dr. Clark has shown us various structural effects by substituents on polymers. This is indeed a timely follow-up for the earlier paper presented to the International Symposium on Advances in Polymer Friction and Wear, in the spring of 1974. (Ref. <u>Advances in Polymer Friction and Wear</u>, Ed. L.H. Lee, Plenum Press, New York, 1974 <u>5A</u>, 241.)

On the Paper by D.E. Williams

G.P. Ceasar (*Xerox Corp.*): Was there any attempt made to correct for edge effects or differences in sintering yield between elements?

D.E. Williams (*Dow Corning Corp.*): We made no attempt to correct for edge effects or for different sputter yields. There should have been no edge effects. The beam current density was

uniform across the entire sample surface as determined by
Faraday cup measurements. However, there may have been related
effects due to the fact that the fracture plane was not completely flat or perpendicular to the sample bar axis. Sputtering of
duplicate fractures at the same voltage gave concordant data so
that such effects do not alter the trends and conclusions of our
work. We are skeptical that sputtering yield differences can be
compensated since such differences will be highly dependent on
the detailed chemical structure of the polymer.

W. Newby (*Aluminum Company of Canada*): Does your data indicate
what form of damage is taking place during sputtering?

D.E. Williams: We suspect that thermally labile functional
groups are released as volatiles within the zone of sputter
damage and subsequently escape. The penetration depth of 0.5 to
2 KV argon ions into polymers has not been determined, but is
probably on the order to tens of angstroms. Within this zone
very high transient temperatures are produced. The voltage
dependence and the appearance of an equilibrium amount of the
elements after prolonged sputtering suggests the sputter zone
is somewhat shallower than the ESCA sampling depth.

On the Paper by D.T. Clark and A. Dilks

C. Turnquist (*USCI (Division of C.R. Bard)*): Is the plasma used
in your studies commerical or homemade?

D.T. Clark (*Durham University, England*): The answer to your
question is both. Most of the work was carried out with a
Tegal Corporation unit capable of up to 100 watts output and
with the pulsing facility which is most important for maintaining a stable plasma at low average power loadings.

On the Paper by J.D. Andrade

C. Turnquist (*USDI (Division of C.R. Bard)*): How was the analyzer pressure of 10^{-9} torr obtained? I didn't notice a separate
pumping system on your schematic.

J.D. Andrade (*University of Utah*): The preparation of chamber
and analyzer chambers are separately ion pumped.

PART II

Infared and Laser Raman Spectroscopy

Introductory Remarks:
Surface Characterization of Polymers by Infrared and Raman Spectroscopy

Lieng-Huang Lee

Wilson Center for Technology
Xerox Corporation
Webster, New York 14580

Infrared[1] and Raman spectroscopy[2] have been used to characterize the surfaces of organic materials. The use of infrared to study the structure of polymers[3-6] was far more extensive than that of Raman spectroscopy especially prior to the incorporation of lasers[7,8] as light sources for the latter. Since the Sixties, several new developments in IR spectroscopy[9,10] have transformed it into a very important analytical tool for studying polymer surfaces. Several papers presented to this Symposium will discuss the potential of these developments.

INTERNAL REFLECTION SPECTROSCOPY

The applications of attenuated total reflection (ATR) or internal reflection spectroscopy (IRS) were first independently reported by Fahrenfort[11] and Harrick.[12] The principles of the technique have been discussed by Harrick.[13] For IRS, the sample, refractive index, n_2, is brought into direct contact with the reflecting surface of a higher refractive index, n_1. The IR beam is introduced into an optically transparent medium at an angle greater than the critical angle. If a material which absorbs one of the wavelengths is placed near the IR beam, the transient wave couples with the material resulting in an attenuated total reflection. Scanning through the wavelengths then gives the spectrum.

IRS has been used to study adhesion, contamination, degradation,[14] fiber structure,[15] lubrication, oxidation, and plasticization of polymers. It has also been employed to determine the spectra of electrons and holes in the semiconductor space-charge region,[16] and of adsorbed molecules on semiconductor and metal surfaces. Dr. Harrick[17] will discuss in detail other applications and experimental techniques related to IRS.

REFLECTION-ABSORPTION SPECTROSCOPY

Another interesting development of infrared spectroscopy came about around the late 1960's. Reflection-absorption spectroscopy (RAS) or external reflection spectroscopy (ERS)[17,18] involves the transmission of an IR beam through ultrathin sample layers on a metal plate and the reflection at a glancing angle. Greenler[19] theorized that the spectral sensitivity at an interface can be enhanced a thousandfold or more by choosing an incident angle between 80 and 88°.

For RAS, the standing wave is somewhat more complex than that for IRS.[20] The sample thickness is normally very thin compared to the wavelength, and it is usually extended from 500Å to a fraction of monolayer. For monolayer adsorption studies, a high vacuum system is needed along with surface preparation equipment. The specialized manipulation also requires nonroutine noise reduction and signal amplification. Therefore, RAS is still not yet suitable for routine analysis.

RAS has been applied to the studies of chemisorption, degradation, oxidation[21] of solid materials, etc. IRS and RAS[22] have also been combined to detect the interaction at the polymer-metal interface. In this part, Allara[23] discusses specifically the experimental techniques of poly(1-butene) studied with RAS.

FOURIER TRANSFORM INFRARED SPECTROSCOPY

The third significant development in infrared spectroscopy is Fourier transformation (FTIR)[24-26] partly due to the availability of commercial equipment in recent years. Since 1965, when Cooley and Tukey[27] published their algorithm, the computation time for Fourier transformation has been reduced by several orders of magnitude. With the introduction of minicomputers and the availability of the Michelson interferometer, it has become possible to obtain on-line, short-time Fourier transformations. In FTIR, all IR signals can be observed simultaneously, and the resultant complex signals (interferogram) are transformed into a standard spectral distribution by Fourier analysis.

FTIR has the advantage of larger energy gathering capabilities, multiple scanning and speed. Furthermore, any two stored spectra can be compared at any time separation between the actual measurement and their substractions. Therfore, it is very convenient to compare the unknown from the known, the unreacted from the reacted, or the unisomerized from the isomerized. Specifically, in polymer technology.[28] FTIR has been employed to study the following:

1. adsorption on polymer surface,[28]

2. chemical modification of polystyrene,[29]

3. hydrolysis of polysiloxanes,[28]

4. irradiation of polyethylene,[30]

5. mechanisms of coupling interaction at the composite interface,[31]

6. oxidation of rubbers,

7. plasticization of polyvinyl chloride.[32]

Since one of the papers on FTIR presented to this Symposium is not included in this volume, the readers should consult directly with some of the cited references.

LASER RAMAN SPECTROSCOPY

In recent years, Raman spectroscopy has also made significant progress. Raman spectroscopy actually is not a new technique. In 1923, Smekal[33] theoretically predicated the "Raman" effect, and the effect was discovered experimentally by Sir. C.V. Raman[34] in 1928. Two years later, Raman was awarded the Nobel Prize in Physics. However, only in the Sixties did Raman spectroscopy become a practical analytical technique. In 1962, Porto and Wood[35] and Stoicheff[36] used pulsed ruby masers to obtain Raman spectra of some strongly scattering liquids. Only with lasers, can good quality spectra be recorded from gases, liquids, single crystals and solids. For polymer, the first spectrum of isotactic polypropylene was obtained by Schaufele[37] in 1967.

Raman and infrared spectroscopy actually complement each other. Unlike IR, Raman spectroscopy can be operated at low frequencies down to 5 cm^{-1} (2000 u). Specifically, Raman spectroscopy[27] offers some advantages over infrared for several functional groups:

1. O-H vibrations are weak in the Raman compared to C-H and N-H vibrations. Therefore, in the presence of O-H, both C-H and N-H vibrations can be better separated

without strong interference in the Raman spectra than in the IR spectra.

2. Vibrations such as C=C, C≡C, S-S, N=N are symmetric or pseudosymmetric. These vibrations are strong in the Raman but weak or absent in the infrared.

3. Both C=S and S-H vibrations are relatively strong in the Raman, whereas the latter is virtually absent in the infrared.

4. The nitrile stretching vibration is always strong in the Raman. However, the intensity of the nitrile group in IR is affected by the adjacent electronegative substituent as in the case of cyanoacrylate.

Though Raman spectroscopy has been employed to study polymers,[38] much research is still needed on polymer surfaces. Recent work by Koenig and his co-workers[39,40] at Case Western Reserve University on the glass surface demonstrates the potential of using laser Raman spectroscopy for surface research. For further study, readers should find the review by Andrews and Hart[41] in this volume to be timely and useful.

REFERENCES

1. M.L. Hair, *Infrared Spectroscopy in Surface Chemistry,* Marcel Dekker, New York (1967).
2. N.B. Colthup, L.H. Daly and S.E. Wiberley, *Introduction to Infrared and Raman Spectroscopy,* Academic Press, New York (1964).
3. R. Zbinden, *Infrared Spectroscopy of High Polymers,* Academic Press, New York (1964).
4. D.O. Hummel and F. Scholl, *IR Analysis of Polymers, Resins and Additives,* II, Verlag Chemie, Weinheim, 1973.
5. H. Takokora, in *Polymer Spectroscopy,* Ed. D.O. Hummel Verlag Chemie, Weinheim, 1974.
6. Chicago Society for Paint Technology, *Intrared Spectroscopy; Its Use in the Coating Industry,* Federation of Societies For Paint Technology, Philadelphia, Pa. (1969).
7. L. Loader, *Basic Laser Raman Spectroscopy,* Heyden/Sadtler, London (1970).
8. M.C. Tobin, *Laser Raman Spectroscopy,* Wiley, New York (1971)
9. K. Holland-Moritz and H.W. Siesler, Appl. Spect. Rev. $\underline{11}$ (1), 1-55 (1976).
10. C.D. Craver, in *Applied Polymer Science,* Ed. J.K. Craver and R.W. Tess, p. 74,Division of Organic Coatings and Plastics Chemistry, American Chemical Society (1975).

11. J. Fahrenfort, Spectrochim. Acta. 17, 698 (1961).
12. N.J. Harrick, Ann. N.Y. Acad. Sci. 101, 928 (1963).
13. N.J. Harrick, *Internal Reflection Spectroscopy*, Wiley-Interscience, New York (1967).
14. M.G. Chan and W.L. Hawkins, Polymer Preprints 9,(2),1638(1968).
15. J. Dechant, Faserforsch. Textiltech. 25, 24 (1974).
16. N.J. Harrick, J. Phys. Chem. Solids 8, 106 (1959).
17. N.J. Harrick, The Proceedings of this Symposium, Vol. 2.
18. S.A. Francis and A.H. Ellison, J. Opt. Soc. Am. 49, 131 (1959).
19. R.G. Greenler, J. Chem. Phys. 44, 310 (1966).
20. H.G. Tompkins, Appl. Spectr. 30 (4), 377 (1976).
21. G.W. Poling, J. Electrochem. Soc. 116, 958 (1969).
22. M.G. Chan and D.L. Allara, Polym. Eng. Sci. 14, 12 (1974); J. Colloid Interface Sci. 47, 697 (1974).
23. D.L. Allara, The Proceedings of this Symposium, Vol. 2.
24. R.J. Bell, *Introductory Fourier Transform Spectroscopy*, Academic Press, New York (1972).
25. A.G. Marshall and M.B. Comisarow, Anal. Chem. 47 (4), 491A (1975).
26. J.B. Bates, Science 191, No. 4222, 31 (1976).
27. J.W. Cooley and J.W. Tukey, Math. Comput. 19, 297 (1965).
28. J.L. Koenig, Appl. Spectr. 29 (4), 293 (1975).
29. T. Matsuda and M.H. Litt, J. Poly. Sci. 12, 489 (1974).
30. D. Tabb, J. Sevich and J.L. Koenig, J. Poly. Sci. 13, 815 (1975).
31. I. Ishida, 22nd. Research Symposium Abstract, Case Western Reserve University, Nov. 9, 1976
32. D.L. Tabb and J.L. Koenig, J. Poly. Sci. in press.
33. A. Smekal, Naturwiss. 11, 873 (1923).
34. C.V. Raman and K.S. Krishnan, Nature 121, 501 (1928).
35. S.P.S. Porto and D.L. Wood, J. Opt. Soc, Am. 52, 251 (1962).
36. B.P. Stoicheff, Tech. Intern. Spectr. Colloq., Uni. of Maryland, June, 1962, Spartan Books, Washington (1963).
37. R.F. Schaufele, J. Opt. Soc. Am. 57, 105 (1967).
38. J.L. Koenig, Macromol. Rev. 6, 59 (1972).
39. J.L. Koenig and P.T.K. Shih, Mat. Sci. & Eng. 20, 127 (1975).
40. P.T.K. Shih and J.L. Koenig, Mat. Sci. & Eng. 20, 137 (1975); 145 (1975).
41. R.D. Andrews and T.R. Hart, The Proceedings of this Symposium, Vol. 2.

Plenary Lecture: Transmission and Reflection Spectroscopy, Nature of the Spectra

N. J. Harrick

*Harrick Scientific Corporation
Ossining, New York 10562*

Transmission and reflection (external, internal and diffuse) are briefly described and spectra compared. It is shown that drastic differences in spectra may occur, not only from the use of different spectroscopy techniques, but also for the same technique, including for transmission, with different sample preparation. These spectra serve to emphasize that care must be exercised both in sample preparation and choice and use of spectroscopy technique. Most of the observed differences in the spectra can be explained by means of optical phenomena also discussed in this paper. Although the spectral differences discussed here were demonstrated only for the infrared, similar effects and explanations will apply to the ultraviolet, visible and far infrared spectral regions.

INTRODUCTION

There are three distinct categories in Reflectance spectroscopy, viz., External Specular, Internal Specular, and Diffuse Reflection. Principles, applications, and relative advantages will be discussed.

External Reflection Spectroscopy (ERS) at near Brewster's angle and using p-polarized light has been employed to study films as thin as a single monolayer on metal surfaces. ERS is a nondestructive method that can be used with advantage in many situations. Via ellipsometry, films as thin as one hundredth of a monolayer can be studied and both index of refraction and extinction coefficient can be measured.

Internal Reflection Spectroscopy (IRS), originally proposed for the study of surfaces, can be employed to study weathering, adhesion, catalysis, adsorbed species on particulate matter, etc. Index of refraction, angle of incidence and polarization control the nature of the spectra and give the method a great deal of flexibility.

The spectra obtained via Diffuse Reflectance are a complex mixture of external and internal reflection and transmission and depend on particle size, shape, polarization, angle of incidence and observation and extinction coefficient; thus very few cases lend themselves to rigorous analysis. Nevertheless, the technique is very important since often no other technique can be used to study many practical cases, e.g., powders, paper, paints, etc., as viewed by the human eye.

It is shown here that spectra for transmission, internal and external reflection depend on a number of parameters such as polarization and angle of incidence and, depending on sample preparation and method of recording spectra, may differ from each other sometimes very dramatically. One must, therefore, use "libraries" of transmission spectra with caution for identifying other spectra. A series of spectra is shown to demonstrate some of the differences that may arise in reflection and transmission spectra. Furthermore, it is shown that many of the differences in the spectra can be explained by the wavelength dependence of the optical constants, the dependence of the reflectivity on polarization and angle of incidence, the occurrence of interfringes and standing waves, and the presence of optical cavities. These spectra serve to emphasize that caution must be exercised both in sample preparation and in the choice and use of spectroscopy techniques. All of these differences and many more may occur simultaneously in diffuse reflectance spectroscopy (DRS). And therefore, in this case (DRS), extreme care must be exercised in the interpretation of spectra.

Reflectance techniques (External, Internal and Diffuse) are very powerful, especially for surface studies and sometimes are the only ones that can be successfully employed for certain studies. Since these techniques are being more widely used, it is important to understand the nature of the spectra.

OPTICAL PHENOMENA*

A. Optical Constants

Optical spectra may be dependent on sample preparation and technique employed to record the spectrum. Therefore, ideally one

---

*These optical phenomena are also discussed in Reference 4g.

would like to determine the optical constants because these specify
the material exactly and are not dependent on sample preparation or
spectroscopic technique. Typical optical constants are shown in
Fig. 1 where the extinction coefficient goes through a maximum and
there is dispersion in the refractive index as the wavelength is
changed through an absorption band.

Since normally the optical constants can not be readily
determined, and since different techniques, which produce different
spectra, must often be employed, it is useful to review a number of
optical phenomena that can be helpful in understanding the differences
in the optical spectra.

B. Reflectivity of an Interface

The reflectivity of an interface is dependent on both the
extinction coefficient and the refractive index and, for normal
incidence, is given by

$$R = \frac{(n-1)^2 + (n^2 k^2)}{(n+1)^2 + n^2 k^2}.$$

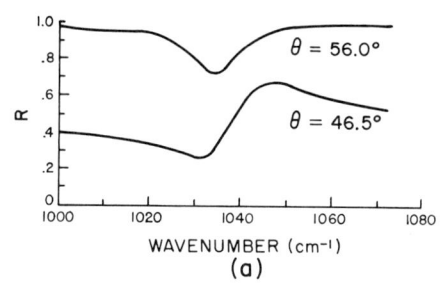

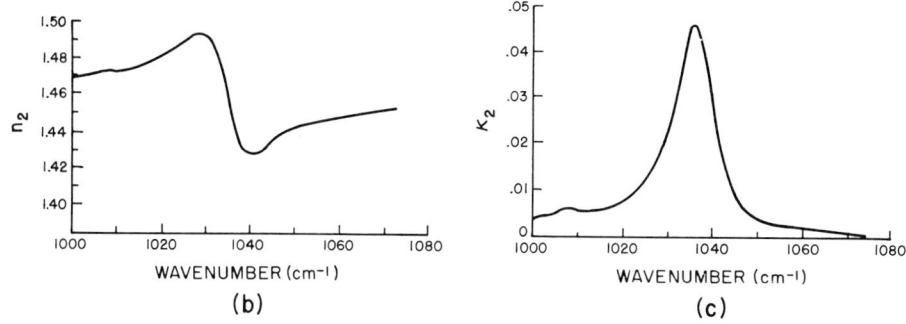

Fig. 1. Optical constants of liquid benzene at 1035 cm^{-1} calculated
from (a) internal reflection measurements, (b) refractive
index, n_2, and (c) extinction coefficient, k_2.

If the wavelength dependence of the optical constants are inserted in this equation, it is evident that the reflectance is more dependent on n than on k and will show a dispersion, qualitatively the same as the dispersion in n and a mirror image of that for internal reflection (light proceeding from dense to less dense medium).

The dependence of the reflectivity on angle of incidence for two different cases is shown in Fig. 2. Note that for external reflection (solid curves) and for parallel polarization the reflectivity is zero at Brewster's (polarizing) angle. There is a similar minimum for internal reflection ($\theta_B + \theta_p = 90°$) and that above the critical angle the reflectivity is 100% for both perpendicular and parallel polarization.

If medium 2 is highly conductive (e.g., metal), the reflection curves resemble the solid curves of Fig. 2 except that at the polarizing angle the reflectivity is not zero. Furthermore, these curves are qualitatively the same whether medium 1 is air (n = 1) or one having a high dielectric constant, i.e., there is no critical angle in this case, even though light proceeds from a medium having a high refractive index to one having a lower one.

C. Interference Fringes

Light striking a film sample is partially reflected at each surface and thus decomposed into many components as shown in Fig. 3a. The components superimpose to interfere constructively or destructively as the wavelength is changed and result in an interference fringe pattern, shown in Fig. 3b, as observed in transmission or reflection. These fringe patterns can be used to calculate film thickness and refractive index but are troublesome in recording spectra because they distort absorption bands and obscure weak ones.

D. Standing Waves

Incoming and reflected outgoing waves superimpose to form a sinusoidal standing wave pattern near the reflecting interface. Even though the light passes through the sample, a weak absorber placed in this wave pattern will interact with it only at the loops and not at the nodes of the electric field amplitude of the standing wave pattern. For a metal, because of its high conductivity, there is an electric field minimum for normal incidence at the surface and there can be little or no interaction with a thin sample on the metal (see Fig. 4a).

For internal reflection, on the other hand, there is, as shown in Fig. 4b, an evanescent wave in the rarer medium and the electric field amplitude is different from zero at the surface and in fact is a maximum at the critical angle resulting in maximum interaction

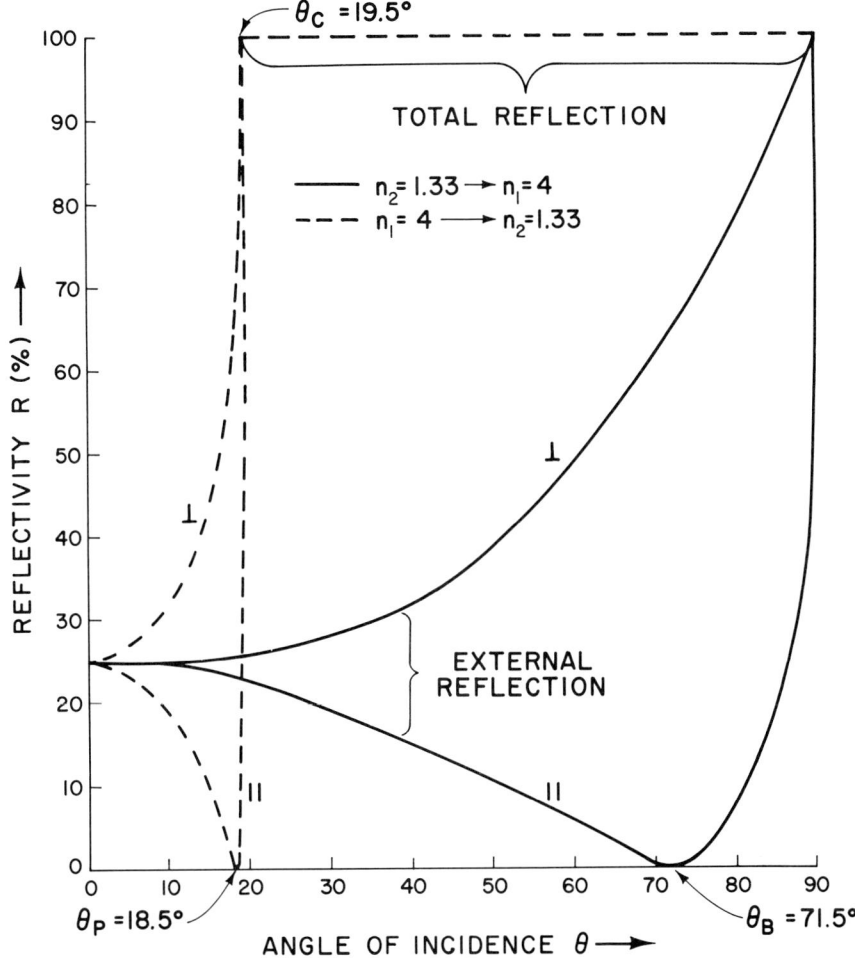

Fig. 2. Reflectivity versus angle of incidence for an interface between media with indices, $n_1 = 4$ and $n_2 = 1.33$, for light polarized perpendicular, $R_\perp$, and parallel, $R_{//}$, to plane of incidence for external reflection (solid lines) and internal reflection (dashed lines). θ_c, θ_B and θ_p are the critical, Brewster's, and principal angles, respectively.

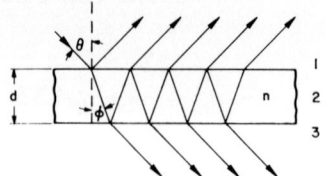

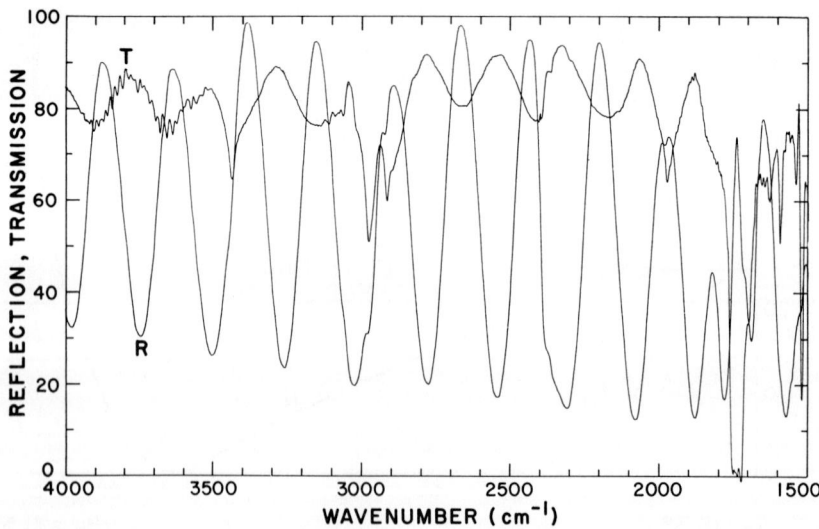

Fig. 3. (a) Decomposition of a beam of light into multiple components by reflection at the front and back surface of a thin film.
(b) Transmission and reflection spectra of a ½-mil Mylar film. Note that better-defined interference fringes with less distortion due to the absorption bands are obtained in reflection.

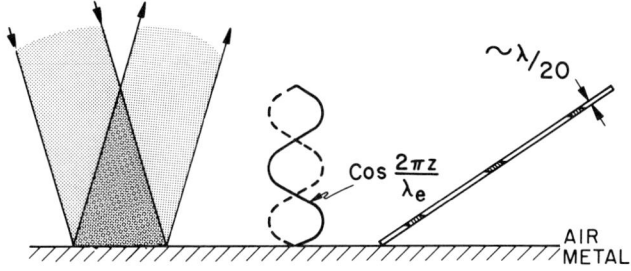
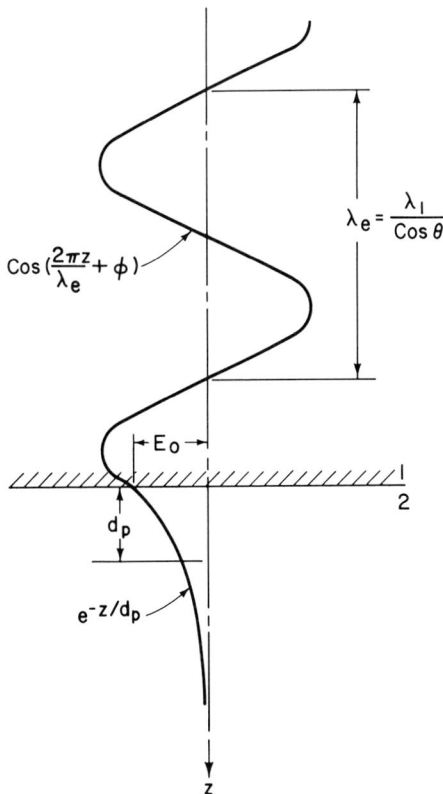

Fig. 4. Standing electric field wave pattern for (a) external reflection from a metal surface, and (b) total internal reflection; (c) electric field vectors in standing wave pattern.

for a thin film right at the surface. Note that in both cases the effective wavelength of the standing wave pattern is dependent on both the wavelength of the incident light and the angle of incidence. Therefore, for a thin film placed parallel to the surface in the sinusoidal wave pattern, the absorption will alternately increase and decrease as either the wavelength or angle of incidence is changed.

E. Electric Field Vectors

It is a known fact that for propagating electromagnetic waves the electric fields can only exist perpendicular to the direction of propagation. Hence, for normal incidence only dipoles that are oriented parallel to the plane of the surface can interact with the electromagnetic radiation for any polarization. It is clear from Fig. 4c that this is also the case for perpendicular polarization at all angles of incidence. For parallel polarization, on the other hand, as θ_1, the angle of incidence, increases E_z becomes finite and at grazing incidence E_z is the only electric field present and only dipoles perpendicular to the surface can interact with the radiation. This is also the case for internal reflection near the critical angle. At grazing incidence and near the critical angle, anisotropic films can thus be studied by simply rotating the plane of polarization and without moving the sample. Calculations of the electric field amplitudes can be found elsewhere[1].

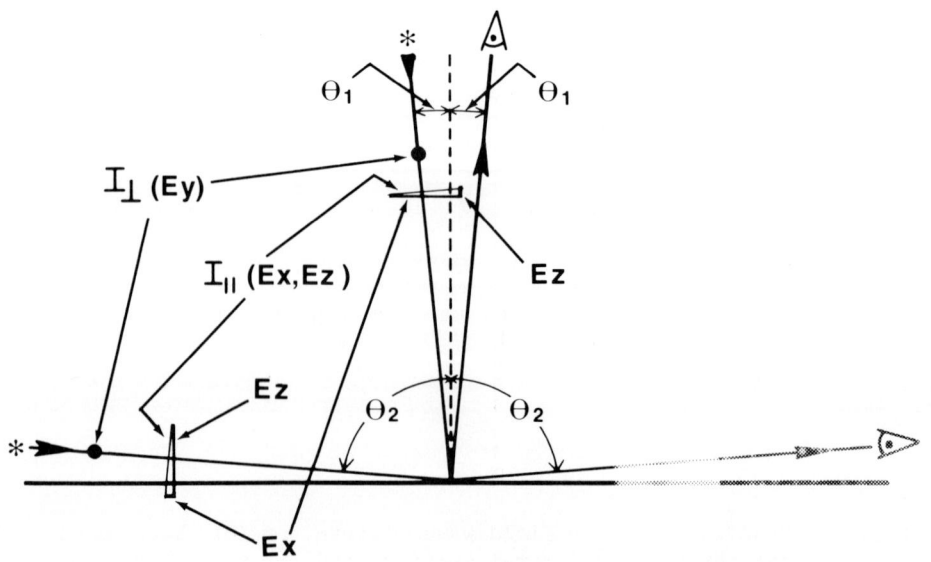

Fig. 4 (c)

F. Optical Cavities

A spectroscopy technique may purposely or sometimes inadvertently be employed in such a way that absorption band intensity and band shape may be highly intensified, diminished or distorted.

For example, a thin film absorber placed on optical prism and examined via Internal Reflection Spectroscopy will yield a spectrum that, except for expected differences, will be comparable to that obtained via Transmission Spectroscopy. If, however, two optically transparent films of certain thickness and refractive index are placed on this prism and then the same thin absorbing film is brought into contact with this structure, there will be total (or highly magnified) absorption for polarized light at a selected wavelength and angle of incidence regardless how weakly absorbing the film may be. What we have here is a resonant structure called an optical cavity. Spectra demonstrating the effect of an optical cavity are shown in Fig. 5. Similar effects may also be observed in External Reflection Spectroscopy.

OPTICAL SPECTROSCOPY TECHNIQUES

Optical spectroscopy is "fingerprinting" (materials). This fingerprint represents the stretching, vibrating, bending of molecules and libraries of spectra are available for indentification by comparison. It is generally assumed that these "spectra are the same day after, regardless where recorded". We will show here that this is not necessarily the case and care must be taken in both sample preparation and use of the technique employed.

Instrumentation for these spectroscopy techniques is discussed briefly in the appendix to this paper.

A. Transmission Spectroscopy

Transmission Spectroscopy (TS) (see insert of Fig. 6a) is the most widely used optical spectroscopy technique; it is also the technique for which the largest collection of reference spectra is available. These spectra are widely used as standards, not only for transmission but also for spectra recorded by other techniques, e.g., external, internal and diffuse reflectance. The following spectra will show that placing too much faith in these libraries can be risky, not only for internal and external reflection spectra, but also for transmission spectra. Examples follow.

Fig. 6a shows that, as predicted by Fig. 2, because R is larger the fringe pattern has greater amplitude at a large angle of incidence. The fringe pattern at $\theta = 80°$ also appears to be more complex; however, on examination with polarized light this

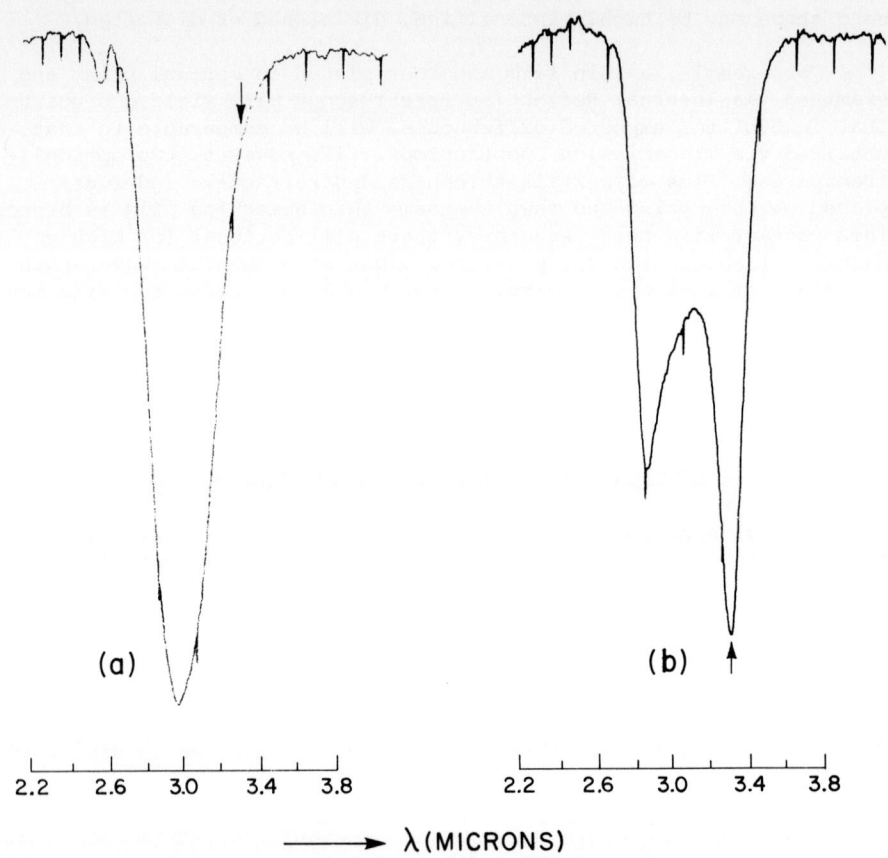

Fig. 5. The O-H absorption band of water at $\theta = 29°$ for $||$-polarization using (a) an uncoated Si prism and (b) the optical $||$-cavity construction with the same prism. The strong amplification for curve (b) at 3.3 leads to the distortion of the broad O-H band.

pattern actually is shown to consist of not a single but of two superimposed fringe patterns (Fig. 6b).

We know from the discussion of Fig. 4c that light with perpendicular polarization has electric field vectors only in the plane of the film, and parallel polarization near grazing incidence for external reflection (or equally near the critical angle for internal

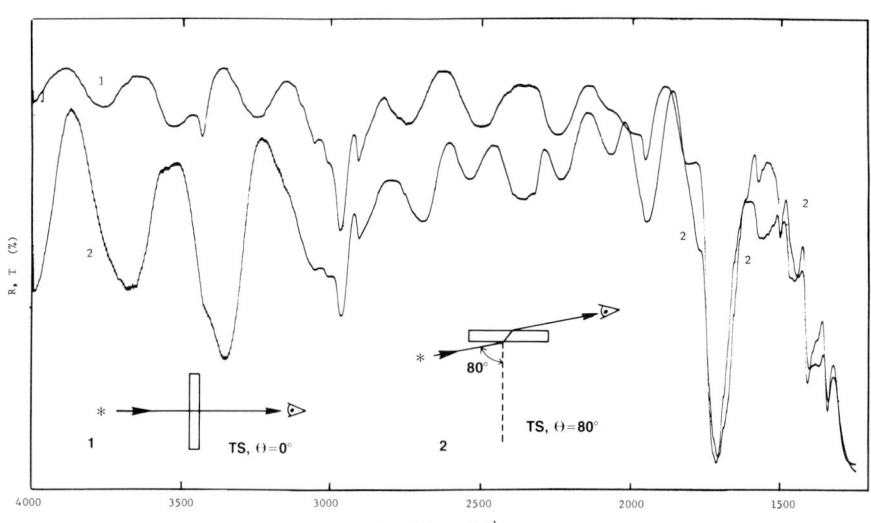

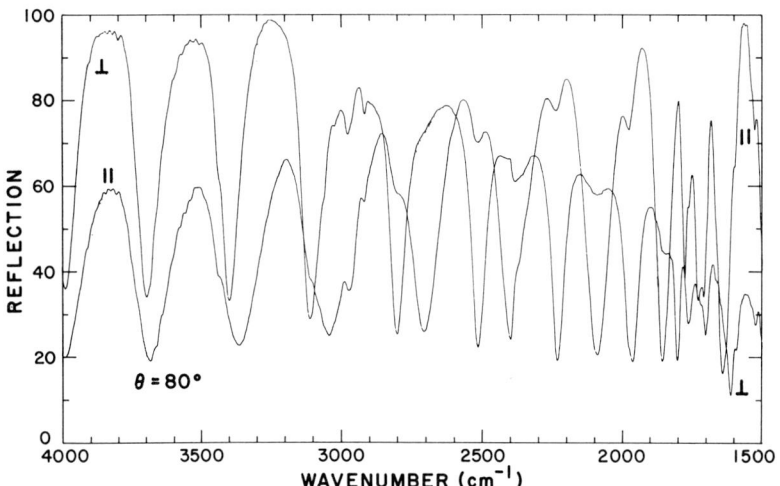

Fig. 6. (a) Transmission spectra of Mylar at θ = 0° and 80°.
(b) Separation of superimposed fringe patterns via external reflection with polarized light at θ = 80°.

reflection) contains only an electric field vector perpendicular to the plane of the film; we can thus immediately conclude that the Mylar film of Fig. 6b is anisotropic. From the fringe patterns[2] in Fig. 6b, we can calculate the refractive index n = 1.652 in the plane of the film and n = 1.603 perpendicular to the plane of the film.

These calculations demonstrate an important use of interference fringes. Sometimes, however, they are a nuisance. Two transmission spectra of one and the same film are shown in Fig. 7. They are different even though the same spectrometer was employed and the sample was untouched between measurements - only the polarization of the beam was changed. Spectrum (1) has strong interference fringes, which obscures weak bands; spectrum (2) has none. The reason there are no fringes in spectrum (2) is that the spectrum was recorded at Brewster's angle where R = 0 (see Fig. 2) and therefore there can be no fringes.* This method of recording transmission spectra should find important use in thin, solid and liquid films and especially where difference spectra are required. If examined carefully, the spectra of Fig. 7 shows more subtle differences in band intensity and shapes as revealed by Fig. 8.

Curve (2) of Fig. 8 shows that it is not readily possible to balance out absorptions of two identical samples exactly because of limited dynamic range of spectrometers. Curve (3), however, shows some real differences in absorption of sample and reference beam where one beam is polarized.

Figs. (6), (7), and (8) show that differences, dependent on angle of incidence and polarization, in transmission spectra may occur. It should be recalled that the light beam is always at least partially polarized in spectrometers (including Fourier transform spectrometers) and that this polarization may not (generally is not) the same in all instruments and hence the phenomena discussed above do play a role in the nature of the transmission spectra.

B. External Reflection Spectroscopy

1. History. The External Specular Reflection Spectroscopy (ERS) method is shown schematically in Fig. 9. Here, one measures the reflected intensity versus wavelength at the desired angle of incidence and polarization. ERS is a very powerful, nondestructive technique that has been in use for many decades but its full potential has not been realized and the method had been highly neglected until recently.

---

*For low reflectivity the interference fringe amplitude is 2R and is therefore zero when R = 0 which is the case for parallel polarization at Brewster's angle.

Transmission and Reflection Spectroscopy   165

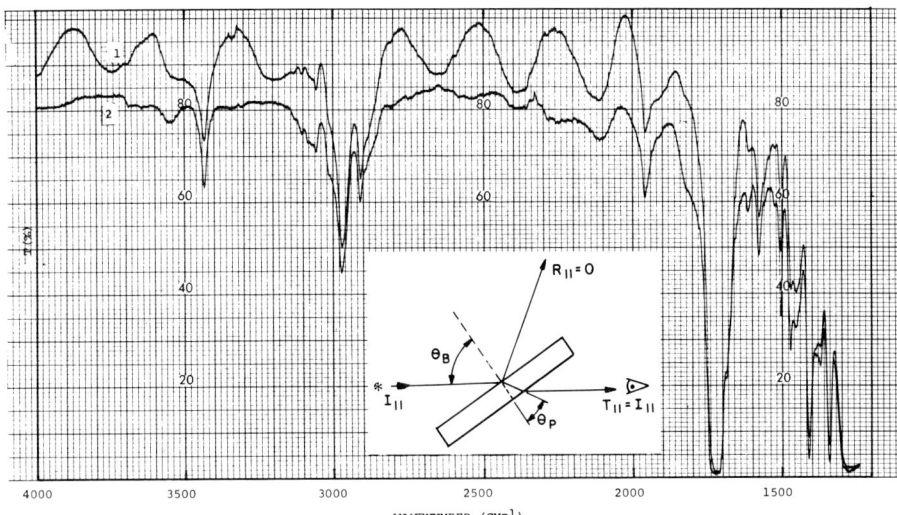

Fig. 7. Transmission spectra of a Mylar film, 6μm thick recorded at θ = 50° with (1) an unpolarized light beam and (2) with light beam polarized parallel to the plane of incidence.

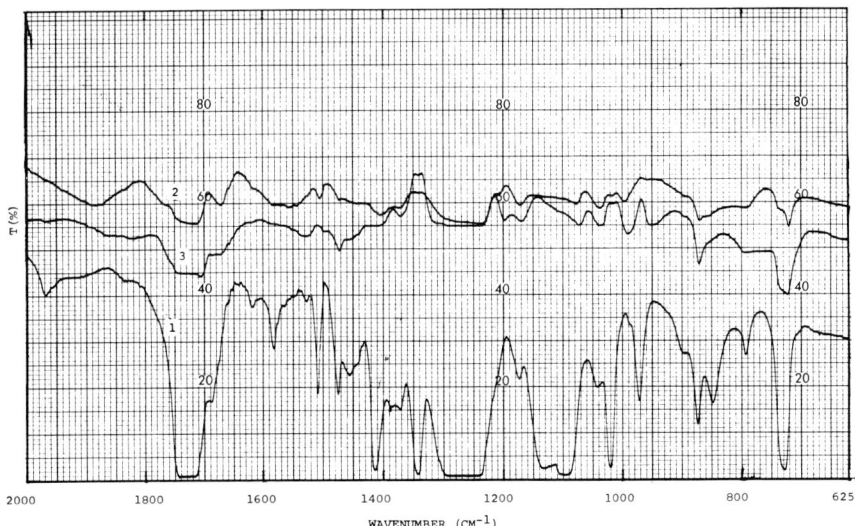

Fig. 8. Spectra of Mylar film, 25μm thick, θ = 50°: (1) in sample beam only; (2) in sample and reference beams; (3) same as (2) except sample beam polarized parallel.

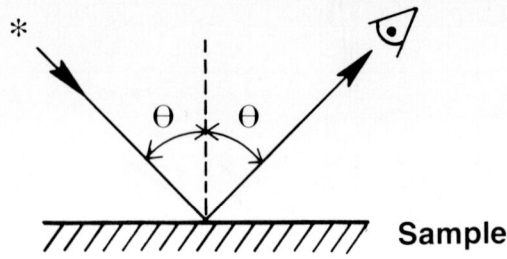

Fig. 9. Schematic diagram depicting external reflection spectroscopy.

Early work includes that of Hagen and Rubens[3a] (1903) on the study of metal optics via ERS. Its current use in the study of adsorbed species on metals begins with the calculations of Francis and Ellison[3b] (1959) in its potential use for the study of catalysis. Their calculations predicted limited sensitivity and effective only with polarized light at large angles of incidence. This thereby discouraged its immediate use until Greenler and coworkers[3c] demonstrated its practical potential for study of monolayer films. The technique has also been employed and analyzed by others[3d]. Currently it is also applied widely to the study of "thick" films[3e].

2. Nature of the Spectra. There are a number of factors that determine the nature of the spectra for ERS and the optical phenomena discussed above must be taken into consideration. Also to be considered are film thickness, nature of the "second" surface, etc. Examples follow.

(a) Thin Films. Films having thickness much less than the wavelength of light are considered to be "thin". Practical problems that require thin film analysis and fall into this category include catalysis, weathering, corrosion, electrochemical reactions, etc. One may conclude from the discussion on standing waves (Fig. 4) that ERS would not be a desirable method to examine thin films. In fact, it would appear that from Fig. 4a the best place to hide a thin film would be to place it on a metal surface since light at or near normal incidence cannot interact with it. Neither is

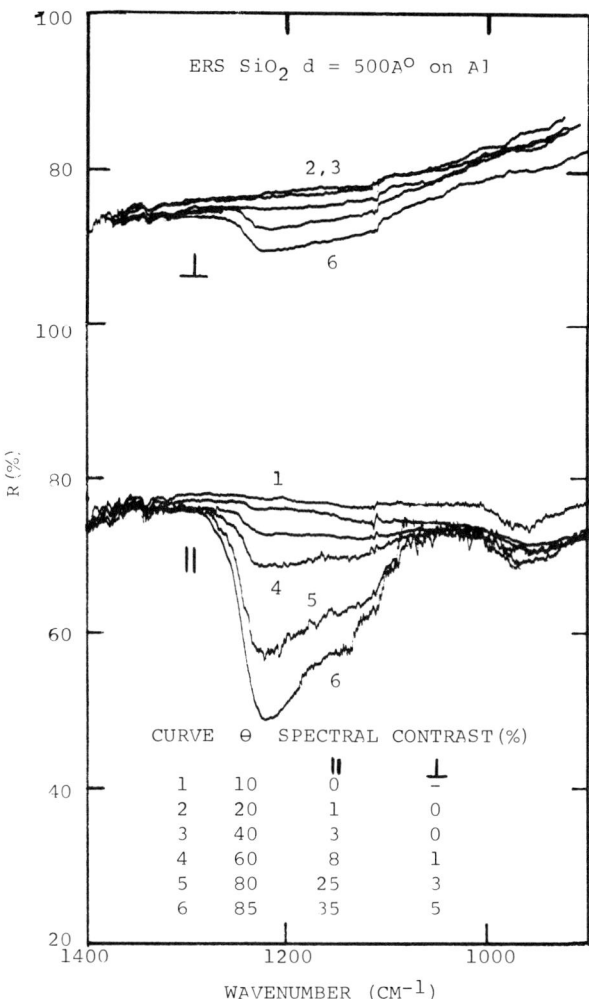

Fig. 10. External reflection spectra of 500 Å of SiO_2 on an Al substrate for perpendicular and parallel polarization at various angles of incidence.

there much interaction with the film using perpendicular polarized light at any angle of incidence. However, if one chooses parallel polarization at large angles of incidence, the film which is undetectable at normal incidence gives a substantial signal as shown in Fig. 10.

One normally associates interference phenomena (fringes, optical cavities) with thick films; i.e., $\lambda/4$, $\lambda/2$, etc. These effects, as shown in Figs. 3-5, can lead to amplification and should not be ruled out even in thin films. The reason for this is that it is not only the physical film thickness that determines its "effective" thickness but also the phase change upon reflection.

We, at least most of us, know that a film of refractive index $n^{\frac{1}{2}}$ and of effective thickness one-quarter wavelength (actual thickness $\lambda/4n$) on a lens of refractive index n acts as an anti-reflective coating. The reason for this is that light reflected at the second surface arrives back at the first surface with exactly the same amplitude and exactly out of phase relative to the component reflected from the first surface, and therefore there can be no reflection or "glare". We all, having blown bubbles at one time or another, are also familiar with "black" soap films; i.e., soap film on a wire loop. Why does this film appear black? It doesn't reflect any light and is therefore undetectable via visual inspection. How can this film act as a quarter wave film since its thickness is zero (only a few monolayers)? This is a free standing film and has the same reflectivity at both the first and second surfaces; therefore the amplitude of the light reflected from the two surfaces will be equal (be matched). How can it have an effective thickness of $\lambda/4$? Phase change explains it all! The light approaching and passing through the film goes from air (low refractive index) to film (high index) to air (low index). A jingle reminds of phase changes involved:

Low to high
Phase change Pi.
High to low,
Phase change? No!

Light from the second surface therefore arrives exactly out of phase relative to that reflected at the first surface and there can be no net reflected light and hence the film is "black" (anti-reflecting).

(b) Thick Films. There are many reasons why one may want to examine thick films via External Reflection Spectroscopy. A few examples are cited below.

• Films, oxides, etc., used to coat metal or other opaque substrates may react over a period of time and there may be a need to study the films to understand the nature of the reaction in order to make changes to prolong life, etc. Many examples may be cited, e.g., coating of telephone cables with polyethylene.

• Non-destructive monitoring in production. The ERS technique is ideal here since moving samples may be studied without inserting them into the sampling compartment of an instrument; furthermore, the light source and detector are on the same side of the sample.

- ERS produces cleaner interference fringes for measuring film thickness and refractive index (see Figs. 3b and 6b).

- ERS may be employed in such a geometry as to eliminate interference fringes.

The above examples imply that the sample studied via ERS may have a variety of configurations. Spectra will be shown here indicating that in each configuration the spectrum of the same sample is different. Furthermore, the spectrum will change with polarization and angle of incidence. Every change can, however, be at least qualitatively explained by the "optical phenomena" already discussed. These spectra will serve to stress the differences that may occur and also demonstrate that terms sometimes used to describe these spectra such as "double absorption" are incorrect and misleading.

The Mylar film on the Al substrate of Fig. 11 appears to show no fringes. This is confirmed by comparing to curve 2 of Fig. 7 which proves that the dips 2600 - 1800 cm^{-1} are indeed absorption bands. Spectra (1) and (2), however, show other differences in addition to the absence of fringes. Note the apparent splitting of the carbonyl band in spectrum (2) at 1725 cm^{-1}. This difference can be understood from examination of curve (1) of Fig. 12. Here, the reflection from the second surface is suppressed and a strong reflection peak is observed at 1725 cm^{-1}. This reflection peak is due to the strong dispersion in n in the vicinity of the absorption band (see Fig. 1).

If the substrate is removed strong and clean interference fringes are observed (curve 2 of Fig. 11) almost free of distortion by the absorption bands, also as shown in Figs. 3b and 6b.

We shall continue to torment a Mylar film by again placing it on an Al substrate and adding some nujol to insure optical contacts and then examine it via ERS at different angles of incidence and different polarizations.

Fig. 13 shows that for two polarizations the spectra are far from the same - band shapes, band intensities, intensity ratios are all different. This is further emphasized by Fig. 14 where all sorts of gyrations occur as the angle of incidence is changed. Attention will be drawn to only a few general trends which can be explained by considering the "optical phenomena" (Figs. 1-5) and with reference to Fig. 14.

- Spectral contrast is less for perpendicular than it is for parallel polarization. We see from Fig. 2 that spectra for perpendicular polarization contains more energy from first surface reflection (R_1 and component (1) of Fig. 14) and which is less sensitive

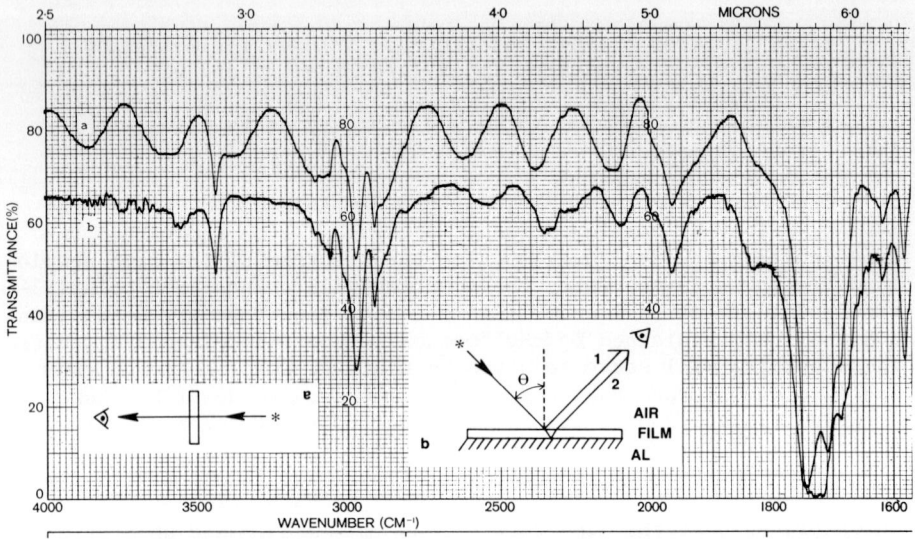

Fig. 11. Comparison of Mylar spectra for (a) transmission θ = 0° and (b) external reflection with Mylar on Al substrate with θ = 12°.

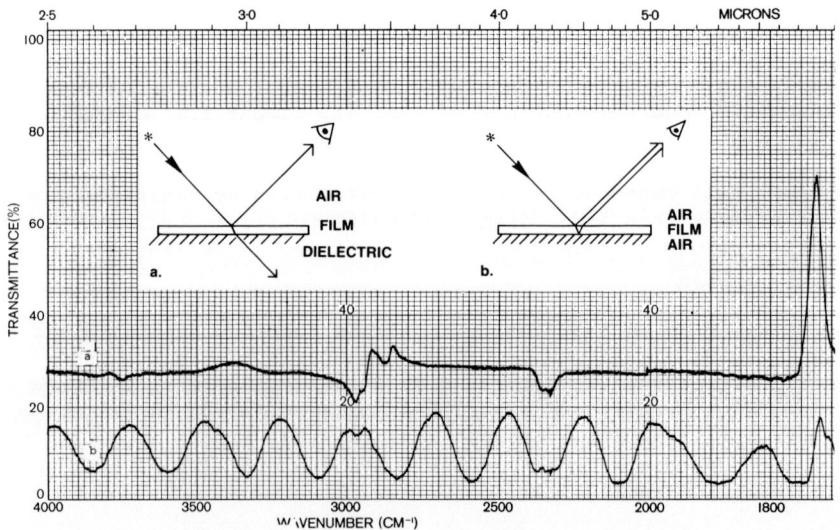

Fig. 12. (a) Spectrum of Mylar optically contacted to dielectric to suppress second surface reflection;
(b) Spectrum of same Mylar film free standing.

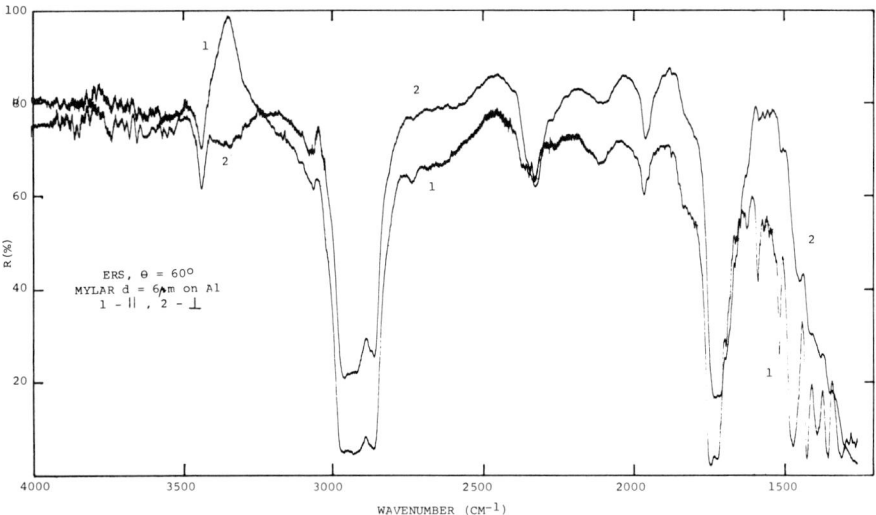

Fig. 13. External reflection spectra of Mylar film on Al substrate at θ = 60° (a) parallel and (b) perpendicular polarization.

to the extinction coefficient, as demonstrated by Fig. 12a, than is component (2) of Fig. 14 and which is more predominant for parallel polarization.

- For large angles (80°) and either polarization, spectral contrast is diminished. Explanation is analogous to that for (a), i.e., see Fig. 2, R_1 is large for both polarizations.

- Band intensity ratios change with angle of incidence. See discussion on standing waves and optical cavities and note in Fig. 14 that the effective wavelength $\lambda e = \dfrac{\lambda}{n \cos \theta}$ is dependent on both angle of incidence and wavelength and the number of "loops" within the film changes.

- Bands appear in "parallel" spectra which are absent in "perpendicular" spectra. Recall that for parallel polarization, especially at large angles, (see also Fig. 6) there is a z-component of the E-field in the film while for perpendicular polarization there is only a y-component. Hence the spectra for perpendicular and parallel polarization at large angles of incidence will be different for anisotropic samples.

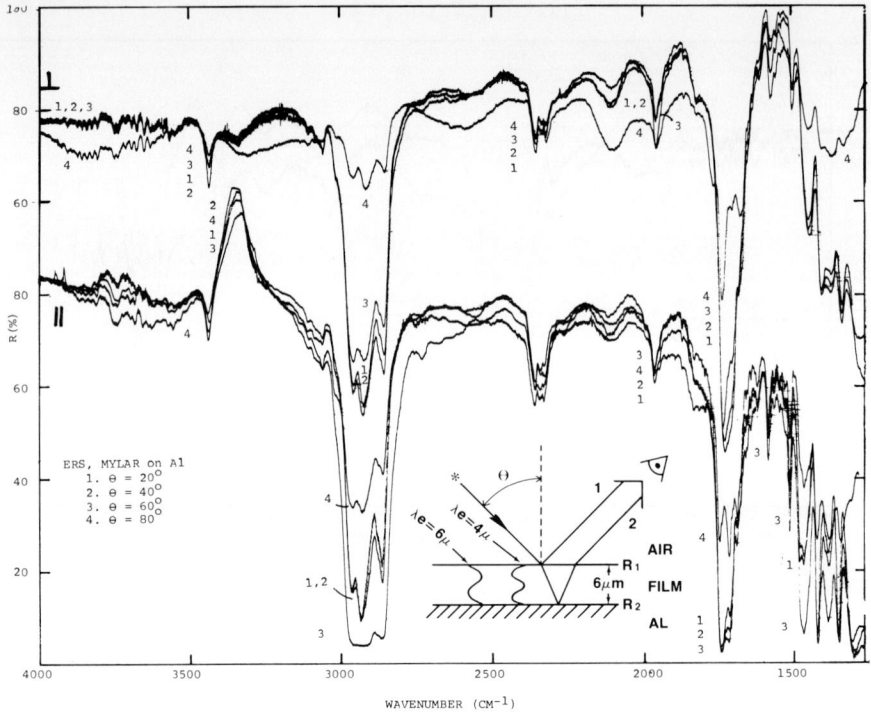

Fig. 14. Spectra of Mylar on Al substrate for perpendicular and parallel polarization at various angles of incidence.

In another form of ERS, called ellipsometry, where phase change upon reflection is measured for polarized light, it is possible to study films as thin as one hundredth of a monolayer. Details of these techniques can be found elsewhere[3f].

C. Internal Reflection Spectroscopy

1. History. The Internal Reflection Spectroscopy (IRS) technique is shown schematically in Fig. 15. Light is introduced into a prism at angles of incidence exceeding the critical angle and the sample material is placed in contact with the prism where it interacts with the evanescent wave (Fig. 4b).

At angles above the critical angle total internal reflection occurs and the surface acts as a perfect mirror and thus hundreds

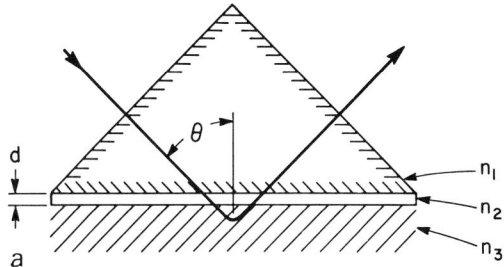

Fig. 15. Schematic diagram depicting the Internal Reflection Spectroscopy technique.

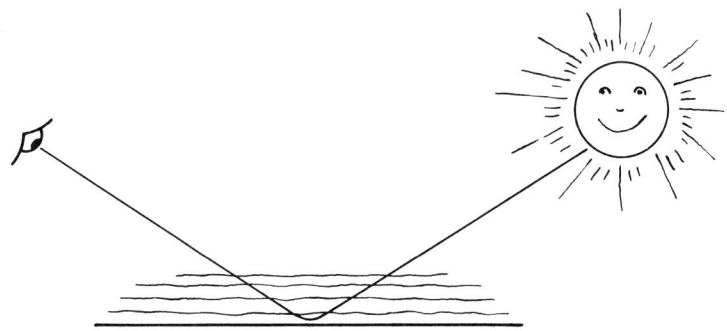

Fig. 16. Mirage - an example of total internal reflection found in nature. Light at grazing incidence is totally reflected by the less dense warm air near the road surface.

of reflections can be employed to detect and study very thin films.

The most common example of total reflection occurring in nature is the mirage (Fig. 16). Here, warm air above the hot road or sand has a lower refractive index and the condition for total internal reflection is met at grazing incidence. The surface in the distance thus appears mirror-like and one sees the sky while looking at the earth.

The presence of the evanescent wave was discovered by Newton more than two and one-half centuries ago and has found many applications such as construction of cold mirrors, beam splitters, light modulators, to mention only a few. Its current wide use in optical spectroscopy stems from the application of multiple internal reflection by Harrick to the study of semiconductor space charge regions[4a] (Fig. 17) and studies of adsorbed species on semiconductor surfaces[4b] (Fig. 18) followed by its application to the study of semiconductor surface states[4c]. Contributions to the early development were also made by Fahrenfort[4d], first to the recording of spectra of highly absorbing samples and then to its use in measuring optical constants[4e], and by Hansen[4f] who was the first to employ these techniques in the visible spectral regions. The IRS technique developed rapidly and found a wide variety of applications which are now well known. These are discussed in a book[4g] devoted to the subject where a complete theory, applications and references can be found. More recent applications and references can be found in Internal Reflection Spectroscopy: Review and Supplement[4h].

It is gratifying that with currently available instrumentation the IRS technique is now capable of achieving the hoped-for sensitivity to study films as thin as 1/100 of a monolayer (see following paper by Jakobsen). IRS thus competes favorably in sensitivity with ESCA, LEED, etc. In applications such as studies in free carrier absorption, surface studies, etc., where it is possible to modulate the absorption, reflectivity changes of one part in 10 million have been achieved.

2. Parameters. In some respects, IRS is mathematically cleaner than ERS since complications such as those shown in Fig. 14 do not occur. Uncertainties do arise when working with solid samples and prisms and when there are intervening films where the nature of the contact is unknown (see optical cavities, Fig. 5).

The IRS technique is a very powerful one where a number of parameters can be varied and which control the nature of the spectrum. These include refractive index of the prisms, angle of incidence, polarization and orientation of the electric fields at the reflecting surface. Fresnel equations predict exactly the interactions involved and in the limiting cases of thick and thin films highly reliable approximations have been derived which give physical insight into the nature of the interactions as has been described before[4g]. Fig. 19 shows the dependence of effective film thickness for thick and thin films on various parameters. Effective thickness determines spectral contrast.

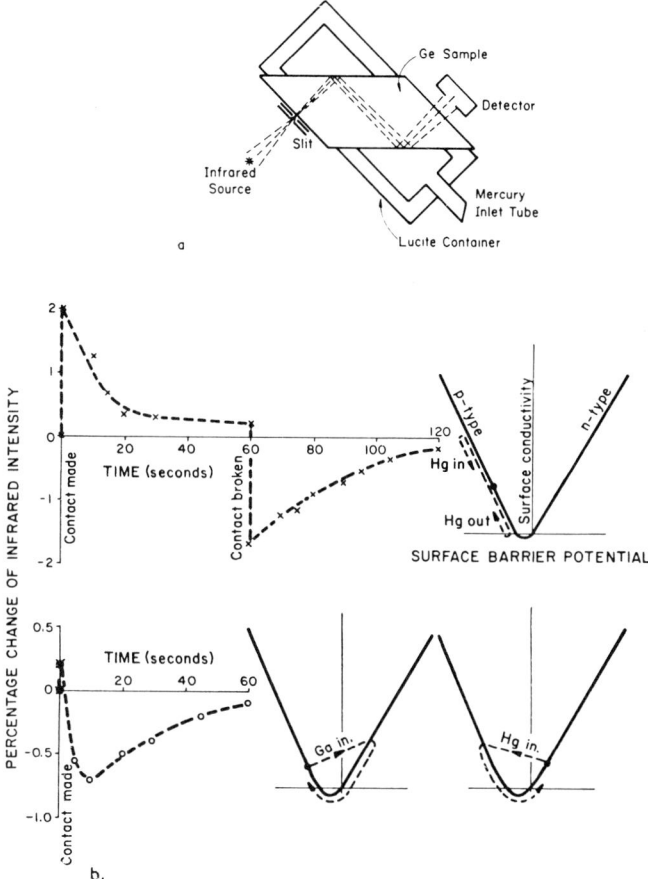

Fig. 17. Experimental arrangement (a) and results (b) for the first experiment involving Internal Reflection Spectroscopy in 1958. This application demonstrated the relaxation phenomena involving free carriers in the semiconductor space charge region and semiconductor surface states.

Still another parameter that can be varied for thin films is the refractive index of the surrounding medium, n_3. Calculations on the effect of varying n_3 are shown in Fig. 20a. It should be noted the strength of interaction (de) may increase, decrease or remain unchanged as n_3 is increased. Fig. 20b gives results showing an increase in spectral contrast when AgCl is pressed against a Ge prism coated with a film of nylon 4. As predicted by theory,

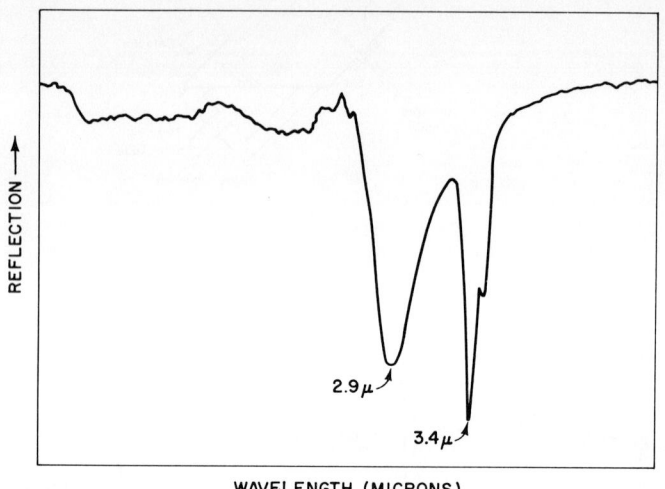

Fig. 18. First Internal Reflection spectrum of naturally occurring adsorbed species on the surface of optical material (silicon internal reflection plate, 165 reflections at $\theta = 45°$).

the increase is greater for parallel than it is for perpendicular polarization.

3. Nature of the spectra. (a) Thin Films. Spectra of thin films recorded via IRS have already been shown in Figs. 18 and 20b. Fig. 21 shows the formation of SiH_4 via bombardment of a Si IRE with atomic hydrogen. When material is removed by means of Argon bombardment, it should be noted that the location of the band for the more deeply penetrating ions is shifted relative to the superficial layers depicting a difference in the nature of the bonding. These spectra demonstrate about a 1/3 monolayer sensitivity. As mentioned above the following paper shows spectra depicting a 1/100 monolayer sensitivity.

(b) Thick Films. A sample of thickness greater than approximately one wavelength can be considered as a thick film. Two spectra are shown to remind us of the nature of internal reflection spectra.

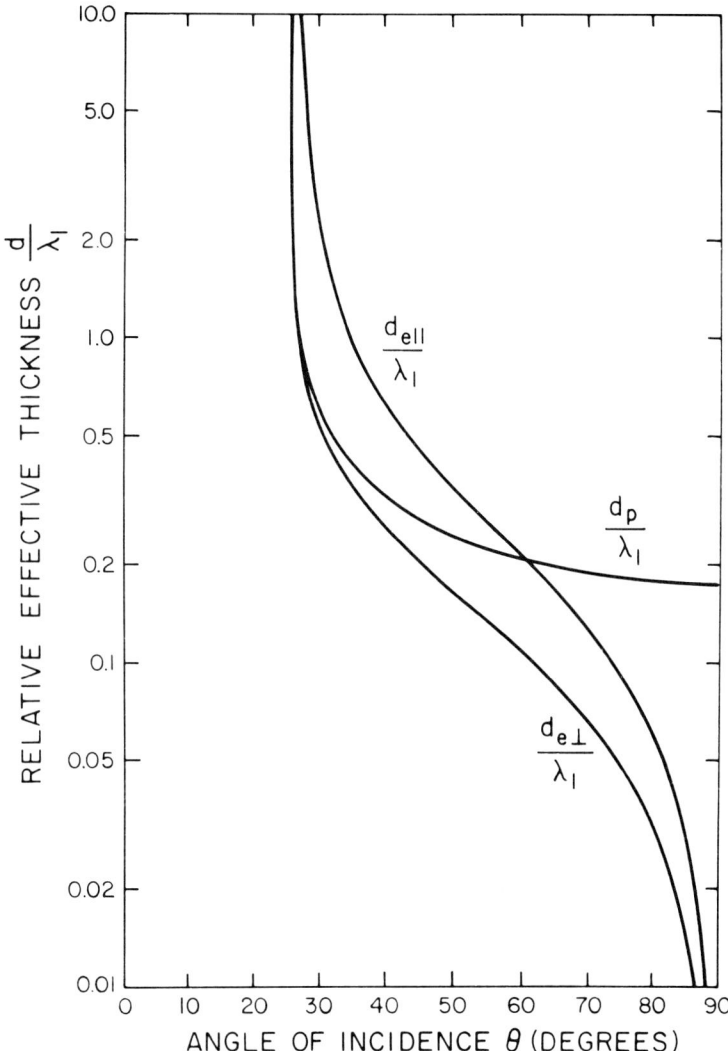

Fig. 19. (a) Relative penetration depth and effective thickness of the evanescent wave versus the angle of incidence for an interface whose refractive index ratio is $n_{21} = 0.423$. Near the critical angle, the effective thicknesses for both polarizations are greater than the penetration depth, and at large angles they are less. (b) Relative effective thickness for a thin film of refractive index, $n_2 = 1.6$, on internal reflection elements (IRE's) of various refractive indices, n_1. Here θ_{ca} is the critical angle of the crystal-air interface.

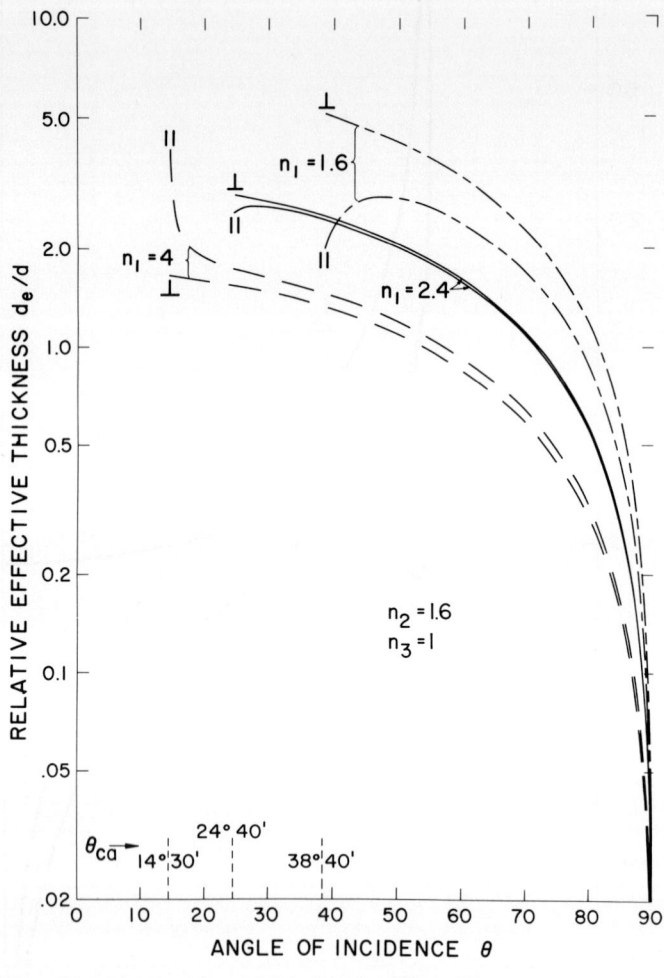

Fig. 19 (b)

The major differences between TS and IRS spectra of thick films are shown in Fig. 22; viz., absence of interference fringes, bands are relatively stronger at longer wavelengths, bands may be broadened on long wavelength side and shifted to longer wavelengths.

Fig. 23 also shows some of the features of Fig. 22 and further shows that the interaction is greater for parallel polarization. Still another feature to be noted is that the spectrum changes from a dispersion-like for measurements near the critical angle to an absorption-like for angles above the critical angle. Advantage

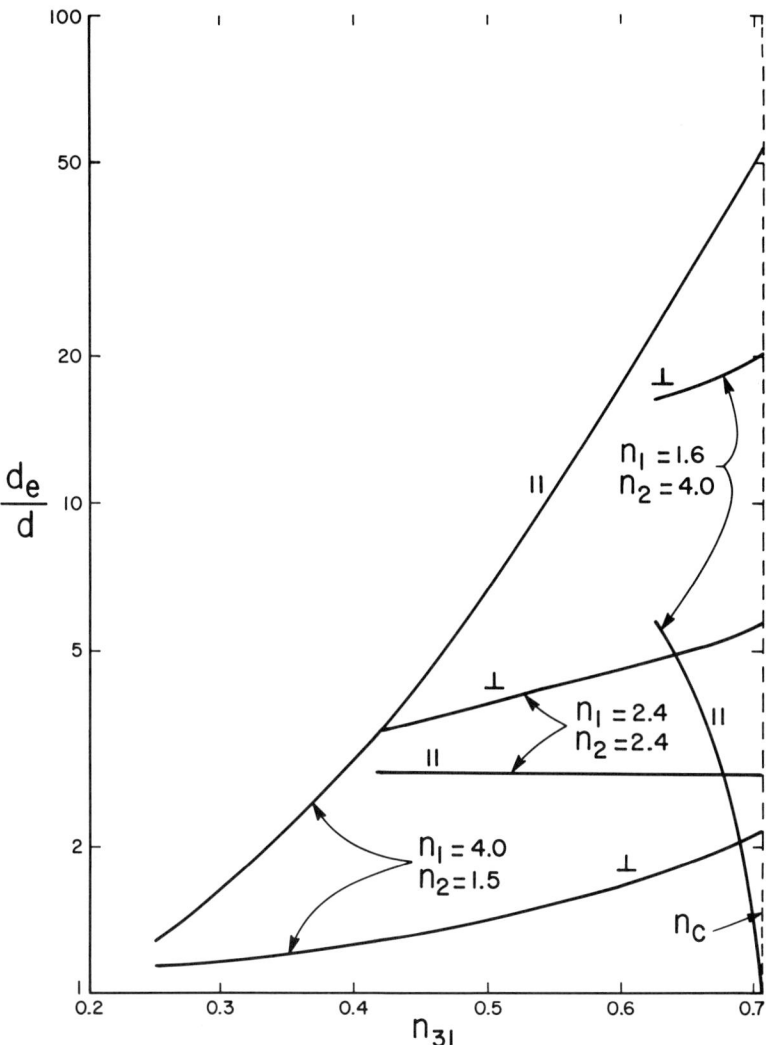

Fig. 20. (a) Effect of immersion showing change in spectral contrast as the refractive index of medium 3 is increased. Note that in general, spectral contrast increases, but in special cases may remain constant or even decrease.

can be taken of this latter effect to determine optical constants (see Fig. 1).

D. Diffuse Reflectance Spectroscopy. There are many instances when by nature the surfaces of a sample may not be smooth or the sample may be of a particulate nature. Transmission and external

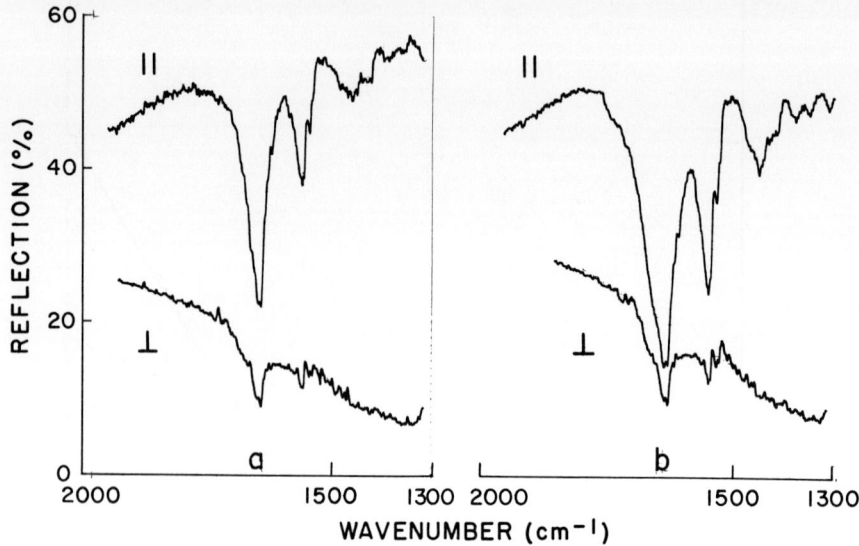

Fig. 20. (a) Effect of immersion showing change in spectral contrast as the refractive index of medium 3 is increased. Note that in general, spectral contrast increases, but in special cases may remain constant or even decrease.
(b) Spectra of 50 Å of Nylon 4 on Ge double-sampling plate (a) before and (b) after bringing AgCl in contact with Nylon covered Ge plate. Note increase in spectral contrast with greater increase for parallel polarization.

reflection cannot be employed without special sample preparation, e.g., pelleting or pressing. Internal reflection has the advantage that even such samples can be examined without special preparation. Still we may wish to examine the samples as seen in nature and thus examine the diffusely reflected light (see Fig. 24).

One may, for example, measure via ERS the cone of the scattered light. An example of this is shown in Fig. 25 where reflection profiles for a mirror (a) and aluminized fabric (b) are compared for angle of observation of 12°. The specular reflection from the mirror shows a maximum at 12° and shows that the angular beam spread is about 6°. The fabric has peak reflection of 5%, however, the integrated (over all angles) reflected light is 22%.

If one were to employ a light collector such as an ellipsoid or integrating sphere, the total reflected light (i.e., the 22%) would be measured immediately, however, information about the angular distribution would be lost. The latter technique is generally known as Diffuse Reflectance Spectroscopy (DRS).

Diffuse reflectance spectra are extremely complex. They

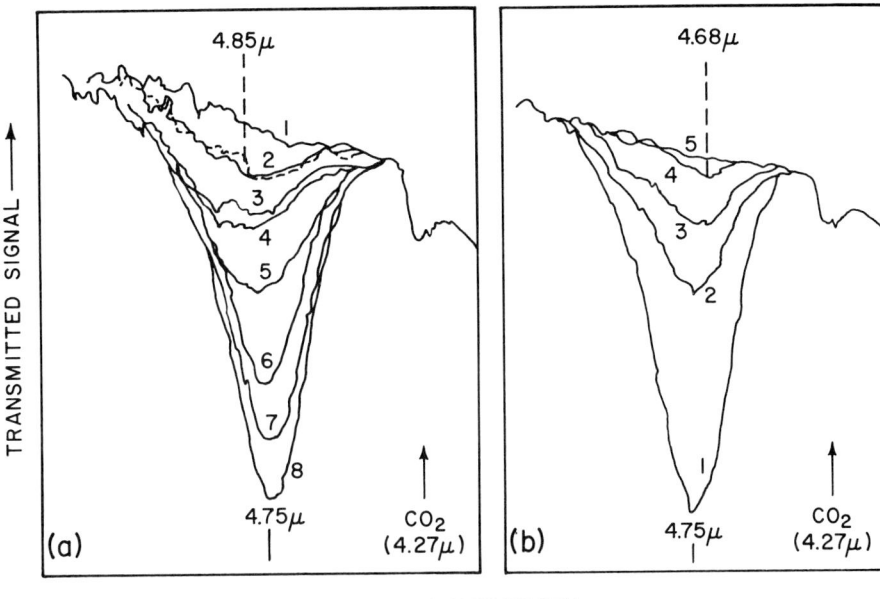

Fig. 21. Internal reflection spectra of a silicon surface, N = 200, θ = 45°. (a) Curve 1 is an initially clean Si surface after sputtering in argon and annealing. The dashed curve nearly superimposed on curve 2 shows the maximum absorption peak resulting from exposure of silicon to atomic hydrogen. Curves 2-8 are due to bombardment with hydrogen ions of increasing energy and for longer times. Note the shift in the absorption peak. (b) Series of spectra showing decrease in strength of silicon hydride band as the silicon is bombarded with argon ions for various lengths of sputtering. Curve 1(b) shows a thick surface layer (same as curve 8(a)). Note the band shift for curve 4(b) relative to curve 2(a).

consist of a mixture of every conceivable spectra that can be obtained via transmission, internal and external reflection as discussed in the preceding pages, all lumped together and further complicated by scattering effects. It is only in special cases that approximate theory is available and the spectra resemble those obtained in transmission. Still the DRS is necessary for the examination of samples such as papers, fabrics, foliage, to name a few.

For more details on DRS, the readers are referred to a number of books on the subject[5a-5d].

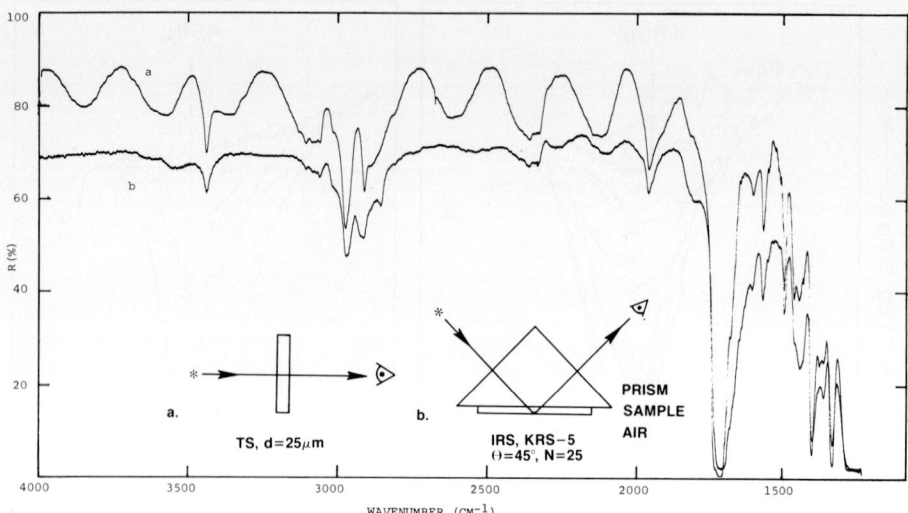

Fig. 22. Comparison of transmission spectrum of Mylar to that recorded via Internal Reflection Spectroscopy of Mylar on KRS-5, θ = 45°, N = 25.

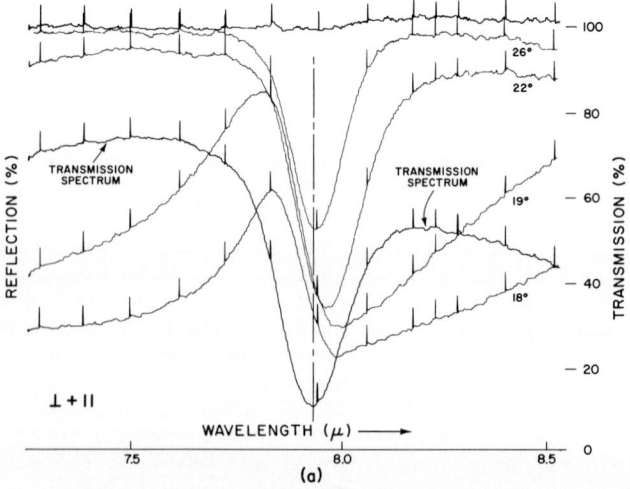

Fig. 23. Silicone lubricant on a Ge Hemicylinder. Spectra of band of silicon lubricant at 7.9 μm. (a) Internal reflection spectra for unpolarized light at various angles of incidence compared to transmission spectrum. (b) Internal reflection spectra at various angles of incidence for ⊥-polarization. (c) Internal reflection spectra at various angles of incidence for ||-polarization.

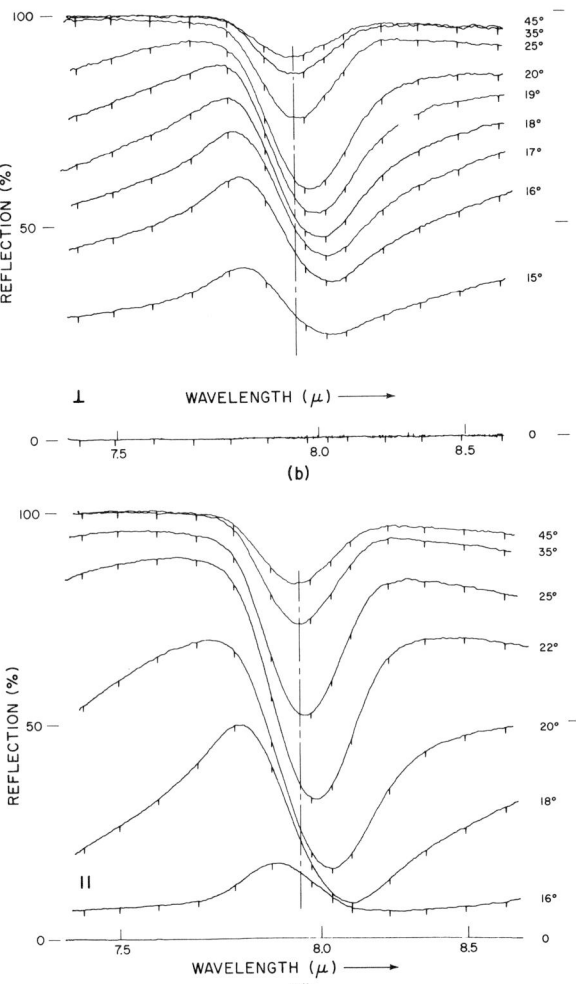

Fig. 23. (a), (b), and (c).

APPENDIX - INSTRUMENTATION

A. Spectrometers

There is a wide variety of spectrometers in use today. They are generally made for simple transmission spectroscopy and often have sample space limitations and general inflexibility for reflection measurements. There is, therefore, a need for attachments and accessories to make them useable for a variety of transmission and reflection studies.

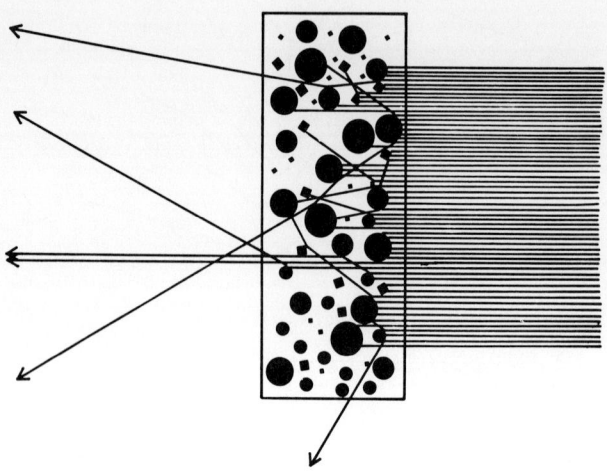

Fig. 24. Diffuse Reflectance.

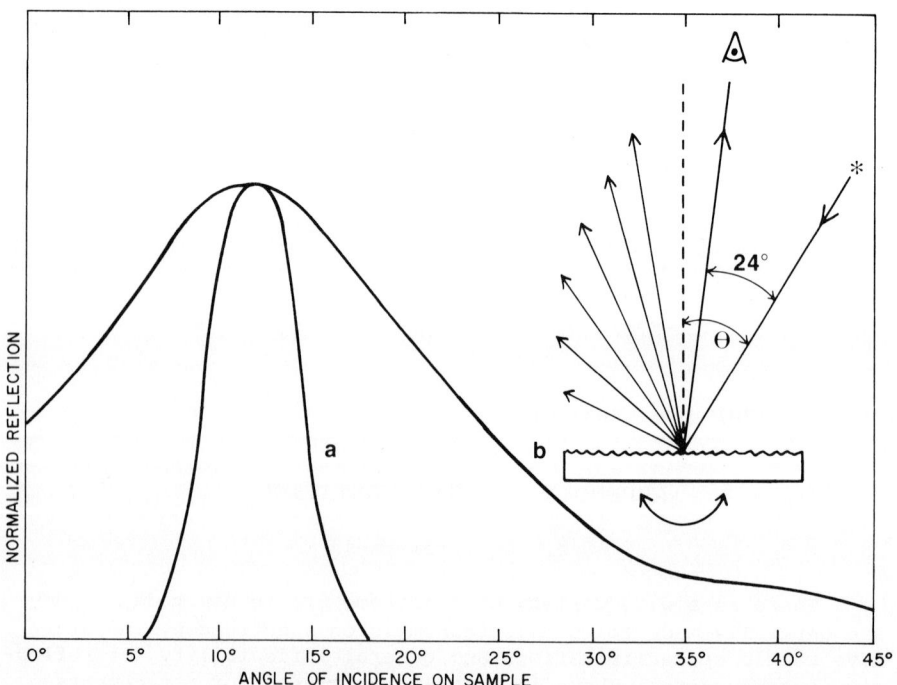

Fig. 25. Cone of reflected light for (a) mirror depicting the angular beam spread of the incoming radiation and (b) aluminized fabric showing extensive light scattering.

A more flexible instrument than conventional spectrometers is the double beam goniometer spectrometer (DBGS) shown in Fig. Al which offers a high degree of versatility for almost all transmission, external and internal reflection measurements, and special attachments are generally not required. The DBGS can be built either with a dispersive (slit) instrument or an interferometer. The source optics are mounted on an arm which pivots at A at the rate of 2θ. The internal reflection elements for the sample and reference beams, as well as the beam chopper and recombiner, are mounted on a platform which is coupled to the 2θ arm and which also pivots about A, but at the rate of θ. This system thus remains in alignment for all angles of incidence. The beam splitter and recombiner are driven synchronously and must move symmetrically relative to AB at the rate of $\tan \theta$ as the angle of incidence is changed. Better than $0.01°$ mechanical accuracy can be obtained on both the θ and 2θ arms. With the ability to control the angle of incidence and the location of the beam focus relative to A, a high degree of flexibility is achieved.

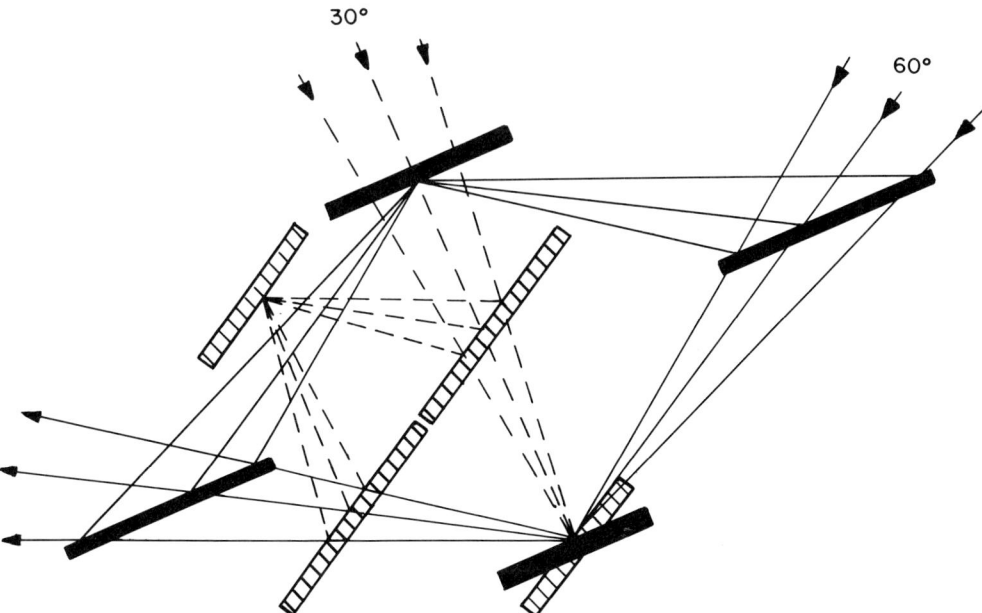

Fig. Al. Double-beam goniometer spectrometer for internal reflection, external reflection, or transmission. Optical layout (a) and photograph (b) show that angle of incidence can be changed continuously without misalignment. IRE's that can be employed include hemicylinders, variable angle double-pass plates, double sampling plates or single-pass plates.

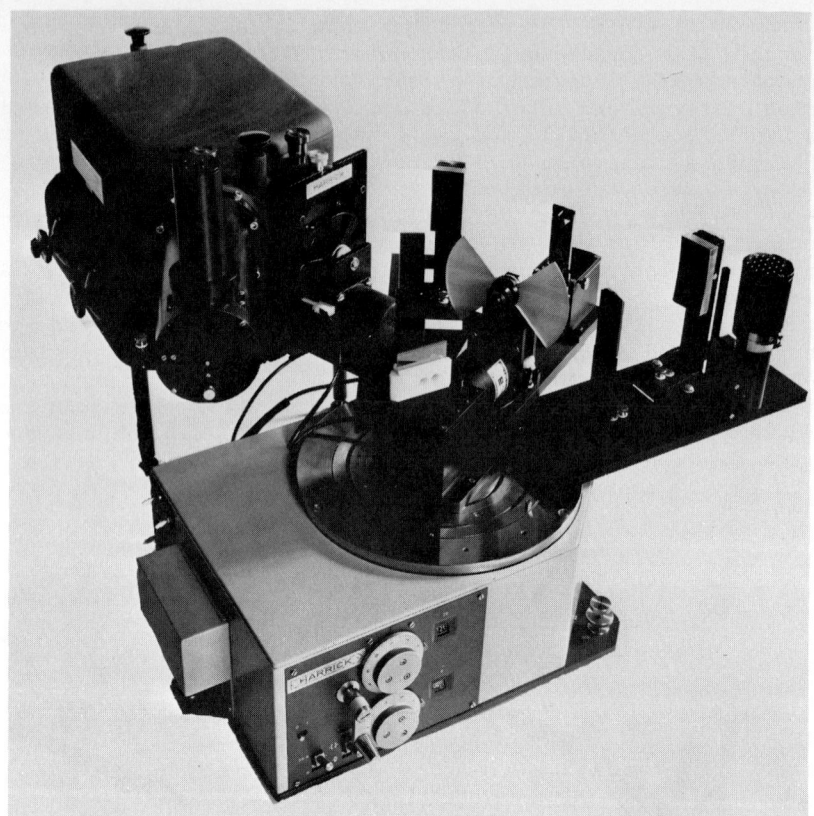

Fig. A1 (b).

By chopping the light beam simultaneously at two frequencies - one via a butterfly mirror, which allows the beam alternately to strike sample point A or reference point B, and a second much faster on-off chopper and use of lock-in amplifiers and ratiometer, the signals can be processed to display percentage absorption or reflection without slit drives or optical nulls.

B. Polarizer

The spectra shown in this paper demonstrate that a polarizer is almost essential for optical spectroscopy, especially since the light beam is partially polarized in all instruments and the state of polarization is generally unknown. A variety of polarizers that cover the entire electromagnetic spectrum is available.

C. Transmission Spectroscopy

In addition to the spectrometer, all that is required for TS in most cases is a sample holder. Figs. 6-8 demonstrate that sometimes it is necessary to change and read the angle of incidence of the light beam stroking the sample. A simple attachment permits this.

D. External (Specular) Reflection Spectroscopy

Figs. 10-14 demonstrate the necessity for changing the angle of incidence for ERS. This can be done very conveniently with the Versatile Reflection Attachment (VRA) and the Retro Mirror Accessory (RMA) shown in Fig. A2. The sample may be placed on either arm RM or MA. The optical layout shows that once this attachment is aligned for one angle of incidence it remains in alignment for all others and hence is ideal for ERS. Multi external reflection can be employed with the VRA by replacing the RMA with a Multi External Reflection Accessory (MESRA) - the boxlike structure shown in Fig. A2. Diffuse reflection measurements such as those shown in Fig. 25 are readily made with the VRA.

The VRA can also be employed for a wide variety of other external reflection studies as well as for Internal Reflection Spectroscopy with double sampling plates.

E. Internal Reflection Spectroscopy

The simplest, and in many respects the best, attachments for Internal Reflection Spectroscopy is the Twin Parallel Mirror Attachment (TPMRA) shown in Fig. A3a and A3b. It consists of the Internal Reflection Element (IRE) and only two plane mirrors. Some of its attractive features are:

- Very efficient and easy to align.
- No unbalance in the atmospheric bands when placed in the spectrometer; note path length in the atmosphere is unchanged.
- By pivoting the IRE about the aperture, A, and allowing M_1 and M_2 to track, a wide range of angles of incidence can be obtained.
- Internal reflection plates of any length can be employed if M_2 is placed on a telescoping arm.
- By replacing the reflection plate with a boxlike structure, the TPMRA can be employed for ERS.

For certain critical measurements involving, e.g., microsamples, vacuum, high or low temperature studies, other more sophisticated attachments are required.

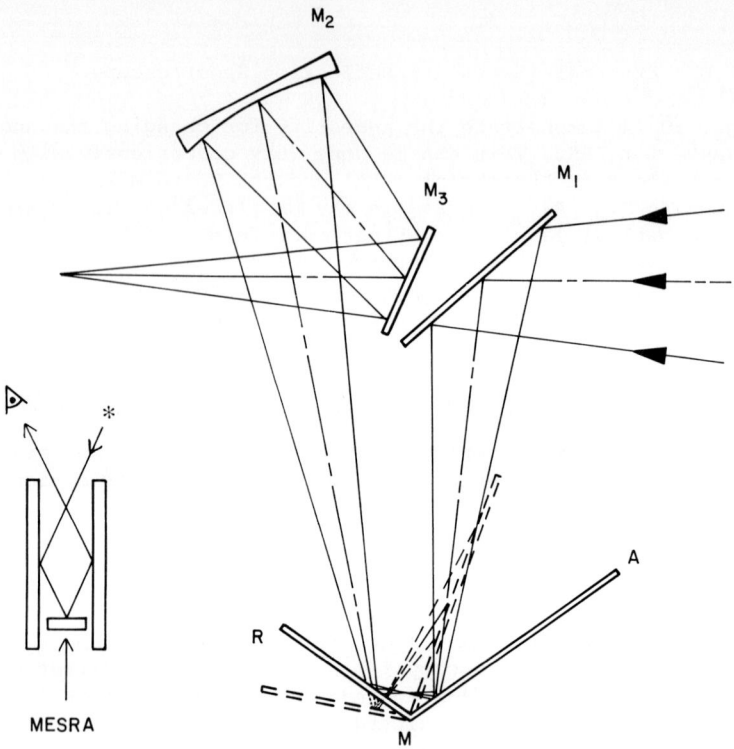

Fig. A2. Versatile Reflection Attachment with Retro-Mirror Accessory (VRA + RMA) for variable angle specular reflection. Once the VRA + RMA is aligned for one angle of incidence, it remains in alignment for all others.

F. Diffuse Reflection Spectroscopy

For Diffuse Reflection Spectroscopy, since light is scattered in all directions, the ideal collector would be a full ellipsoid (Fig. A4a) where the sample is placed and illuminated at one focus and the detector is placed at the other focus. Suitable large ellipsoids are not readily available and integrating spheres are (Fig. A4b) most often employed.

Fig. A2b.

The integrating sphere is a spherical enclosure whose interior is coated with a white scatterer, i.e., material which scatters light uniformly over the spectral range of interest and has no absorption bands. Monochromatic light enters the sphere through a port, strikes the sample, and eventually exits to strike the detector.

Integrating spheres work satisfactorily in the visible spectral range where suitable white scatterers (e.g., MgO) and sensitive large area detectors (photomultiplier tubes) are available. Both of these are unavailable for the infrared, and therefore there is a need for ellipsoid type collectors.

REFERENCES

1. N. J. Harrick, J. Opt. Soc. Am., 55, 851 (1965); N. J. Harrick, and F. K. du Pré, Appl. Opt., 5, 1739 (1966); W. N. Hansen, J. Opt. Soc. Am., 58, 380 (1968); also see Reference 4g.
2. N. J. Harrick, Appl. Opt., 10, 2344 (1971).
3. External Reflectance Spectroscopy
   (a) Hagen and Rubens, Ann. Physik, 11, 873 (1903).

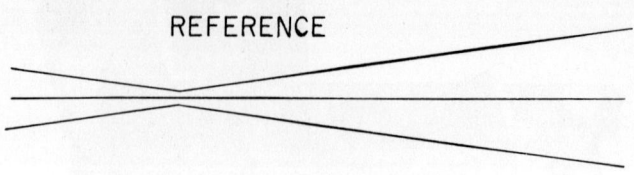

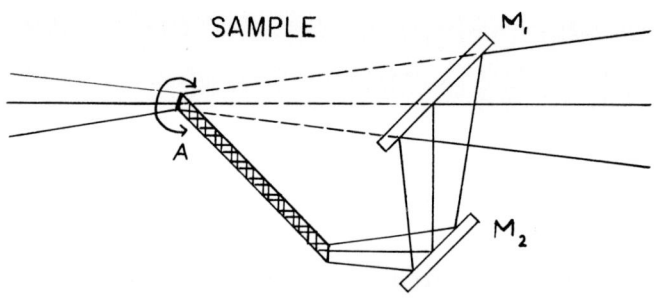

Fig. A3(a). The TPMRA for Internal Reflection Accessory. Note no unbalance in the atmospheric bands when attachment is placed in the spectrometer.

(b) S. A. Francis, and A. H. Ellison, J. Opt. Soc. Am., 49, 131 (1959).
(c) R. G. Greenler, J. Chem. Phys., 44, 310 (1966); H. G. Tompkins and R. G. Greenler, Surface Science, 28, 194 (1971); M. Kottke and R. G. Greenler, Rev. Sci. Instrum., 42, 1235 (1971).
(d) R. W. Hannah, Appl. Spectry., 17, 23 (1963); G. W. Poling, J. Colloid & Interface Sci., 34, 365 (1970); D. J. Drmaj and K. E. Hayes, J. Catal., 19, 154 (1970); J. D. E. McIntyre and D. E. Aspnes, Surface Sci., 24, 417 (1971); H. G. Tompkins and D. L. Allara, Rev. Sci. Instrum., 45,

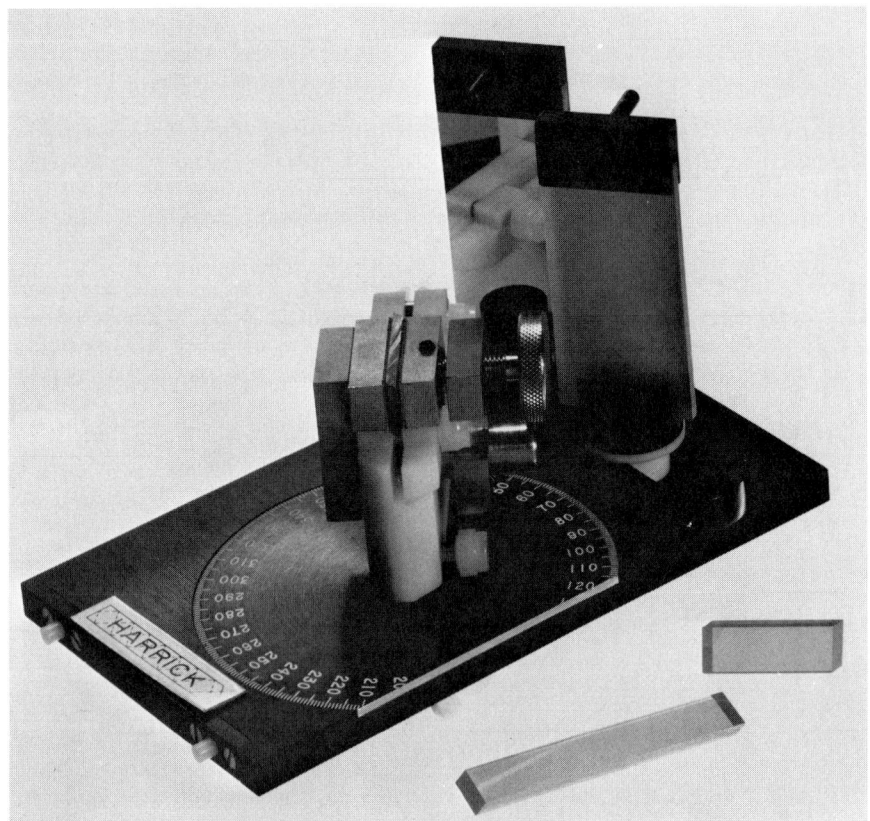

Fig. A3(b).

        1221 (1974); K. E. Hayes, Can. J. Spectry., 20, 57 (1975).
- (e) M. G. Chan and W. L. Hawkins, Polymer Reprints, 2, 1639 (1968); M. G. Chan and D. L. Allara, J. Colloid Interface Sci., 47, 697 (1974).
- (f) R. J. Archer, J. Opt. Soc. Am., 52, 970 (1962).

4. Internal Reflection Spectroscopy
   - (a) N. J. Harrick, J. Chem. Phys. Solids, 8, 106 (1959).
   - (b) N. J. Harrick, Phys. Rev. Letters, 4, 224 (1960).
   - (c) N. J. Harrick, Phys. Rev., 125, 1165 (1962).
   - (d) J. Fahrenfort, Spectrochim. Acta, 17, 698 (1961).
   - (e) J. Fahrenfort and W. M. Visser, Spectrochim. Acta, 18, 1103 (1962).
   - (f) W. N. Hansen, Anal. Chem., 35, 765 (1963).
   - (g) N. J. Harrick, *Internal Reflection Spectroscopy*, Wiley, Interscience, New York (1967).
   - (h) N. J. Harrick, *Internal Reflection Spectroscopy: Review and Supplement* (in preparation).

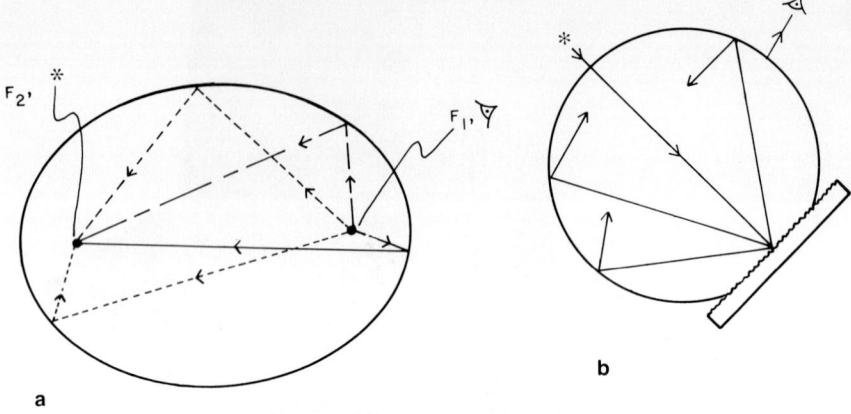

Fig. A4. Radiation collectors for Diffuse Reflectance (a) Ellipsoid. (b) Integrating Sphere.

5. Diffuse Reflectance Spectroscopy
   (a) W. W. Wendlandt and H. G. Hecht, *Reflectance Spectroscopy*, Interscience, New York (1966).
   (b) W. W. Wendlandt, Ed., *Modern Aspects of Reflectance Spectroscopy*, Plenum Press, New York (1968).
   (c) G. Kortüm, *Reflectance Spectroscopy*, Springer-Verlag, New York (1969).
   (d) R. W. Frei and J. D. MacNeil, *Diffuse Reflectance Spectroscopy in Environmental Problem-Solving*, CRC Press, Cleveland, Ohio (1973).

The Study of Thin Polymer Films on Metal Surfaces Using Reflection-Absorption Spectroscopy

OXIDATION OF POLY(1-BUTENE) ON GOLD AND COPPER

D. L. Allara

Bell Laboratories
Murray Hill, New Jersey 07974

Reflection-Absorption (R-A) infrared spectroscopy can be of considerable value in the study of chemical changes in thin polymer films on reflective metal surfaces. Recent advances in experimental techniques and theoretical understanding have facilitated the general use of the R-A technique for thin film and interface studies. This paper briefly reviews the current state of the technique and presents specific data on the oxidation (with O_2) of thin (<2000Å) poly(1-butene) films on gold and copper at 60 and 90°C. These data indicate that on gold, the oxidation proceeds via processes similar or the same as bulk oxidation whereas on copper, the mechanisms appear almost entirely controlled by interfacial chemistry. These observations are consistent with known autoxidation mechanisms and significantly add to previous, less rigorous studies of oxidation at the copper, polyolefin interface.

INTRODUCTION

Polymer-metal composite systems are of considerable practical importance, and a great need exists to understand the chemical characteristics of the polymer-metal interface. The number of techniques for examining the interface and the adjacent surface regions is quite limited. Most of these techniques involve the interaction of the composite with ionic or neutral particles and/or photons of various energies. It is desirable, whenever possible, to examine the interface without inducing physical damage. This

requirement significantly restricts the range of energies which
can be used to probe the interface and obviously excludes physical
separation of the interface prior to any analyses. One method
which appears promising for direct, nondestructive analysis of
chemical groups is reflection-absorption (R-A) spectroscopy in
the infrared energy region. The technique has seen only limited
application hitherto but holds the potential to become a significant
tool in polymer-metal interfacial studies. It is the purpose of
this paper to describe the major experimental and theoretical
features of the R-A method and to illustrate the usefulness with
recent experimental results.

GENERAL DESCRIPTION OF THE METHOD

Infrared radiation reflected at glancing angles from a reflective metal surface generates a standing wave with a nonzero value of the electric field at the metal surface. Only the component of the electric field of the incident radiation which is polarized parallel to the plane of incidence (where the plane of incidence is defined as the plane perpendicular to the reflecting surface and parallel to the incident and reflected beams) will give rise to a significant surface field. The latter can, in turn, interact with appropriate chemical groups at or near the surface with loss of beam power at the absorbing frequencies, which gives rise to typical absorption bands. Absorbing groups can be detected when they are located anywhere between the metal and a depth equal to a substantial fraction of the wavelength of the incident light (typically several thousand angstroms for several μm radiation). A recent and useful review of the method has been given by Tompkins[1]. When thin polymers films (e.g., < $\sim 10^3$ Å) are deposited on reflective metal surfaces, R-A spectroscopy can be used to characterize chemical groups originally present in the film and, of particular interest, can potentially provide information on chemical reactions between the metal surface and the polymer. Other typically used modes of infrared spectroscopy are transmission, internal reflection[2] (IRS), and emission[3]. Of these, only the latter two are generally applicable for metal substrates and can provide useful complements or alternatives to R-A spectroscopy. However, the emission method requires heating the sample, which may be a disadvantage. The IRS method requires covering the sample face with a reflection element, which thus precludes access of reaction gases or external radiation to the surface for in situ studies. However, it is possible in some cases to obtain spectra of metal-polymer interfaces by covering the surface of an internal reflection element with a very thin (e.g., < several hundred angstroms in order to be partially transparent) metal film and coating the exposed metal surface with the desired polymer film. This method may not be feasible if the metal is consumed (chemically altered) by interfacial reactions.

EXPERIMENTAL CONSIDERATIONS

General Considerations

Samples are prepared using optically flat metal surfaces (polished, evaporated, sputtered, etc.) of such metals as Au, Cu, Ni, Pd, Ag, Si, etc. Polymer films can be deposited by such methods as spin coating, solution dipping, and spray coating. The spectra are obtained using standard infrared instruments fitted with commercially available optics for reflection. Reaction chambers for direct studies of gas-film interactions have been described in the literature[4]. One or more reflections may be used. The single reflection mode generally gives better base lines, makes variations of the reflection angles easier, and generally allows better access to the surface for external radiation where desired. In multiple reflection, two parallel reflecting plates are placed a small distance apart (perhaps several mm) to cause the indicent beam to reflect alternately off each plate during passage between the plates. Multiple reflection can give higher sensitivity than single reflection up to an optimum number of reflections, unique for each metal where other conditions are fixed[1,5]. In order to maximize absorption, angles of incidence near 90° are necessary for thin films (< several hundred angstroms). As thickness increases, the angle of incidence can be decreased.

Calibrations

Band positions are usually, but not necessarily, identical to those in transmission. Unlike transmission, the refractive index of the absorbing material can influence band positions (e.g., 28 cm^{-1} shift for the Cu_2O 609 cm^{-1} band[6]). On the other hand, band intensities do not show any simple, general linear relationship between the film thickness and transmission extinction coefficients (e.g., a Beer's law-type relation) and calibrations must be done for quantitative studies. Plots of band intensities (absorbances or fraction of power absorbed) against measured film thickness must be done for each unique combination of a particular metal substrate, polymer film, angle of incidence, absorption band, and temperature (for variations of more than $\sim 50°$). Such plots can be calculated fairly well from theory (see below) when all the optical constants of the system are accurately known. An example of an experimental plot is given in Figure 1 for poly(1-butene) films on copper.

THEORY

Theoretical relationships between the ratio of the power of the incident and reflected beams and the optical constants of the metal, film, and ambient environment are well known from classical theory (e.g., see References 7 and 8 and references therein to earlier work). The physical description of the experiment is shown in Figure 2, where $\hat{n}_i$ is the complex refractive index of the ith phase, n_i and k_i are the corresponding real and imaginary parts, ϕ is the angle of incidence, and the E's are the amplitudes of the electric fields of the incoming (+) and outgoing (-) waves in the various phases. By specifying the refractive indices, ϕ, the wavelength and state of polarization of the incoming radiation, and film thickness, the optical boundary equations of the system can be specified[7,8]. Values of the fraction of amplitude reflected, E_1^-/E_1^+, can be obtained from the equations and the corresponding reflectance (R) calculated. The equations are best solved by machine computation. Exact calculations for hypothetical films on copper were carried out (on a Honeywell 6000 computer) using $n_2=1.5$, $n_3=2.4$, $k_3=28.0$ and varying k_2. The results are shown in Figure 1, where good agreement exists between the experimental and calculated curve shapes (using k_2 (2915 cm^{-1}) = 0.17 and k_2(1465 cm^{-1}) = 0.11-0.12 for the calculation).

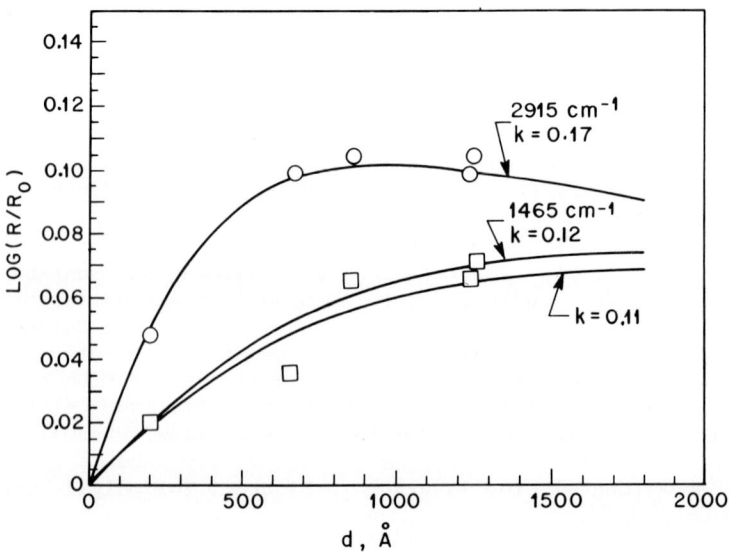

Fig. 1. Calibration plot for the 2915 and 1465 cm^{-1} bands of poly(1-butene) on Cu; single reflection ϕ=85ρ,
  O - 2915 cm^{-1} band
  □ - 1465 cm^{-1} band

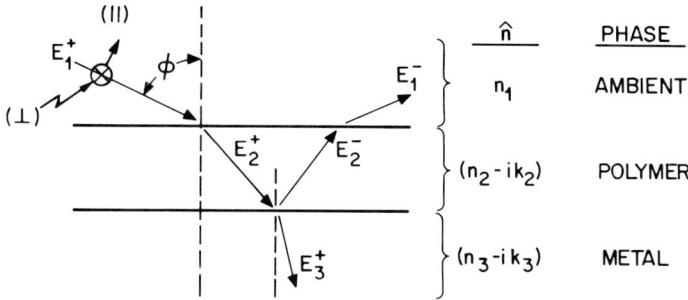

Fig. 2. Description of the single reflection experiment.

Approximate analytical solutions have been derived for cases where the ratio of film thickness to incident wavelength is quite small[8,9] ($<10^{-3}$). Thus for a 6 μm band, a film thickness of less than 60Å is required for an approximate solution. Such restrictions on the linear approximation range can be seen easily in Figure 1. Interesting variations of intensity with distance occur with films considerably thicker than 0.2 μm and considerable caution is required in interpreting spectra of such films[10].

AN EXPERIMENTAL APPLICATION - THE INTERACTION OF POLY(1-BUTENE) AND COPPER(OXIDE) IN THE PRESENCE OF O_2

Background

Copper metal surfaces are known to be catalytically active, in polyolefin oxidation, under commonly encountered conditions of temperature and exposure to air. Previous studies[11,12] of the mechanisms involved have indicated that a complex combination of chemical[11] and physical processes[12] in both the bulk and interface regions are involved. The main conclusions for the interfacial mechanisms were obtained from laminate experiments in which thick polyethylene films (∿0.006 inch) were physically separated from copper surfaces after oxidation and examined by internal reflection spectroscopy[11]. This study suffered from two limitations: 1) the penetration depth of the probing radiation in the internal reflection method was of the magnitude of 1 μm and thus was predominantly an analysis of bulk polyethylene chemistry rather than the interface region chemistry, and 2) the repeated physical separation of the interface left doubt as to the completeness[12] and reproducibility of the polymer, metal contact.

The present experiments were designed to overcome these problems. Very thin (<2000Å) films of poly(1-butene) were prepared on copper and gold surfaces, exposed to O_2 at 60° and 90° and R-A spectra accumulated during the oxidation. These experiments thus provided constant and complete interfacial contact and also emphasized interfacial chemistry relative to bulk because of the very small film thickness. Thus, these experiments could serve as close models for the interfacial chemistry of thick films on copper. Poly(1-butene) was chosen as the polyolefin because of the considerable ease in preparing thin films as compared to polyethylene used in previous studies[11,12]. Although detailed differences exist between the oxidative behavior of the two polyolefins (mainly absolute rate data) the major features of the Cu-polymer interfacial oxidation are expected to be the same, thus allowing overall mechanistic conclusions obtained with one polyolefin to be applicable to another.

EXPERIMENTAL

Poly(1-butene) was purified by dissolving in hot chlorobenzene and precipitating with methanol. Films between 100 and 2000Å were spun from hot chlorobenzene onto Cu and Au films evaporated on polished Si wafer substrates and the polymer films dried in vacuo. The intensities of the bands at 2913 cm^{-1} (C-H stretch) and 1465 cm^{-1} (-CH$_2$-deformation) were calibrated using films with known thicknesses determined by ellipsometry at 1.959 eV. Because of errors associated partly with the variation in substrate quality, the determined film thicknesses are conservatively estimated as good to ±20 percent (probably much better). The films showed no detectable initial absorption due to C-O or C=O bonds. The films were heated at 60 and 90° in 1 atm of O_2 for various lengths of time and examined at selected intervals by single-reflection R-A spectroscopy on a Perkin Elmer 621 spectrometer equipped with optical accessories supplied by Harrick Scientific. All spectra were taken using an 85° angle of incidence, and only the component of the radiation parallel to the plane of incidence was analyzed.

RESULTS AND DISCUSSION

The initial spectra for two films on copper [∼150 and 1250Å of poly-1-butene)] are shown in Figure 3. The spectra for films on gold show band positions identical to those of the copper films and intensities very similar to equivalent thickness copper samples. The approximate band intensity, film depth calibration curve for copper is given in Figure 1. The major bands are the C-H stretch (maximum at 2913 cm^{-1}) and C-H deformations (1465 and 1380 cm^{-1}), as observed in transmission spectra. There is no initial absorption observable in the 1500-1800 cm^{-1} region where the base line is quite flat. Bands with -log R/Ro $\sim$ 0.003 are observable provided that half widths are <50 cm^{-1}.

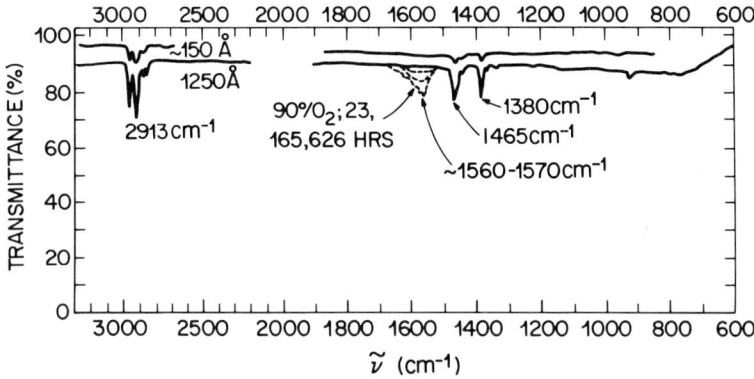

Fig. 3. Poly(1-butene) films on Cu; single reflection, $\phi=85°$.

After 20 hours at 90°, the Cu films showed the appearance of a broad band at 1560-1570 cm^{-1}; at 60° $\sim$30 to 100 hours were required. After $\sim$150 hours at 90° and $\sim$400 hours at 60°, the 1570 cm^{-1} band leveled off to a nearly steady value for the 90° run but increased after several thousand hours at 60°. These data are summarized in Figures 4 and 5. The growth of the 1570 cm^{-1} band at 90° is also shown in Figure 3, and complete spectra for 165 hours reaction at 90° are shown in Figure 6. It is interesting that at both temperatures the thinnest poly(1-butene) films show the greatest absorption for the 1570 cm^{-1} band. None of the copper samples gave spectra with absorption between 1700 and 1750 cm^{-1} (see Figure 6). The presence of volatile, corrosive oxidation products is suggested from the observed growth of the 1570 cm^{-1} band on a bare and initially clean copper film placed in the 90° reaction vessel with the oxidizing poly(1-butene) films (see Figure 4).

In contrast, spectra of the gold samples showed no absorption between 1600-1700 cm^{-1}, but after 200 and 400 hours bands at $\sim$1715 cm^{-1} could be detected for the $\sim$150 and $\sim$1500Å films, respectively. Again, the thinnest film gave the strongest absorption (see above

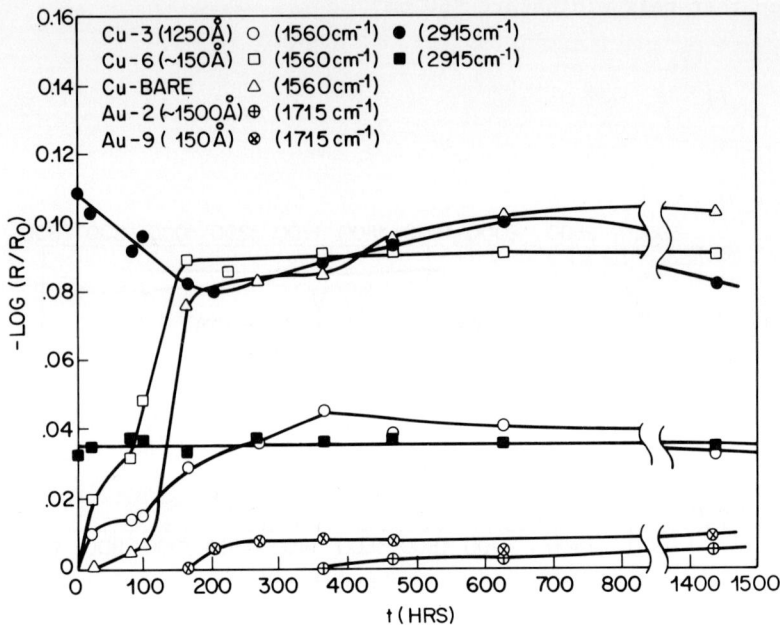

Fig. 4. Oxidation of poly(1-butene) films at 90°C; single reflection, $\phi=85°$;

for the 1570 cm^{-1} band). Figure 7 shows the IRS spectrum of an oxidized bulk poly(1-butene) film ($\sim$0.006 inch). The broad absorption at $\sim$1715 cm^{-1} is thus characteristic of a bulk oxidation.

In Figure 8, a direct comparison of the R-A and IRS spectra is made for the 1500Å film after 165 hours at 90° (no 1715 cm^{-1} band detectable yet). The C-H stretching and deformation bands were also measured for the copper films at 90°. The 2915 cm^{-1} band (see Figure 4) appears invariant with time in the 150Å film but appears to cycle in intensity for the 1250Å film with an average value of about 0.09. The latter phenomenon is at present unexplained, particularly in view of the steady value of the thin film which suggests that experimental errors in the measurements are slight. The C-H deformation bands at 1380 and 1465 cm^{-1} become progressively broadened and obscured as the oxidation proceeds due to the presence of new absorption in this region (e.g., see Figure 6). Therefore, quantitative measurements of these bands were not attempted.

Oxidation of Poly (1-Butene)   201

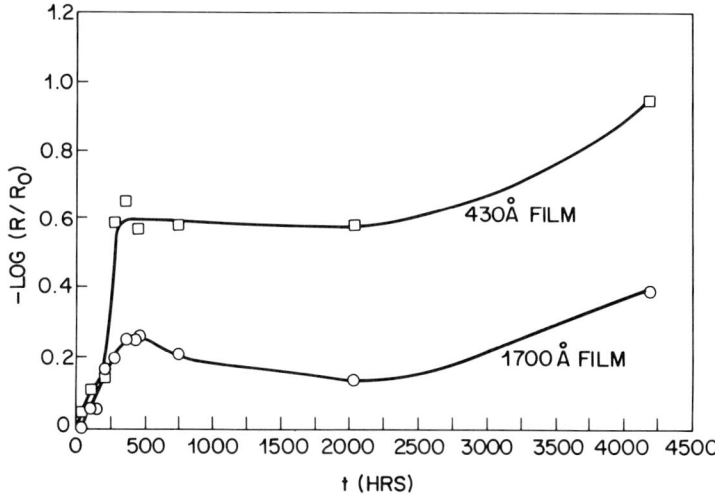

Fig. 5. Oxidation of poly(1-butene) films on Cu at 60°C; single reflection, $\phi=85°$, 1570-1600 cm^{-1} band.

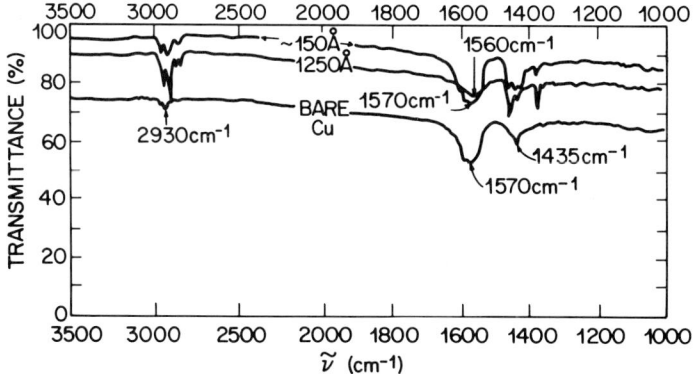

Fig. 6. Oxidized poly(1-butene) films on Cu; 165 hours/90°C; single reflection, $\phi=85°$.

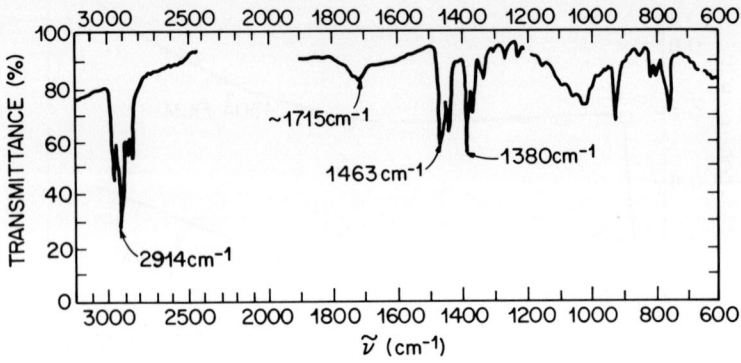

Fig. 7. Oxidized bulk poly(1-butene) film, 230 hours/90°C; IRS spectra, KRS5 element, $\phi=45°$, 25 reflections.

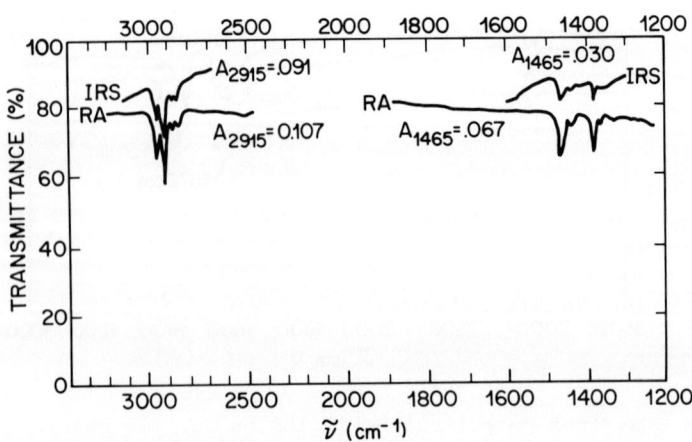

Fig. 8. Oxidized poly(1-butene) film on Au, 165 hrs/90°C; single reflection ($\phi=85°$) and IRS (KRS5, 45°, N=25) spectra.

The qualitative interpretation of these gold results appears straightforward. The chemical changes which occur during the oxidation of the polyolefin film over a gold surface appear to be quite similar to those during bulk oxidation (see Figures 7 and 8) of thick films. After an induction period of ~150-250 hours, absorption appears at 1710-1730 cm^{-1} and, although weak, is definitely observed even in very thin (~150Å) films. This absorption is typical of carbonyl groups in carboxylic acids, aldehydes, and ketones, and is normally observed in bulk polyolefin oxidation[13]. The oxidation, as monitored by carbonyl absorption, levels off after several hundred hours, a phenomenon typically observed in polyolefin oxidation[14]. Thus, we can conclude that the most reasonable mechanisms for the oxidation of thin surface films of poly(1-butene) over inert surfaces such as gold are the typical mechanisms applicable to bulk oxidation. A proposed, simplified reaction scheme, based on the extensive literature in autoxidation, is shown in Equations 1-10. More complex[15] and quantitative[16] schemes have been proposed but for purposes of this discussion can be ignored. A typical laboratory sample of a polyolefin inevitably will contain small amounts of oxygenated products as shown below because of normally unavoidable oxidation during processing. Usually, the trace hydroperoxide is assumed to be responsible for the initial occurrence of Equation 1. For reference a poly(1-butene) structure is shown below also:

(1) $RO_2H \rightarrow RO^\bullet + HO^\bullet$

(2) $RO^\bullet, HO^\bullet + RH \rightarrow \underline{ROH}, \underline{H_2O} + R^\bullet$

(3) $R^\bullet + O_2 \rightarrow RO_2^\bullet$

(4) $RO_2^\bullet + RH \rightarrow RO_2H + R(2°)^\bullet$

(5) $\searrow R(3°)^\bullet$

(6) $R(2°)O_2H \rightarrow \underline{ketone} + H_2O$

(7) $R(2°)O^\bullet \rightarrow \underline{aldehyde} + R'(2°)^\bullet$

(8) $R(3°)O^\bullet \rightarrow \underline{ketone} + R'(1°)^\bullet$

(9) $R'(1°)^\bullet + O_2 \xrightarrow[\text{several steps}]{}$ aldehydes, acids, low MW organic products (volatile), CO, CO_2, H_2O

(10) Aldehydes + $O_2 \xrightarrow[\text{several steps}]{}$ products similar to Reaction 9

A comparison of the characteristics of the oxidation of thin poly(1-butene) films over copper and gold clearly implies different mechanisms. The lack of typical carbonyl absorption (within an experimental limit) suggests that the carbonyl forming reactions (6-10), which contribute to product over gold, must be significantly by-passed over copper and/or any carbonyl product formed is unstable towards further oxidation with conversion to other noncarbonyl products. The major feature of the copper substrate samples, of course, is the rapid appearance of the relatively strong 1560-1570 cm^{-1} absorption in the very early stages of the oxidation. This absorption, most reasonably, can be attributed to the anti-symmetric vibration of the carboxylate ion[17], RCO_2^- where R is an alkyl group of variable length. The carboxylate ion is clearly a product of the reaction of a carboxylic acid with the copper(oxide) surface[4] and Reaction 11 (which will consist of several steps) is given as a likely possibility.

(11) $4RCO_2H + Cu_2O + 1/2\ O_2 \rightarrow 2(RCO_2)_2Cu + 2H_2O$

A previous study[18] of the gas-solid interactions of acetic acid and Cu_2O/Cu surfaces has shown that a carboxylic acid will readily (within hours) chemisorb on copper(oxide) at ambient temperatures to form multilayer structures of carboxylate ions in the absence of oxygen. In the presence of oxygen, at long reaction times, and temperatures above ambient (60 and 90°), copper-salt formation, as shown in Equation 11, is expected. It is not clear whether the copper salts are dispersed throughout the polymer film or localized right at the interface. The distribution characteristics will, of course, affect the measured carboxylate band intensities somewhat since the further the carboxylate groups are located from the surface, the weaker the absorption (see calibration curve in Figure 1, as an example). It is possible that the observed phenomenon of more intense absorption in the thinner films (see Figures 4 and 5, ∿1560 cm^{-1} curves) could be due to the localization of salt near the interface in the thin films.

Copper stearate, a typical carboxylate salt, previously has been shown[12] to be an active catalyst for the autoxidation of bulk polyethylene at 100°. The most likely role of the copper salts is the catalytic decomposition of hydroperoxides which may occur as given in Equations 12 and 13 (which will consist of several steps).

(12) $Cu(RCO_2)_2 + RO_2H \rightarrow Cu(RCO_2) + RCO_2H + RO_2\cdot$

(13) $Cu(RCO_2) + RO_2H + RCO_2H \rightarrow Cu(RCO_2)_2 + H_2O + RO\cdot$

More detailed descriptions of the chemistry in Reactions 12 and 13 have been given by Richardson[19]. Other reactions involving copper salts are, of course, possible; but at present no evidence exists to allow specific proposals. The major influence of Reactions 12 and 13 is to accelerate the decomposition of hydroperoxides to radicals (normally accomplished by Reactions 1 and 6). This will accelerate the autoxidation rate of the polymer and correspondingly will increase the contribution of the initiation reactions to overall product formation. Thus, reactions involving RO· radicals will increase Equations 7, 8, and eventually 9·, and this will lead to extensive production of carboxylic acid, chain scission, and volatile product formation. The former will lead to increased copper carboxylate formation, and the observed initial, accelerated build-up of salt (see Figures 4 and 5) is consistent with this proposal. The proposed chain scission leading to volatile products is consistent with the observation of chemisorption of volatile acidic or acid precursor products on the bare copper mirror as determined by the appearance of strong RCO_2^- absorption (Figure 4).

CONCLUSIONS

From these data, it can be concluded that the only major product of oxidation in a <2000Å film of polybutene on Cu is carboxylate ion. Further, this product is formed quite early in the oxidation process during the initiation stage. Thus, the copper-polymer interfacial chemistry completely controls the degradation of the polymer and the results indicate extensive chain scission of the polymer (roughly each acid group formed will result from at least one chain scission if oxidation of the side chains is negligible) and salt formation. It is not clear whether the salt is dispersed throughout the polymer matrix or localized at the interface, but there is no question that the initial nature of the metal polymer interface has been dramatically altered. Such findings are in agreement with earlier studies[11,12] using vastly thicker films (∿0.006 inch), in which the interface was not analyzed directly but rather after physical separation of the two surfaces. The present study avoids the uncertainties of such separation and allows study of only the portion of the polyolefin film which is close enough to the metal surface to be significantly affected by the surface chemistry.

REFERENCES

1. H. G. Tompkins in *Methods of Surface Analysis*, edited by A. W. Czanderna, Elsevier Publishing Company, Amsterdam, 1975, Chapter 10.
2. For example see N. J. Harrick, *Internal Reflection Spectroscopy*, Interscience, New York, 1967.
3. For example see I. Coleman and M. J. D. Low, Spectrochim Acta 23, 1293 (1966); M. J. D. Low and I. Coleman, Appl. Opt. 5, 1453 (1966) and references therein.
4. H. G. Tompkins and D. L. Allara, Rev. Sci. Instrum. 45, 1221 (1974).
5. R. G. Greenler, J. Chem. Phys. 50, 1963 (1969).
6. R. G. Greenler, R. R. Rahn, and J. P. Schwartz, J. Catalysis 23, 42 (1971).
7. R. G. Greenler, J. Chem. Phys. 44, 310 (1966).
8. J. D. E. McIntyre and D. E. Aspnes, Surface Sci. 24, 417 (1971).
9. S. A. Francis and A. H. Ellison, J. Opt. Soc. Am. 49, 131 (1959).
10. D. L. Allara, unpublished results.
11. M. G. Chan and D. L. Allara, J. Colloid and Interface Sci. 47, 697 (1974).
12. D. L. Allara, C. W. White, R. L. Meek, and T. H. Briggs, J. Polymer Sci., 14, 93 (1976).
13. J. P. Luongo, J. Polymer Sci. 42, 139 (1960); J. H. Adams, J. Polymer Sci. Part A-1 8, 1077 (1970); M. G. Chan and W. L. Hawkins, Polymer Preprints Amer. Chem. Soc., Div. Polymer Chem. 9, 1639 (1968); D. J. Carlsson and D. M. Wiles, Macromolecules 2, 587 (1969).
14. R. H. Hansen in *Thermal Stability of Polymers*, Vol. I. ed. by R. T. Conley, Marcel Dekker, New York, 1970, Chapter 6.
15. For example J. Adams, J. Polymer Sci. Part A-1 8, 1077 (1970).
16. E. Niki, C. Decker, and F. R. Mayo, J. Polymer Sci., Polymer Chem. Ed. 11, 2813 (1973), and references therein; D. L. Allara and D. Edelson, Rubber Chem. and Tech. 45, 437 (1972).
17. For example, see: L. J. Bellamy, *The Infrared Spectra of Complex Molecules*, John Wiley, New York, 1958, Chapter 10.
18. H. G. Tompkins and D. L. Allara, J. Colloid Interface Sci. 49, 410 (1974).
19. W. H. Richardson, J. Amer. Chem. Soc. 88, 975 (1966).

Use of Laser Raman Techniques in the Study of Polymers

R. D. Andrews and T. R. Hart

*Department of Chemistry and Chemical Engineering
and Department of Physics
Stevens Institute of Technology
Hoboken, New Jersey 07030*

The laser Raman technique employs a laser beam as an activating source rather than the mercury light used in the conventional Raman method. This has the advantage of a much higher light intensity so that measurements can be made on a time-scale of minutes rather than hours or days. The laser beam is also very concentrated and localized which allows the use of very small samples, such as milligrams of liquid in a capillary tube, or inspection of a very localized area of a surface. Although the Raman technique is closely related to the infrared method, the Raman technique makes possible the observation of very low-frequency modes which are beyond the range of the usual infrared spectrometers. This feature has recently made possible the observation of hydrogen-bonding modes in ice, near 4 cm^{-1}. The intensity of the laser beam also allows the observation of Raman spectra by reflection from an opaque surface. Another advantage of the laser is that it is possible to use different incident frequencies and tunable dye lasers to selectively observe and emphasize certain regions of the Raman spectrum. Applications and potential applications of the laser Raman technique to polymers are reviewed and discussed.

CONTENTS

INTRODUCTION
EXPERIMENTAL METHOD
APPLICATIONS
  A. Biological Polymers
  B. Amorphous Polymers
  C. Crystalline Polymers

D. Polymerization and Degradation Processes
E. Molecular Complexes with Polymers
F. Surface Layers and Adsorbed Molecules

INTRODUCTION

Following the first observation of the Raman effect in 1928, most of the studies carried out in subsequent years made use of high-intensity mercury arcs for the irradiating light source. However, with these arc lamps it normally required hours or days to obtain a satisfactory Raman spectrum using photographic recording, and the method remained a relatively inconvenient one. The information obtained from such measurements was similar to that obtained from infrared spectra, and while the infrared method was extensively developed and applied because of its generally greater convenience, the Raman method fell into a state of neglect. This situation was drastically changed by the development of the laser, since by use of this extremely intense light source (and additional related instrumentation) very high quality Raman spectra could be obtained in a matter of minutes. The applicability of the method to a number of different types of systems was explored by various investigators, and in the polymer field it was discovered that conformational transitions in polypeptides (both natural and synthetic) could be seen in a very clear-cut way in laser Raman spectra. This led to a very active use of the laser Raman technique in the biochemical area, which still continues to expand and develop at the present time. This is only one of many applications to the polymer field, however, and it is the purpose of this review to give a general picture of some of the recent successful applications of the method to polymer problems, so that the reader will obtain a general idea as to where the method has been found to be particularly useful thus far. Since the method is in its infancy, in some respects, no definitive description can be given at the present time of the ultimate limitations of the method, since some of the most serious past and present limitations seem in the process of being overcome. Undoubtedly some of the greatest successes of the method are still to come, and a decade from now, a much more adequate description will be possible of the potentials and limitations of the method. The present review makes no attempt to be comprehensive or complete (the number of publications in the field is already appreciable), but rather might be described as an "illustrative" review which focuses attention on successful applications and current trends in the polymer area.

The first Raman spectrum of a polymer (polystyrene) was measured[1] in 1932, and a general review of the limited studies of Raman spectra of polymers in the pre-laser period was given in 1964 by Nielsen[2]. At the end of his review, Nielsen states that "it is too early to make any predictions as to what role lasers may come to play in Raman spectroscopy; however, the role is likely to be important". Three years later, in 1967, the first high-quality laser Raman spectrum of a polymer (isotactic polypropylene) was published by Schaufele[3]. The results were so attractive that the method was quickly taken up by a number of workers. A review paper giving laser Raman spectra of several polymers was published in 1969 by Hendra[4], as well as another more general review paper[5] on the laser Raman method, including some comments on polymers. A review paper on polymers, with particular attention to hydrocarbon and polyolefin types, was published by Schaufele[6] in 1970. The experimental problems in obtaining high-quality laser Raman spectra from polymers were discussed in an article by Gall, Hendra, Watson and Peacock[7] in 1971. Koenig has also published two fairly extensive review articles on the results of laser Raman studies of polymers--one in 1971 on synthetic polymers[8], and the other in 1972 on biological polymers[9]. Raman data were discussed in a general article on inelastic laser light scattering from biological and synthetic polymers by Peticolas[10] in 1972. Review papers on the laser Raman method as applied to polymers, for a more general audience, have also appeared[11-13]. The use of the method in studies in the biological field has been discussed recently in a book chapter by Spiro[14].

Infrared and Raman spectroscopy both look at the vibrational spectra of molecular systems, and consequently there is a great deal of overlap in the applicability of the two techniques. For research work, however, the differences can be significant. Scanning IR spectrometers have a long wavelength limit of approximately 20-40μ (500-250 cm^{-1}). This limitation means that low-frequency vibrations of molecules will be missed by normal IR spectroscopy. Raman scattering techniques have been developed to operate down to 5 cm^{-1} (2000μ) to investigate very low frequency modes associated with vibrations of polymer chain backbones and lattice modes of polymer crystals.

There is a complementary nature between IR and Raman scattering, since for molecular systems that have inversion symmetry the vibrations can be either Raman-active, or infrared-active, or silent (not excited by incident radiation). Polymer systems do not have complete inversion symmetry, so the vibrations are generally both Raman and IR active; however, a line which is weak in one type of spectrum will often be strong in the other. This is illustrated in Fig. 1, which shows corresponding IR and Raman spectra[15] of an amorphous acrylic polymer (polymethyl acrylate). The IR and Raman-active vibrational modes are also often described as corresponding to vibrations in which the atomic displacements produce a change of dipole moment (polarization) or of bond polarizability, respectively. One consequence of this is that bonds between different types of atoms (a C-H or C-Cl bond, for example) will generally be IR-active, while a bond between atoms of the same type (such as the C-C bond, which occurs in most chain backbones and is of special importance in polymers) will show up particularly in the Raman spectrum.

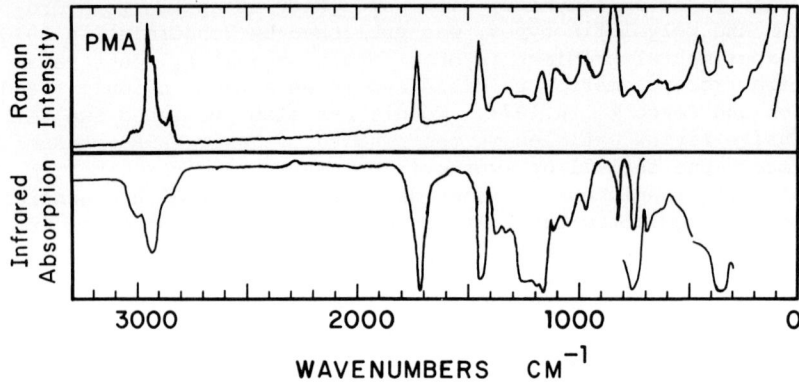

Fig. 1. Raman and infrared spectra of polymethyl acrylate at room temperature (from Ref. 15).

The spectral resolution is comparable in the two cases, but one advantage of the Raman method is that in cases where a vibrational mode is so strongly infrared-active that it saturates a certain region of the IR spectrum and makes it difficult or impossible to locate other weak lines in the vicinity of such a strong line, a Raman spectrum can often be taken which will reveal the details of the spectral structure in such a region clearly. This type of problem would arise in particularly critical form, for example, in observing the weak spectral lines from isotopic substitutions which are slightly shifted from their normal positions. The Raman-active lines will be polarized (in the same sense as the incoming laser beam) or depolarized, according to their symmetry. The equivalent effect in infrared spectra is the optical dichroism (parallel vs. perpendicular). One important area of difference between IR and Raman measurements is that IR is almost always done in transmission, whereas laser Raman scattering is easily done by reflection from surfaces. This means that thin coatings on opaque backings can be studied using Raman scattering. The problems of special sample preparation are, in general, much less with the Raman method than in the case of IR measurements. Another practical feature of great importance is that glass and water show only slight amounts of Raman scattering, which means that glass sample cells can easily be used (e.g., in dealing with liquid samples), and that samples containing absorbed water or even in aqueous solution can easily be measured (in contrast to IR, where the presence of water usually causes severe difficulties because of its strong and broad IR absorption).

The general nature of the Raman effect is illustrated schematically in Fig. 2, in terms of an energy level diagram. Whereas the IR effect involves simply the absorption of a photon of the correct incident frequency, the Raman effect involves the loss or gain of a quantum of energy from an incident photon, which is observed as a change of frequency, $\Delta\nu$, of the photon when it is scattered. Because quanta can be either lost or gained during scattering (the scattering is referred to as "inelastic" when this happens), the Raman spectrum actually consists of two identical sets of frequencies, one set higher and one set lower than the incident frequency (as mirror image spectra around the incident frequency). The lower frequency set is called the Stokes spectrum, and the higher frequency set the Anti-Stokes spectrum. The Stokes spectrum is the more intense, and it is therefore in almost all cases the one shown; this is normally evident from the fact that $\Delta\nu$ values are shown as increasing toward the left on the abscissa scale for Raman spectra. Most incident photons are simply scattered elastically, and show no frequency change (so-called Raleigh scattering), but about one photon in a million undergoes a Raman frequency shift. The Raman-scattered light is therefore extremely weak, and a basic experimental problem is to detect and record this very weak spectrum, which must be separated from the background of intense Raleigh-scattered light. The techniques used for doing this are indicated in the following section on Experimental Method.

Resonance Raman scattering has become an important technique for selecting specific features of molecular systems[16]. With tunable dye lasers operating throughout the visible and into the near ultraviolet, resonance spectra can now conveniently be obtained. The resonance phenomenon occurs because the Raman scattering tensor,

$$(\alpha_{ij})_{mn} = \frac{1}{h} \sum_e \left[\frac{(M_j)_{me}(M_i)_{en}}{\nu_e - \nu_o} + \frac{(M_i)_{me}(M_j)_{en}}{\nu_e + \nu_s} \right] \quad (1)$$

which arises from second order perturbation theory[17,18], contains denominator terms of a resonance type. Near resonance, when ν_e is approximately equal to ν_o, there is a strong enhancement of the scattering which is prevented from becoming infinite by inclusion of a damping constant in the formulas. The frequency ν_o is the incident laser frequency and ν_e is the frequency corresponding to the transition energy to a state of energy E_3 which is a virtual intermediate state, as indicated in Fig. 2. The frequency of the scattered light is ν_s. The electric dipole moments $(M_j)_{me}$ and

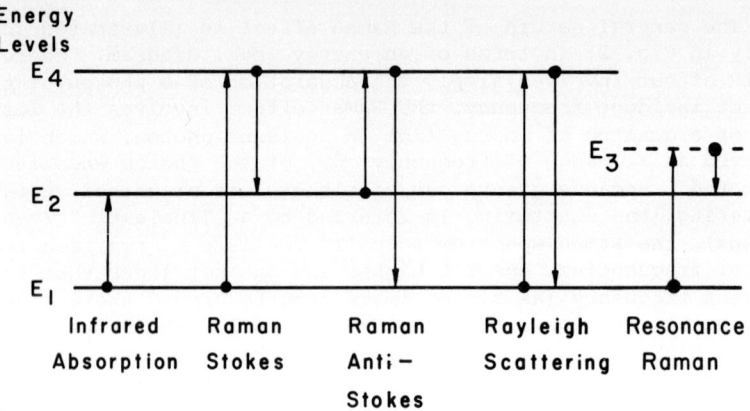

Fig. 2. Energy level diagram illustrating the comparative nature of infrared absorption, Rayleigh and Raman (Stokes and Anti-Stokes) scattering, and the resonance Raman effect. E_1 and E_2 represent the ground state and excited state for a given vibration, while E_3 and E_4 represent virtual intermediate states. The resonance Raman effect occurs when the photon energy is equal to that of a low-lying excited state E_3.

$(M_i)_{en}$ connect the initial state "m" and the final state "n" to the virtual intermediate excited states "e" which are summed over. The subscripts "i" and "j" give the polarizations of the incoming and outgoing (scattered) radiation and are responsible for the polarization selection rules of Raman scattering.

The total intensity of a Raman line for randomly oriented molecules is given by:

$$I = \frac{2^7 \pi^5}{3^2 c^4} I_o \nu_s^4 \sum_{ij} |\alpha_{ij}|^2 \qquad (2)$$

where I_o is the intensity of the incoming laser light. It is to be noted that there is an overall fourth-power frequency dependence, ν_s^4, so that it is advantageous to use higher frequency (shorter wavelength) lasers, in order to obtain greater intensity in the Raman spectrum.

The laser Raman field is now in a state of rapid development, and during the past nine years of polymer studies much of the work has been exploratory. This exploratory activity will certainly continue, including the development of increasingly sophisticated experimental techniques which can minimize or eliminate some of the problems which have caused trouble up to the present time, such as fluorescence and excessive sample heating. The development of tunable dye lasers will make possible a greatly increased use of the resonance Raman (RR) method, which has already proved its value in several important cases. Details of equipment and procedure will be discussed in the section on Experimental Method, which follows. Applications will be discussed in a section after that, under several general topic headings, concluding with a section on the application of the laser Raman technique to the investigation of surface coatings and molecules adsorbed on surfaces, since this relates to the special interest of the present Symposium.

EXPERIMENTAL METHOD

Since Raman scattering gives the vibration frequencies of the normal modes of the sample material by measuring the shift in frequency of the scattered light relative to the incident light, it is desirable to have an intense, monochromatic light source. Numerous lasers operating at different frequencies are now available and two of the best for Raman scattering are the argon-ion laser which operates at 488.0 nm (blue) and 514.5 nm (green), and the helium-neon laser operating at 632.8 nm (red).* Laser powers of 50 mW to 2 watts in continuous operation are available. In addition to the principal laser radiation, there is considerable fluorescence from the gas discharge which must be filtered out with a narrow band (10Å) blocked interference filter if a clean Raman spectrum is to be obtained.

The laser beam is focused onto the sample and the reflected and transmitted light is caught in a beam trap and dumped. The Raman scattered light is frequently collected at angles from 30° to 90° relative to the incident laser beam. Several necessary precautions should be noted. If a powerful laser beam is focused to a small spot with a condensing lens, most samples will be damaged by the heat. However, a cylindrical condensing lens can be used which distributes the laser energy over a long elliptical-shaped region that can be oriented parallel to the slits of the spectrometer, and this greatly decreases the heating effect. Other methods which have been employed to decrease sample heating are the use of a rotating sample cell[19], and a method in which the incident beam scans the sample surface in a rotating circular path[20].

---

*These wavelengths can be expressed in Ångstrom Units, rather than nanometers, as 4880, 5145 and 6328Å.

Special cells allowing controlled temperature variation of the sample have also been developed[21].

Another serious problem encountered in many Raman measurements is fluorescence from the sample, which in unfavorable cases can completely obscure the Raman spectrum. Methods of bringing this problem under control vary from case to case. In some cases, the fluorescence can be decreased to a low enough level by prolonged irradiation of the sample with the laser beam. In other cases, special purification or cleaning treatments may be required, as when metal oxide surfaces are heated in oxygen at high temperatures[22] to burn off traces of adsorbed hydrocarbons on the surface which generate fluorescent compounds when exposed to the laser beam. Another possible method of handling the fluorescence problem, which has potential for the future, is to use a pulsed laser and then to discriminate the Raman signal from the fluorescence on the basis of time-delay (Raman scattering is essentially instantaneous, while fluorescence is characterized by a slight time delay).

It should also be noted that most high-power lasers are polarized and that the Raman scattering is, in general, not isotropic. In an unfavorable configuration, Raman lines can be missed because the scattering is polarization-forbidden. For oriented samples the polarization of the beam can be used to advantage to inter the spectrum, and to draw conclusions regarding the nature of the orientation (an example of this is discussed in Section B on Amorphous Polymers). In cases where an unpolarized spectrum is desired from an oriented sample, a polarization scrambler can be used in the laser beam. However, unwanted polarization scrambling produced by birefringence and turbidity in the sample can also be a source of difficulty when polarized Raman measurements are being made.

The Raman-scattered light is weak, and it is desirable to collect as much light as possible using a short focal length (f/1) lens. For this light to be detected, however, it is necessary to use a second lens which focuses the collected light onto the slit of the spectrometer and matches the optical aperture of the spectrometer (often around f/7). A grating spectrometer should be selected which has high contrast between elastically scattered laser light and Raman scattered light. Imperfections in gratings produce a diffuse background ("grating ghosts") which can bury weak Raman lines in the noise. A double grating spectrometer largely eliminates the diffuse scattering problem, except near the laser line (within 100 cm^{-1}). Recently developed holographic gratings further reduce the background problem so that it is now possible to work within about 1 cm^{-1} of the laser line (which corresponds to extremely low frequency modes of vibration). Other methods used to reduce background (particularly near the incident frequency) are triple-grating monochromators, and an iodine-vapor absorption cell (which can only be used with the 514.5 nm line of the argon-ion laser). A method of making measurements very close to the exciting line by use of a vibrating glass plate at the exit slit of the monochromator has also been reported[23,24]. This method actually involves measurement of the first derivative of Raman intensity with respect to frequency, which allows a better discrimination of signal against background and background noise.

If the exciting laser radiation is in the visible, the Raman-scattered light can be detected with a photomultiplier (e.g., one with an S-20 spectral response). Various systems have been used to process the electrical output from the photomultipliers. The simplest is to use a D.C. micro-microammeter with a recorder output and measure the total anode current. Alternately, the laser beam can be chopped and a lock-in amplifier used; or a photon-counting system can be used. At most these electronic systems differ by only a factor of two in the signal-to-noise ratio in the final spectrum, and are selected on the basis of the type of experiment to be performed. By use of signal-averaging with a computer, the signal-to-noise ratio can be enhanced in proportion to the square root of the number of spectra being averaged.

A typical experimental arrangement for laser Raman measurements is shown in Fig. 3.

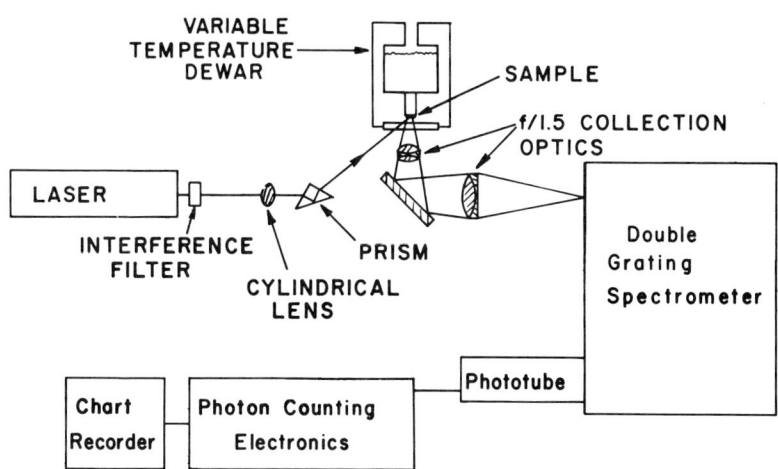

Fig. 3. One type of experimental arrangement for laser Raman measurements. This shows a system for taking measurements in back reflection from an opaque sample, including the use of a cylindrical lens to minimize local heating of the sample and a temperature control system for the sample.

APPLICATIONS

A. Biological Polymers

A field in which the laser Raman technique has found important applications has been the area of biochemistry, and in particular synthetic and natural polypeptides. Interest in this area was stimulated by the early discovery that conformational transitions of such polypeptides (both in solution and in the solid state) could be clearly observed in the laser Raman spectrum. Such transitions have been most commonly identified and studied by means of optical rotatory dispersion (ORD) measurements; however, optical rotation represents a summation effect of the molecule as a whole, whereas the Raman spectrum by giving information on individual molecular groups allows a more detailed and specific picture of the conformational transition to be deduced.

To illustrate this particular application, an important early study by Small and Peticolas[25] may be cited. They found several features of the Raman spectrum of polypeptides to be sensitive to conformation changes. They observed that there was a reduction of intensity of certain ring vibrations of the polynucleotide bases when "stacking" or ordering occurs (e.g., in helix formation); they named this phenomenon "Raman hypochromism". (The theory of this effect has recently been discussed by Painter and Koenig[26].) Since the frequencies are different for each type of base, this provides a method of estimating the average amount of order of each type of base in partially ordered helical systems.

When two synthetic polypeptides, poly-A (polyadenylic acid) and poly-U (polyuridylic acid), were mixed in aqueous solution to form a double-stranded helical complex, there was also a very large increase in intensity of a sharp, strongly polarized band at about 815 cm^{-1}. This band was assigned to a vibration of the phosphate-sugar portions of the molecule, and is related in a sensitive way to the backbone conformation of the polymer. When this poly A-poly U complex undergoes a conformational transition at 59°C, which can be described as a cooperative melting of the complex, or a breakdown of the double-stranded helix structure, the intensity of this band drops suddenly to about half its original intensity (see Fig. 4). A very similar effect is seen in RNA (Fig. 4), but the decrease with temperature is more gradual, indicating a non-cooperative conformational transition in this polymer. When the transition takes place in poly A-poly U, there is also a marked increase in intensity of the adenine band at 730 cm^{-1} and the uracil bands at 1236 and 1660 cm^{-1} (the hypochromic effect taking place in reverse as temperature increases).

In addition to the above, frequency changes are seen in the 1600-1700 cm^{-1} region of the spectrum where carbonyl group

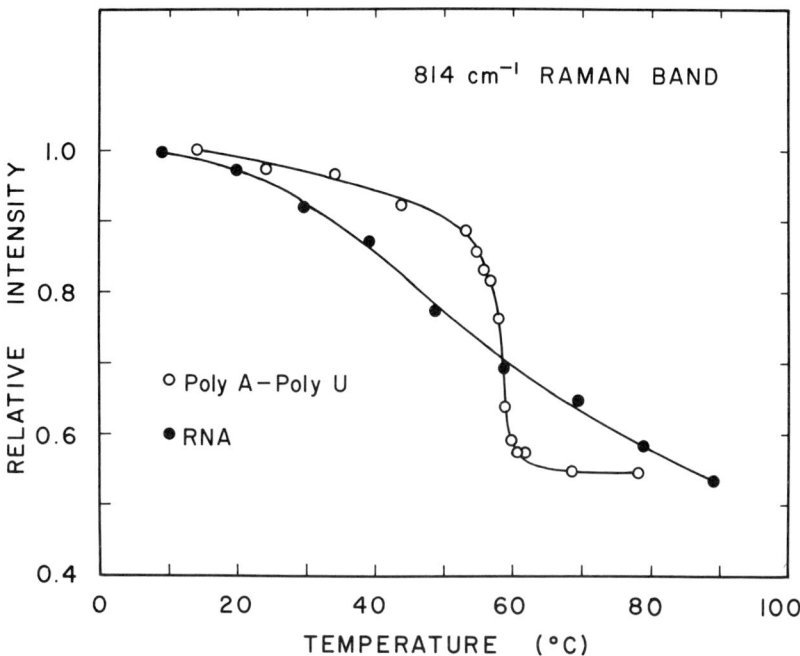

Fig. 4. Effect of conformational transition on relative intensity of the 814 cm^{-1} Raman line of a helical complex of two synthetic polypeptides (poly-A and poly-U) and of yeast RNA, the reference temperatures being taken as 15° and 10°C, respectively (data re-plotted from Ref. 25).

stretching is found. These are related to changes in hydrogen-bonding of the carbonyl group when conformational changes take place. Conformational transitions can take place not only by temperature change, but also by changes of pH, ionic strength, and solvent composition, and these can always be observed in the Raman spectrum[27,28]. The conformational changes (α-helix formation in solution, for example) produced by changes of ionic strength due to the presence of specific salts can also be studied[29], as well as conformation changes produced by mechanical deformation[30]. Conformation changes are also, of course, directly involved in the process of protein denaturation, and Raman studies[31-33] have made important contributions to the understanding of this phenomenon. One of these studies[33] led to the proposal of a new mechanism for the heat denaturation of bovine serum albumin, involving a stepwise unfolding of α-helices followed by aggregation. The conformations and conformational changes associated with copolymerization and sequence isomerism can also be observed[34,35]; this type of information is needed in order to understand the behavior of such complex copolymers as RNA.

The above conformation-sensitive changes can be described as changes in intensity or frequency related to whether a particular small group in the molecule finds itself in a helical or non-helical environment. However, in addition to this, low-frequency Raman lines can be seen which are associated with longitudinal vibrational modes of the helix itself. Observation of these frequencies has been reported, for example, in an article by Lewis and Scheraga[27], who investigated the laser Raman spectra of random copolymers of hydroxybutylglutamine (HBG) and glycine in the solid state. Poly-HBG was considered to be a host polymer, and different amounts of glycine were introduced as guest residues. The fluorescence of the samples fortunately decayed without sample degradation on prolonged exposure to the laser beam, and a 10-cm iodine cell containing iodine vapor at 65°C was used to permit measurements down to within 5 cm^{-1} of the exciting line (the 5145Å line of an argon-ion laser). A number of lines were observed which could be assigned to helix vibrational modes: at 46, 55, 68, 74, 89, 111, 205, 260, 313 and 382 cm^{-1}. These would all be lines corresponding to a poly-HBG helix. The glycine was always present in limited quantity and was believed to be present in helices always as isolated single residues. Glycine residues in coil and in helix environments could be identified by specific lines at 1005 and 1040 cm^{-1}, respectively, associated with a skeletal stretching mode of the glycine residue in the two types of environment. Many differences were also observed in the various frequencies of the amide group between the helix and coil structures, in agreement with the observations of other workers. They used optical rotatory dispersion measurements as an independent method of assessing the helical content of their polymers.

Some care must always be exercised in assigning low-frequency modes to helical structures, as shown in a recent paper by Genzel and coworkers[24], who measured Raman spectra from aqueous solutions of lysozyme (a protein enzyme) and two crystalline forms (orthorhombic and triclinic) of lysozyme (from hen egg white). Their measurements extended down to 5 cm^{-1}, and while they found that the solution and crystalline spectra were similar above 100 cm^{-1}, a low-frequency band at 25 cm^{-1} observed in the crystalline state was not observed for the polymer in solution, indicating that this low-frequency line corresponds to a crystal lattice mode rather than an intramolecular vibration. Assignment of such frequencies can often be aided by a normal mode analysis. A discussion of helical vibration modes is included in a review paper by Koenig[9] mentioned earlier.

In addition to studies of the behavior of polypeptide molecules as a whole, it is also possible to study specific sites on the molecule which are of interest because of their biological activity. Some interesting examples of this type of study are provided by the recent investigations by Spiro and coworkers[36] of the heme group (iron-porphyrin ring complex) in heme proteins, in which use was made of the resonance Raman effect, which greatly intensified the Raman lines arising from vibrations in the porphyrin ring. The most familiar example of such a heme protein is, of course, hemoglobin, which can exist either in an oxidized or resting state (i.e. with or without bound oxygen) and can be "poisoned" by complexing of the central Fe ion with carbon monoxide or cyanide ion. These reactions of the heme group are accompanied by changes of geometrical form of the porphyrin ring (in particular, a deviation from planarity of the ring, called "doming"), as well as changes of oxidation state and spin state of the central Fe ion. Data from several heme proteins have been correlated and interpreted[36], and Spiro has published a general review article[16] on the resonance Raman effect with particular application to the study of heme proteins. These studies provide a new experimental approach to the details of structure of the heme group, and the structural changes related to its specific chemical reactivity.

A related study which is also of interest is an investigation[37] of "blue" copper proteins, which also have a powerful chromophore group with an electronic transition which can be excited by the laser beam (a necessity in producing the resonance Raman effect). This study also included use of a tunable dye (Rhodamine 6G) laser, which is valuable not only for its ability to improve or optimize the resonance effect, but also because data on the frequency-dependence of the resonance effect are valuable in elucidating the theoretical nature of the spectrum obtained. Of the three proteins studied, the one which gave the best spectrum was plastocyanin (extracted from spinach leaves); this molecule plays a role in the photosynthesis process, by making possible an intramolecular electron transfer. The laser Raman data, when combined with electron spin resonance data and arguments based on ligand field theory, allow a fairly precise structure for the "blue" copper center to be postulated, based also on the assumption that the electronic transition giving rise to the resonance effect is a charge transfer transition between the central Cu and a sulfur ligand. The structure deduced is that there are five ligand groups around a central Cu, in a geometrical arrangement which is a slightly distorted trigonal bipyramid. There are three "strong" ligands in the equatorial plane, one a cysteine mercaptide sulfur, and the other two presumably nitrogen ligands (but not imidazole). The axial positions are occupied by less strongly bound ligands which might be nitrogen or oxygen.

Other colored compounds of biological interest have been studied by the resonance Raman method, including carotenoid pigments[38], vitamin B_{12} (containing a Co^{+2} ion in the center of a corrin ring) and derivatives[39], and ferredoxin[40].

B. Amorphous Polymers

Although amorphous polymers do not form a separate class of materials in regard to applications of the laser Raman method, this heading is adopted for convenience, since this section will be concerned with the polarization and depolarization characteristics of Raman spectra which can be related to molecular orientation and conformation, and the applications of this type of analysis have mostly involved amorphous polymers, since they constitute the simplest case (less complicated than the 2-phase structure of a semicrystalline polymer).

We will first discuss a fairly detailed study of molecular orientation in uniaxailly oriented polymethyl methacrylate (PMMA) by Purvis and Bower[41]. They utilized an experimental arrangement in which the plane of polarization of the incident laser beam could have two positions (vertical and horizontal), and the polarization axis of the collection optics could have the same two types of orientation, thus giving four different configurations. Samples with three different degrees of uniaxial orientation, oriented at three different temperatures, were used; and the orientation axis of the sample could be set at different angles with respect to the propagation axis of the laser beam. A considerable amount of data was gathered. Four different Raman lines were included in each measurement: the lines at 486, 562, 604 and 1732 cm^{-1}. The first, third and fourth frequencies are assigned to the $C - C - O$ part of the ester side group, while the second frequency is assigned to a C-C-C skeletal mode of the backbone carbon chain. Typical results obtained for one of the samples, using one of the Raman lines, are shown in Fig. 5; these are polar plots of Raman intensity for the four different polarization configurations as a function of rotational position of the orientation axis of the sample (expressed by an angle β).

To interpret the data, mathematical expressions had to be developed, which involved some assumptions at the molecular level. It was assumed that the basic unit being oriented was a chain segment which could be described by a single Raman scattering tensor. It was also assumed that the ester group was a rigid planar structure. Both planar zig-zag and helical conformations were considered for the chain backbone. The assumption was also made of the additivity of Raman tensors for individual bonds. Five parameters were derived which could express the scattering intensity for any combination of scattering geometry and polarization axes. Values of $\overline{\cos^2\theta}$ and $\overline{\cos^4\theta}$ could be deduced for the structural units (chain segments) of the molecules, where θ is the angle between the axis of the individual chain segment and the overall

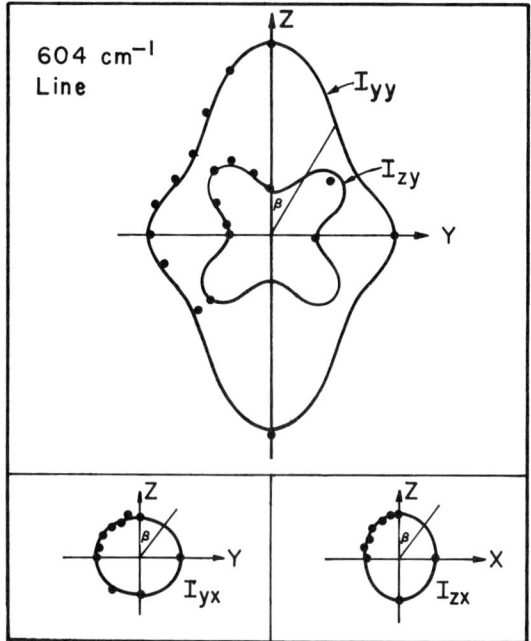

Fig. 5. Polar plots of the angular dependence of intensity of the 604 cm^{-1} Raman line in uniaxially oriented polymethyl methacrylate for each of the four different configurations of incident and scattered polarization axes, as a function of angle (β) of the orientation axis of the sample. Curves are least squares fits to the experimental points and are plotted on a common arbitrary intensity scale. (Data from Ref. 41, with the original authors' subscript notation of 1,2,3 changed to x,y,z.)

uniaxial orientation axis of the sample. These values could be compared with values derived from broad-line NMR measurements, and good agreement was obtained. However, the nature of the change in numerical values with degree of orientation indicated "either that the nature of the structural units changes on drawing or that there are significant interactions between them which affect the values of their Raman tensors and which depend on their degree of orientation". The authors point out that further study on a greater variety of oriented samples is necessary to resolve this question, such as samples with the same degree of birefringence, but produced by orienting at different temperatures, since they feel that the Raman tensors for the individual modes of vibration probably do not change with the degree of orientation and that it is therefore possible that the changes of the effective Raman tensors observed for PMMA may be due to conformational changes in the molecules.

Measurements of the Raman depolarization ratio, ρ, of the 1002 cm^{-1} band of polystyrene have been used by Speak and Shepherd[42] to follow conformational changes of polystyrene dissolved in three different solvents (cyclohexane, methyl ethyl ketone and deuterobenzene) as a function of temperature in the range 20-60°C. The largest changes were observed in the poorest solvent (cyclohexane), while in the best solvent (deuterobenzene) no change at all was seen in ρ. Changes in the depolarization ratio when the conformation changes are due to the fact that gauche and trans isomers in the chain have different polarizability properties and thus affect the polarization of the scattered light through their effect on the elements of the Raman tensor. They conclude that the Raman tensor for a trans isomer has a smaller anisotropy (off-diagonal terms are smaller relative to the diagonal terms) than the Raman tensor for a gauche isomer. They compare their depolarization data with conventional light-scattering data which gives the change in end-to-end distance of the molecule vs. temperature. The two types of data give consistent results, and allow a satisfactory calculation of the relative populations of trans and gauche structure as a function of temperature.

Related investigations have been carried out on dimethyl siloxane polymers in the solid state by Shepherd and coworkers[43,44]. They measured the depolarization ratios for two Raman bands, one at 490.9 cm^{-1} which is an Si-O skeletal mode, and the other at 2906.8 cm^{-1} which is the symmetric stretching mode of the methyl group. These are both strong, highly polarized bands. In the first study[43], the depolarization ratio was measured as a function of strain for samples of radiation-crosslinked polymer, and also as a function of temperature (between 30°-105°C) in an uncrosslinked sample. From the measurements vs. temperature, it could be calculated that the energy of a gauche conformation is higher than that of the trans by approximately 1000 ± 140 cals/mole. For the unstretched rubber at room temperature, it is calculated that the molecules will be about 3/4 trans. They were also able to derive a value of the constant required in a theoretical equation giving the relative isomer population as a function of strain (which also involves the degree of crosslinking). In the second study[44] the depolarization measurements vs. temperature were repeated, and were carried out over a more extended temperature range, from -45° to 100°C. The higher-temperature data gave them revised values for the trans-gauche energy difference of 790 ± 100 and 820 ± 50 cals/mole from the 2907 cm^{-1} and 491 cm^{-1} bands, respectively. A new observation was made in the temperature region below room temperature: values of the depolarization ratio for both Raman lines went through a minimum near 0°C, and then started to increase as temperature was lowered further. This was interpreted as due to the presence of a helical type of conformation beginning to appear as temperature was lowered, which constitutes a third isomeric form (in addition to trans and gauche) and also represents a pre-crystalline structure, since this polymer crystallizes in helical form. A conformational energy of helix formation was estimated as 3300 ± 800 and 3100 ± 1000 cals/mole from the 2907 cm^{-1} and 491 cm^{-1} Raman bands. A maximum limit on ΔS for helix formation was also calculated as 16 ± 3 e.u. Measurement of the Raman spectrum of the crystallized polymer showed a new band at 468 cm^{-1}. The relative intensities of this band and the 491 cm^{-1} band observed in the amorphous polymer provides a convenient method of estimating relative crystalline and amorphous fractions in this polymer.

In addition to conformation, the configuration of molecules is also a question of importance, and Raman data can be helpful in connection with this problem. The differentiation of cis and trans structures around a double bond is relatively easy; an illustration of this as applied to an unsaturated **polyes**ter resin is given later in Section D. The structure of polybutadiene rubber has also been very successfully studied by Raman measurements[45]. The application to the stereo structure (isotactic, syndiotactic, atactic) and conformation of vinyl polymers has been discussed by Koenig[8]. This is a more difficult problem, as can be seen from a study of polyacrylonitrile[46]; this area of application has not been followed up very extensively, probably because of the widespread and very successful use of NMR measurements for this purpose.

Some observations have been made[15] of the Raman spectrum of an amorphous polymer (polymethyl acrylate) above and below its glass transition temperature of 6°C. No differences in the Raman spectrum were observed in the glassy and rubbery states, in accordance with what would be expected from the generally accepted theories of the nature of the glass transition.

C. Crystalline Polymers

A number of laser Raman studies have been carried out on crystalline polymers. Raman spectra were measured on polyethylene (PE) even before the advent of the laser[2]; however, the purpose of those studies was usually to aid in the elucidation of the infrared spectrum of PE, so that proper assignments could be given to the IR-active frequencies. More recently, laser Raman spectra have been obtained on PE by several authors, but there have been some conflicting assignments of frequencies, although the data have largely been in agreement. In an attempt to resolve the problems of the Raman spectrum of PE in a definitive way, Gall, Hendra,

Peacock, Cudby and Willis[47] have measured laser Raman spectra on a wide variety of polyethylenes in the range of 700 - 3200 cm^{-1}, including: low-density PE, high-density PE of different molecular weights, quenched transparent PE, PE crystallized under pressure to give extended-chain crystals, single crystal mats, a low molecular weight paraffin hydrocarbon, and also a PE melt. They observed several crystallinity-sensitive bands, and also bands in the 700 - 900 cm^{-1} range (particularly a band at 725 cm^{-1}) indicating <u>gauche</u> methylene groups, as well as weak features in the 1600 - 1700 cm^{-1} region due to vinyl and vinylidene groups which are present as impurities or defects in the structure. Using these results as well as data on oriented PE samples and the data published by other workers, these same authors[48] have attempted to give a complete and definitive interpretation of the entire Raman spectrum of PE between 700 and 3200 wavenumbers. Related to this work is a study[49] of the spectrum of PE under high hydrostatic pressure (2 kbar), which shows "correlation splitting" of certain of the lines under pressure (due to the fact that a molecular chain in the unit cell of the crystal can vibrate either in-phase or out-of-phase with its neighbor). This is a predicted effect and provides additional evidence for the correctness of some of the vibrational assignments, particularly when polarization measurements are made on the split components.

One of the phenomena related to the Raman spectrum of polyethylene which has received a great deal of attention is the presence of very low-frequency longitudinal acoustic modes (sometimes referred to as "accordion modes") of a completely extended zig-zag, all-<u>trans</u>, polymethylene chain. These low-frequency modes were observed in both liquid and solid paraffin hydrocarbons in early pre-laser Raman studies, and their nature was correctly interpreted by Mizushima and Shimanouchi[50] in 1949. In his 1970 review article, Schaufele[6] pointed out the relevance of this work to the case of polyethylene, and shortly thereafter laser Raman data on PE single crystals were reported by three different research groups[51-55]. In the paper by Olf, Peterlin and Peticolas[53], they observe this low-frequency mode both in single crystal mats and in bulk-crystallized PE, and in both linear and branched PE. The mode is more clearly seen in linear PE, however, as is evident from Fig. 6, which shows data on single crystal mats of the two types of PE in the low-frequency region near the incident laser frequency. The use of an iodine-vapor filter to eliminate the unshifted scattered light (Tyndall and Raleigh scattering) allowed measurements to be made down to within 5 cm^{-1} of the incident argon laser line (5145Å). In contrast to the related spectra observed for long-chain paraffin hydrocarbons[56], where many overtones can be seen, the unique feature of the PE spectrum is the fact that only the fundamental frequency is observed, with no overtones. This low-frequency line corresponds to the longitudinal "accordion" vibration of a linear all-<u>trans</u> sequence of methylene groups, and clearly corresponds to the rigid, linear sections of chain between chain-folds in the PE crystal lamellae. The lamellar thickness can be measured by low-angle x-ray

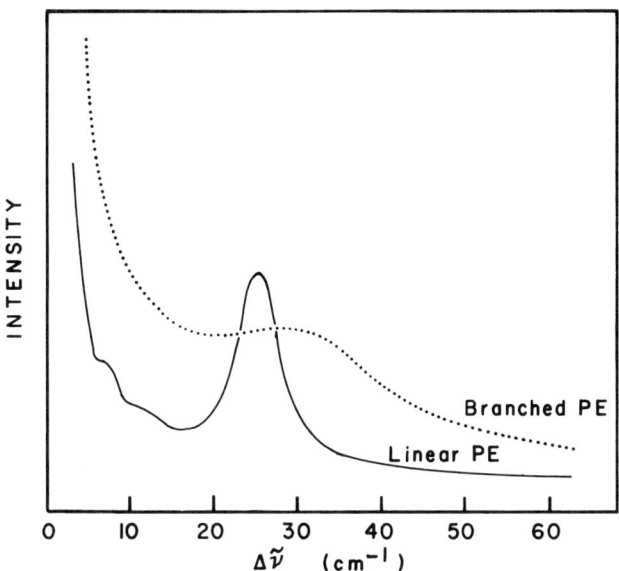

Fig. 6. Raman spectra in the low-wavenumber region for single crystal mats of linear and branched polyethylene (from Ref. 53).

scattering, and by applying a correction for the tilt of these chain segments in the lamellae (since they are not exactly perpendicular to the lamellar surfaces), the length of the rigid chain segments can be calculated. From the peak frequency in the Raman spectrum, a segment length can also be calculated, and this is found to agree extremely well with the x-ray values. Since lamellar thickness can be changed by annealing, measurements were made on both unannealed and annealed single crystals. The results for segment length from unannealed PE were: 126Å from x-ray, 119Å from Raman; and from annealed single crystals: 222Å from x-ray, 218Å from Raman. The authors feel that the absence of overtones is due to the presence of chain folds in the lamellae, and they feel that this result tends to support the "loose loop" rather than the "tight fold" picture of lamellar structure. They attempted to observe this low-frequency mode in other crystalline polymers (isotactic polypropylene, two crystal forms of polyvinylidene fluoride, polyoxymethylene, and nylon 66), but without success.

However, Folkes et al.[55] were able to observe this type of mode in a second polymer: polyester 10,10 (polydecamethylene sebacate), also prepared in the form of a single crystal mat. This polymer has polymethylene sequences in the backbone, interrupted by periodic ester groups. Just as the present article was being submitted for publication, the observation of this type of acoustic mode for a third polymer (polyethylene oxide) was reported[56a]. In both the polyester 10,10 and PEO, the rigid chain segment whose accordion frequency was observed was that extending between the chain folds at the lamellar surfaces, and therefore included ester groups or ether oxygen atoms as well as methylene groups.

The use of these low-frequency acoustic modes has not been restricted to high polymers. A recent study[57] has involved their measurement in surfactants of the sulfonated hydrocarbon type, both in the crystalline solid state and in solution. It is concluded that the hydrocarbon chains are in the fully-extended zig-zag conformation in the crystalline solid, but are more coiled up in solution, since the lowest-frequency mode observable in the solid can no longer be seen in solution. It should be pointed out that these low-frequency modes, when observed in the Raman spectrum of solid polymers, do not always represent intramolecular vibrational modes; they can also represent crystal lattice modes, as indicated in a recent paper[58] on the polarized Raman spectrum of the straight-chain hydrocarbon n-$C_{36}H_{74}$ in the form of single crystals. Further information relevant to these interpretations is contained in a paper[59] which has just appeared on the Raman and IR spectra of a series of isotactic poly(1-alkyl ethylenes), where the alkyl side group varies in length from 1-20 carbon atoms. The isotactic structure tends to make the chain backbone adopt a helical conformation of one type or another, and low-frequency modes can be seen related to the vibrations of these helices when the length of the side group is short. When the side groups are of intermediate length, they assume an extended zig-zag conformation and show accordion mode spectral lines corresponding to this structure. When the side groups become still longer, they crystallize into the type of lattice characteristic of polyethylene, and corresponding crystal lattice modes are seen in the low-frequency region of the spectra.

Melveger[60] has measured laser Raman spectra from polyethylene terephthalate (PET), and found that crystallization of the polymer produced a very marked narrowing of the band corresponding to the Raman-active C=O stretching frequency at 1096 cm^{-1}. The change of crystallinity could also be observed by density changes of the polymer, and this was used as an index of degree of crystallinity. Although McGraw[61] had used the band at 632 cm^{-1} as an internal intensity reference for the 1096 cm^{-1} band, in attempting to use the latter band as an index of percent crystallinity, Melveger found that this procedure was not satisfactory because the proposed reference band was affected by molecular orientation. Melveger concluded that rather than using band intensity, it was more satisfactory to use band width (of the 1096 cm^{-1} band) as the crystallinity index. The effect of crystallinity on this band is shown in Fig. 7. This carbonyl group is part of the ester group in the polymer backbone, and in the crystal lattice of this polymer, the ester group, including the carbonyl, is a planar structure, whose planarity is stabilized by resonance effects. The broadening of the carbonyl band is attributed to the various degrees of out-of-plane rotational position which the carbonyl group can assume in the amorphous state. The half-width of this band correlates well with density values for different PET samples, and was found not to be influenced by molecular orientation.

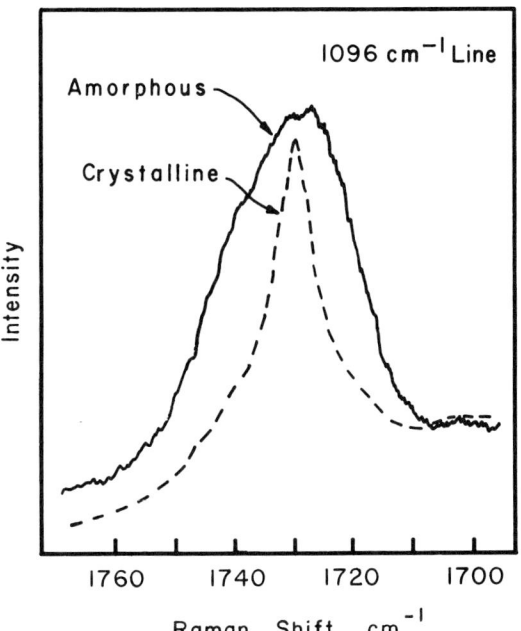

Fig. 7. Appearance of the 1096 cm^{-1} carbonyl stretching band of quenched and crystallized polyethylene terephthalate (from Ref. 60).

Another feature of polymer crystallinity of great interest is the fact that some polymers show more than one crystalline form. One example of this is polyvinylidene fluoride (PVDF), which has been recognized as having two different crystalline forms, and perhaps a third. A very detailed analysis of both the laser Raman and the IR spectra of this polymer in its various crystalline forms has been carried out by Kobayashi et al.[62], including a calculation of normal modes for both the molecules and the crystal lattice structures. They confirm from their analysis that Form III is a definitely different crystal form from Form I, even though the molecular chains are in the fully-extended zig-zag conformation (all-<u>trans</u>) in both cases. Form II is easily distinguishable from the other two by the fact that the molecular conformation is different (alternating <u>trans</u> and <u>gauche</u>). Vibrational lattice modes are calculated and observed for the three crystalline forms in the region of 50-85 cm^{-1}. Raman spectra have also just been reported for the two crystalline modifications of poly (trans-1,4 butadiene) by Evans and Woodward[63], measured from single crystal mats. Form I is stable at room temperature and atmospheric pressure, while Form II is stable (at 1 atm.) at temperatures between 71°C and the melting point. They observe an increase in Raman scattering intensity above the transition at 71°, apparently due to an increase in molecular mobility and a greater variety of chain conformations in the Form II crystal. They also observe crystal lattice vibrational bands in the region 90-120 cm^{-1}.

D. Polymerization and Degradation Processes

The course of the polymerization process and the final structure of thermoset resins can be studied using the laser Raman technique. Infrared measurements have been used for the same purpose, but a decided advantage of the Raman method is its lower sensitivity to water, which is usually a product of the condensation process, and which tends to obscure infrared spectra. A further complication with infrared measurements is that they must be carried out on thin films, and there are some questions as to whether the reactions are the same in thin films and in bulk.

An example of this type of study is the investigation by Koenig and Shih[64] of the polymerization and structure of a low molecular weight unsaturated polyester resin ($\overline{M}_n$ = 1300) of the type which can then be copolymerized with styrene (by addition polymerization) to give a thermoset resin such as that used to make fiberglass-reinforced plastics. It is actually possible to study such a thermoset resin directly in the form of a composite containing glass fibers or spheres, because glass does not seriously interfere with the Raman measurements. However, the study here was carried out on unfilled polymer. The unsaturated polyester was made by polymerization of maleic anhydride with a mixture of propylene and diethylene glycols, and it was found that several of the Raman lines showed major changes during the polymerization process. The necessary assignment of the spectral lines was carried out for both the monomers and the final polymers.

Fig. 8 (taken from their publication) shows the behavior of several spectral lines (with intensities normalized by comparison to the intensities of lines which showed no change as a result of the polymerization) during the first two hours of the polymerization process. One of the significant things that happens during the polymerization is the partial isomerization of the carbon-carbon double bond from the _cis_ form (maleic anhydride or maleic ester) to the _trans_ form (fumaric ester). The spectral lines at 1168 and 1645 cm^{-1} are assigned to the =CH- bending and C=C stretching vibrations, respectively, in the _cis_ form of the C-C=C-C group. The line at 1662 cm^{-1} is assigned to the C=C stretching in the _trans_ form of C-C=C-C, the line at 1280 cm^{-1} is assigned to the =C-O- stretching which couples with the same _trans_ structure, and the line at 892 cm^{-1} is assigned to the stretching of the C-C-O group related to this _trans_ C-C=C-C. It is evident from Fig. 8 that the three lines related to the _trans_, or fumaric, structure (at 892, 1280 and 1662 cm^{-1}) show the same sort of general shape: a rapid initial increase followed by a leveling-off. The two lines related to the _cis_, or maleic, structure (1168 and 1645 cm^{-1}) both show, in contrast to the above, a gradual decrease during the same time period. The line at 1213 cm^{-1} is assigned to the stretching vibration of the C-O-C bond of the glycol-ester linkage which is produced during the esterification reaction. This line can therefore be used as a quantitative measure of the extent of reaction and shows a linear increase with time during the same two-hour period (a result to be expected from the theory of the kinetics of such reactions). The Raman spectrum measured at the end of the polymerization reaction indicated that the final polymer contained 55% fumarate and 45% maleate unsaturation, compared to 100% maleate in the original monomer. An additional piece of conformational information was that the glycol fragments were in the _gauche_ isomeric form in the chain.

Raman measurements were also made during the copolymerization of this polyester with styrene. In the reaction mixture (containing 30-34% styrene), the styrene spectrum predominated. The reaction of styrene could be followed by the disappearance of the bands at 1632 and 1662 cm^{-1} corresponding to the C=C stretching of the styrene vinyl group. It was found that the styrene has a stronger tendency to copolymerize with fumarate than with maleate unsaturation: there was a reduction of concentration of the unreacted _trans_ C=C bond at 1662 cm^{-1} by about 41%, while the concentration of the _cis_ C=C bond at 1645 cm^{-1} decreased by only 5.5%. Lines associated with the vinyl group of the styrene monomer weaken or disappear completely during the copolymerization, but the characteristic lines of polystyrene (of high molecular weight) are not seen in the thermoset product. It is concluded that the styrene forms

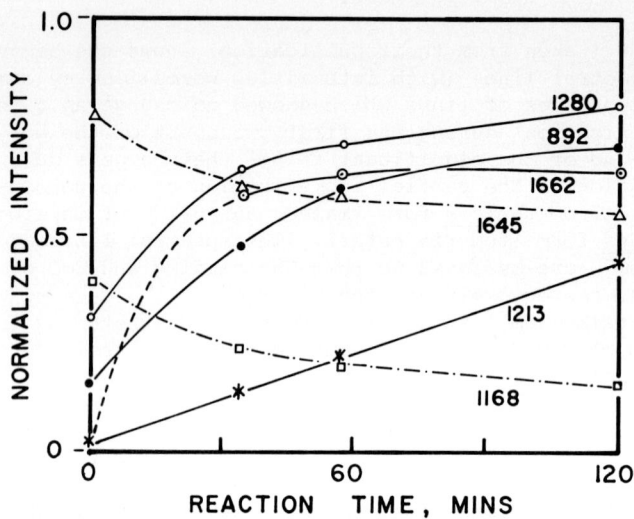

Fig. 8. Change of intensity of several Raman lines during polymerization of a polyester made from maleic anhydride and propylene and diethylene glycols (from Ref. 64).

very short crosslinks (containing about two styrene units) which tie the polyester chains together. There are 6-8 chemical repeat units in each polyester chain, of which 4-5 are fumarate double bonds, the remainder being maleate. After creation of the final thermoset structure, about two fumarate double bonds of each polyester chain have, on the average, reacted with styrene.

Another study of a similar type has been carried out by Chow and Chow[65] on the polymerization reaction involved in the formation of water-soluble phenol-formaldehyde resins called "resols". Analogous polymers can also be made using p-cresol instead of phenol, in which case the polymers consist of linear, rather than branched, chains. The polymerization reaction is carried out in water solution, and this prevents infrared measurements being made directly on the reaction mixture. Raman spectra can, however, be measured. In order to interpret the spectra obtained from the reaction mixture, spectra were also measured on simple model compounds. A number of bands were found which could be used to interpret the changes occurring during the polymerization process: phenol 998 and 814 cm^{-1}; 2-substituted phenol 1042 and 784 cm^{-1};

4-substituted phenol 850 and 645 cm^{-1}; 2,4-disubstituted phenol 570 cm^{-1}; and a superimposed line from 2-substituted, 2,4-disubstituted and 2,4,6-trisubstituted phenol at 784 cm^{-1}. This last composite line showed a linear increase vs. molecular weight of the polymer, and therefore provided a simple and convenient measure of the extent of reaction.

Changes in intensity of four of these peaks as a function of molecular weight during the early stages of the condensation reaction are shown in Fig. 9. Peak intensities here are expressed relative to the intensity of a line in the spectrum at 998 cm^{-1}, which did not change significantly during the polymerization. Phenol (814 cm^{-1}) disappears rapidly at the beginning, followed by the transient appearance of appreciable amounts of 2-substituted (1042 cm^{-1}) and 4-substituted (850 cm^{-1}) phenol. It is evident that the rate of formation is more rapid for the 2-monosubstituted phenol. The appearance of 2,4-disubstituted phenol (570 cm^{-1}) is first observed at a molecular weight of about 300, followed by a slow further increase. There is a second sharp decrease in phenol concentration when molecular weight reaches about 550; the explanation for this is not clear. These Raman results are consistent with the results of investigations on the same system using paper chromatography, but the Raman data give more detailed information on the early stages of the reaction. In the polymerization with p-cresol, the Raman spectrum changes very rapidly at the beginning of the reaction, indicating a very rapid condensation taking place during the first 10 minutes. After that, the spectrum shows very little change, and is very similar in appearance at 10 min. and 180 min. of reaction.

The use of laser Raman measurements in the study of polymer degradation is illustrated by a recent study of the structure of thermally degraded PVC by Gerrard and Maddams[66]. They find, in agreement with previous work of Liebman et al.[67,68], that strong bands at 1124 and 1511 cm^{-1} appear in the spectrum of thermally degraded PVC, which was heated for different times at 150-190°C. These bands are associated with the formation of conjugated sequences of double bonds in the degraded polymer. Although these conjugated sequences are present in very low concentration (less than 0.01%), they give strong spectral lines due to a resonance Raman effect; the presence of this resonance effect was demonstrated by measuring spectra with different frequencies of incident light, using argon-ion and krypton-ion lasers. The authors show that different frequencies of incident light excite resonance in polyenes of different sequence lengths, and they point out how useful tunable lasers could be in this type of study. They associate the two strong lines at 1511 and 1124 cm^{-1} with conjugated sequences of 13 double bonds, and 16-17 double bonds, respectively, in the degraded structure.

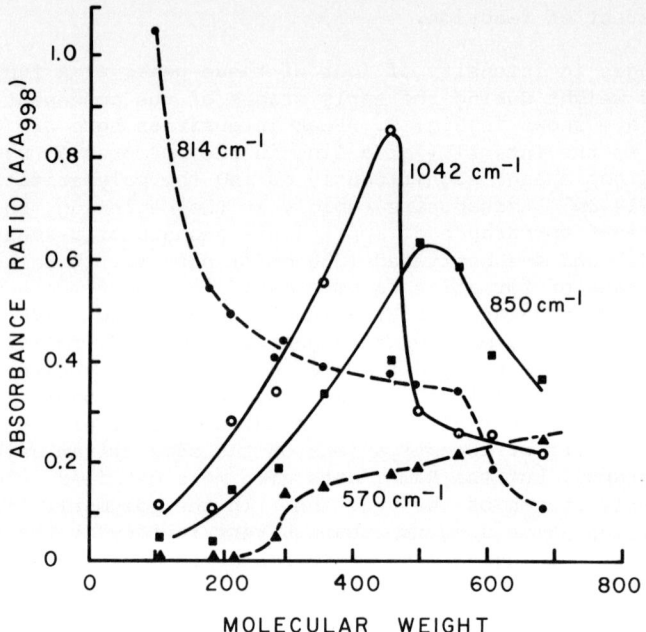

Fig. 9. Change of intensity of four Raman lines during polymerization of a "resol" resin made from phenol and formaldehyde (from Ref. 65).

E. Molecular Complexes with Polymers

An area of polymer behavior of interest both in biochemistry and to polymer studies in general, is the area of complex formation. One interesting case of complex formation already studied by the laser Raman method is that of the complexes formed between triiodide ion and both starch (amylose) and polyvinyl alcohol (PVA), which show a resonance Raman effect. These complexes both have a characteristic blue color, and seem similar in many respects. (It may be of interest to note that Polaroid polarizers are made by forming this complex in a highly oriented PVA sheet.)

The Raman spectrum of the two complexes in the low wavenumber region shows very clearly the three characteristic vibrational frequencies of the triiodide ion, as seen in Fig. 10, which shows the spectrum of the I_3^- complex with PVA, obtained with a He-Ne laser, as reported by Inagaki et al.[69]. The clearly evident peaks at 45, 106 and 155 cm^{-1} correspond, respectively, to the bending mode, the symmetrical stretching mode and the unsymmetrical stretching mode of the triiodide ion. The peaks at 265 and 310 cm^{-1} are apparently a combination mode (106 + 155 = 265) and a first overtone (2 x 155 = 310). The 155 cm^{-1} line is the predominant one in the spectrum, as a result of the resonance effect, as is also the case in the starch-iodine complex[70] (where this peak appears at 163 cm^{-1}). This contrasts with the spectrum of the simple I_3^- ion in aqueous solution, where the <u>symmetrical</u> stretching mode (observed at 114 cm^{-1}) is the predominant line in the spectrum, the unsymmetrical mode only appearing as a weak shoulder on this peak, at 148 cm^{-1}. Other interesting effects are also seen -- for example, a maximum in intensity of the resonance-enhanced band at 155-163 cm^{-1} as a function of concentration (the maximum occurs at about 10^{-4} molar iodine concentration in both complexes). This is a concentration much lower than that which could be used for the measurement of an ordinary non-resonance Raman spectrum, and shows the effectiveness of resonance enhancement of Raman spectral lines in increasing the sensitivity of the spectrum for the detection of very low concentrations. This phenomenon of a maximum in intensity vs. concentration has been discussed in a recent publication[71]. The blue complex disappears when heated to about 60°C, and the Raman lines above also disappear at the same time, indicating that they are specifically associated with the complex.

The spectra obtained are the same in nature for the starch-iodine complex when He-Ne and Ar$^+$ ion lasers are used. This was true also for one PVA specimen in the PVA-iodine complex, but another PVA sample showed a difference in spectrum for the two lasers, as a result of an absorption band at 4900Å very close to the wavelength of the Ar$^+$ laser (4880Å). This absorption produced a powerful enhancement of the symmetrical stretching band at 108 cm^{-1} which then dominated the whole spectrum and gave observable overtone bands up to the sixth overtone as the only feature of the spectrum in this wavelength region. A summary of the resonance effects observed in the two complexes and in the free I_3^- ion in aqueous solution are shown in Table 1 (reproduced from Ref. 69). The yellow solutions in the case of the higher molecular weight PVA (sample B, with D.P. = 2000) are solutions where the concentration of boric acid (also added to the solutions) was too low for the complex to form. A corresponding red color is obtained at low boric acid concentration for the lower molecular weight PVA (sample A, with D.P. = 500). The authors conclude that their results are consistent with the generally accepted picture of these complexes in which triiodide ions are held in the axial cavity of helical regions of the starch and PVA molecules.

Very similar Raman data on both of these complexes have been reported by Heyde et al.[72], including some polarization

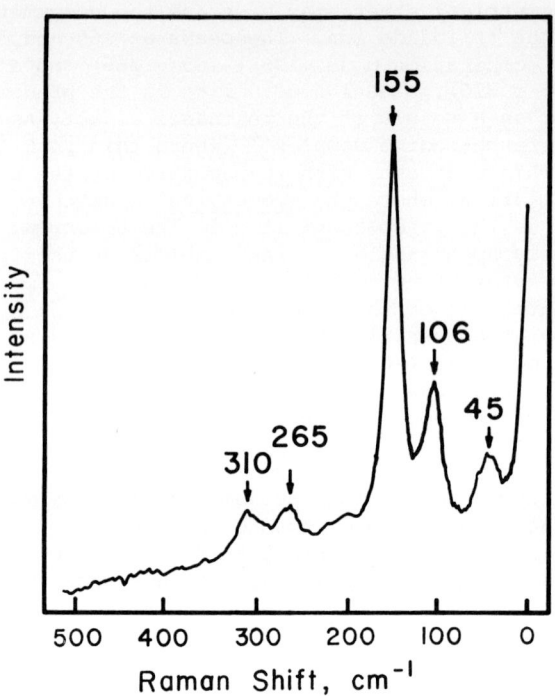

Fig. 10. Resonance Raman spectrum in the low-frequency region of tri-iodide ion complexed with polyvinyl alcohol, obtained using a He-Ne laser (from Ref. 69).

measurements. These authors also observe a low-frequency mode in these complexes not reported by the authors above (e.g., at 27 cm^{-1} in the amylose complex) which they attribute to the longitudinal vibration of the whole sequence of iodine atoms in the polymer helix. From the value of this frequency, using normal mode calculations, they obtain a value of 28 atoms for the iodine sequence length in both the amylose and PVA complexes.

An example of complex formation in the biological field is provided by a study of complex formation between haptens and a series of antibodies reported by Carey, Froese and Schneider[73], using the resonance Raman method. It was found that many spectral changes accompanied the complex formation. Two conclusions drawn from these changes were that complex formation caused twisting of aromatic rings around the azo group of the hapten, producing

Table 1

Resonance Enhancement Effects

I_3^- Ion in PVA and Starch Complexes and in Free State

Exciting Line	PVA - A Red	PVA - A Blue	PVA - B Yellow	PVA - B Blue	Starch Complex	Free I_3^-
632.8 nm	s	s,A	none	s,A	s,A	s
488.0 nm	S	S	none	s,A	s,A	s

The letters indicate the <u>symmetric</u> and <u>antisymmetric</u> lines of the I_3^- ion. A capital letter indicates a great enhancement of intensity, while a small letter indicates a small resonance enhancement.

different conformational states, and that there was a particular type of charge distribution at the antibody complexing site which produced a decrease in double bond character of the N-O bonds in substituent NO_2 groups in the 2 and 4 positions of the phenyl ring.

This also illustrates another feature of the use of the laser Raman method which is often true, but seems particularly evident in biological problems. This is that in answering complex questions, a combination of different types of experimental data is usually required, and often a good deal of imagination and intuition added to that, to arrive at a reasonable hypothesis, let alone a final answer. The laser Raman method is in many cases able to provide certain types of information that cannot be obtained by other methods, which may be of critical importance, but this method (like IR, NMR, and other methods) has its special areas of usefulness, and the specific information which it supplies must often be built into a larger pattern of experimental data and theory.

A resonance Raman study of methyl orange bound to bovine serum albumin has also been reported[74].

F. Surface Layers and Adsorbed Molecules

Only a limited amount of work has been done on the study of thin layers on surfaces and of adsorbed molecules on solid substrates; and most of the work to date has been motivated by an interest in the phenomena associated with heterogeneous catalysis at solid surfaces, and has involved the study of small molecules. A recent paper[75], however, has reported measurements on a thin film of a polymer (polystyrene) on a solid substrate, and it can be expected that further work on this type of system will be reported in the future. Familiarizing oneself with the laser Raman work carried out to date on surfaces and adsorbed molecules has been made much easier by the appearance recently of two review articles in this area: one by Egerton and Hardin[76], and the other by Cooney, Curthoys and Tam[77]. Although there is some duplication and overlap between these two articles, the former gives a more detailed discussion of experimental data, while the latter gives more attention to the theory and experimental method (and in fact provides an interesting and readable introduction to the laser Raman method in general).

A serious problem in this type of measurement, of course, is the question of sensitivity, since the amount of sample being observed is extremely small. In addition, there is often a background fluorescence from the solid substrate which tends to obscure the signal from the adsorbed layer. Usually it is difficult to obtain a measurable Raman spectrum from a monolayer, but there are occasional exceptions to this. Hendra and Loader[78] have reported measurements of adsorbed Br_2 molecules on silica gel, in which a Raman spectrum could be obtained from the Br_2 (which is a strong Raman scatterer) at a coverage of only 6% of a monolayer. But usually at least a few molecular layers are required. This paper discusses the spectra of a number of different adsorbed molecules on silica and silica gel, and a paper by Buechler and Turkevich[22] discusses a number of different adsorbed molecules, mainly on a substrate of Vycor glass.

Hendra et al.[79] discuss the adsorption of pyridine and the competitive adsorption of pyridine and piperidine on oxide surfaces. These are useful as "probe" molecules to identify various types of sites on the surface related to catalytic activity. It seems fairly easy to identify pyridine molecules adsorbed at Lewis acid sites on the surface and to distinguish these from hydrogen-bonded pyridine attached to hydroxyl groups on the surface and simple physically adsorbed pyridine, but it is very difficult to identify pyridine molecules specifically adsorbed at Brönsted sites on the surface, even when piperidine (which would be expected to displace pyridine from Brönsted sites) was used in combination with the pyridine. Angell[80] has measured Raman spectra of zeolites (silica-alumina minerals containing water and cations) and of small molecules (acetonitrile, propylene, acrolein, carbon dioxide) adsorbed on them. The small molecules showed spectra similar to (but not identical with) the spectra observed for the corresponding liquids, indicating that these molecules were only physically adsorbed. Egerton et al.[81] show that when pyridine is adsorbed on a Y-zeolite containing a series of cations, the frequency of the ν_1 Raman band from the ring-breathing mode of pyridine gives an excellent linear correlation with the charge/radius ratio of the cation, strongly suggesting that the main adsorption is at cation sites.

Greenler and Slager[82] have discussed a method for obtaining the spectrum of a thin coating layer on a reflecting metal surface (by a multiple reflection technique), and in applying the method to benzoic acid on a silver film were able to observe Raman lines from the benzoic acid at a coating thickness of 50Å. Fleischmann, Hendra and McQuillan[83] measured Raman spectra from pyridine adsorbed on the surface of a silver electrode from water solution at different potentials and were able to draw conclusions regarding the interaction of pyridine and water molecules in the double layer at the electrode surface. A recent publication by Howe, Watters and Greenler[75] discusses the sensitivity of Raman measurements made on CCl_4 adsorbed on a microporous silica gel, and on polystyrene deposited from benzene solution on the surfaces of a pair of silver mirrors used in the multiple-reflection method. They conclude that the lowest limits of detection are at about 4% of a monolayer for the CCl_4, and at a thickness of 50Å for the polystyrene. They state that the Raman reflection technique, as used here, is about an order of magnitude less sensitive than is needed to study monolayers, but that this might still be possible by using molecules which produce a resonance Raman effect.

Another recent publication[84] describes the successful measurement of Raman spectra on a monolayer of an interaction product of a cationic surfactant and an anionic azo dye (methyl orange) deposited at the interface between water and carbon tetrachloride. The methyl orange shows a resonance Raman effect, and the measurements were made using a new method of total reflection of the exciting light at the interface, in which the laser beam enters the glass or acrylic sample cell through the CCl_4 layer at a high incidence angle (84°, compared to the critical Brewster's angle of 67°). The beam is totally reflected at the interface, and is then reflected back through the cell along the original path, thus increasing the intensity of the spectrum. Polarization measurements could be made conveniently, and from these it was calculated that the long axes of the methyl orange molecules in the monolayer were oriented at an average angle of inclination of about 55° to the axis normal to the interface. The frequencies of the methyl orange lines in the monolayer were also shifted from those observed in aqueous solution toward the values observed for solid methyl orange, suggesting that the molecules in the monolayer are in an environment somewhat similar to a crystal field.

REFERENCES

1. R. Signer and J. Weiler, Helv. Chim. Acta, **15**, 649 (1932).
2. J. R. Nielsen, J. Polymer Sci., **C7**, 19 (1964).
3. R. F. Schaufele, J. Opt. Soc. Am., **57**, 105 (1967).
4. P. J. Hendra, Fortschr. Hochpolym.-Forsch. (Adv. Polymer Sci.), **6**, 151 (1969).
5. P. J. Hendra and P. M. Stratton, Chem. Rev., **69**, 325 (1969).
6. R. F. Schaufele, J. Polymer Sci., **D4**, 67 (1970).
7. M. J. Gall, P. J. Hendra, D. S. Watson and C. J. Peacock, Appl. Spectr., **25**, 423 (1971).
8. J. L. Koenig, Appl. Spectr. Rev., **4**, 233 (1971).
9. J. L. Koenig, J. Polymer Sci., **D6**, 59 (1972).
10. W. L. Peticolas, Fortshr. Hochpolym.-Forsch. (Adv. Polymer Sci.), **9**, 285 (1972).
11. M. E. A. Cudby, H. A. Willis, P. J. Hendra and C. J. Peacock, Chem. Ind. (London) 1971, p. 531.
12. J. L. Koenig, Chem. Tech., **2**, 226 (1972).
13. J. L. Koenig, Chem. Tech., **2**, 411 (1972).
14. T. G. Spiro, Chap. 2 in *Chemical and Biochemical Applications of Lasers*, Ed. C. B. Moore, Academic Press, New York (1974).
15. S. B. Adams, T. R. Hart and R. D. Andrews, A.C.S. Preprints, Div. Org. Coatings and Plastics Chem., **32**, 117 (1972).
16. T. G. Spiro, Accts. Chem. Res., **7**, 339 (1974).
17. G. Placzek, UCRL Translation No. 526L from *Handbuch der Radiologie*, Vol. 2, Ed. E. Marx, Akademische Verlagsgesellschaft, Leipzig, p. 209.
18. E. B. Wilson, J. C. Decius and P. C. Cross, *Molecular Vibrations*, McGraw Hill, New York (1955).
19. W. Kiefer and H. J. Bernstein, Appl. Spectr., **25**, 500, 609 (1971).
20. N. Zimmerer and W. Kiefer, Appl. Spectr., **28**, 279 (1974).
21. K. Holland-Moritz and I. Modric, Progr. Colloid and Polymer Sci., **57**, 212 (1975).
22. E. Buechler and J. Turkevich, J. Phys. Chem., **76**, 2325 (1972).
23. Y. Yacoby and A. Linz, Phys. Rev., **B9**, 2723 (1974).
24. L. Genzel, F. Keilmann, T. P. Martin, G. Winterling, Y. Yacoby, H. Fröhlich, and M. W. Makinen, Biopolymers, **15**, 219 (1976).
25. E. W. Small and W. L. Peticolas, Biopolymers, **10**, 1377 (1971).
26. P. C. Painter and J. L. Koenig, Biopolymers, **15**, 241 (1976).
27. J. L. Koenig and B. Frushour, Biopolymers, **11**, 1871 (1972).
28. T-J. Hu, J. L. Lippert and W. L. Peticolas, Biopolymers, **12**, 2161 (1973).
29. P. C. Painter and J. L. Koenig, Biopolymers, **15**, 229 (1976).
30. B. G. Frushour and J. L. Koenig, Biopolymers, **13**, 455 (1974).
31. J. L. Koenig and B. G. Frushour, Biopolymers, **11**, 2505 (1972).
32. B. G. Frushour and J. L. Koenig, Biopolymers, **13**, 1809 (1974).
33. V. J. C. Lin and J. L. Koenig, Biopolymers, **15**, 203 (1976).
34. B. Prescott, R. Gamache, J. Livramento and G. J. Thomas, Jr., Biopolymers, **13**, 1821 (1974).
35. A. Lewis and H. A. Scheraga, Macromolecules, **5**, 450 (1972).

36. T. G. Spiro and T. C. Strekas, J. Am. Chem. Soc., 96, 338 (1974).
37. V. Miskowski, S-P. W. Tang, T. G. Spiro, E. Shapiro and T. H. Moss, Biochemistry, 14, 1244 (1975).
38. L. Rimai, R. G. Kilponen and D. Gill, J. Am. Chem. Soc., 92, 3824 (1970).
39. W. T. Wozniak and T. G. Spiro, J. Am. Chem. Soc., 95, 3402 (1973).
40. S-P. W. Tang, T. G. Spiro, C. Antanaitis, T. H. Moss, R. H. Holm, T. Herskovitz and L. E. Mortensen, Biochem. Biophys. Res. Comm., 62, 1 (1975).
41. J. Purvis and D. I. Bower, Polymer, 15, 645 (1974).
42. R. Speak and I. W. Shepherd, J. Polymer Sci.-Symp., 44, 209 (1974).
43. J. Maxfield and I. W. Shepherd, Chem. Phys. Lett., 19, 541 (1973).
44. A. J. Hartley and I. W. Shepherd, J. Polymer Sci. - Phys., 14, 643 (1976).
45. S. W. Cornell and J. L. Koenig, Macromolecules, 2, 540 (1969).
46. Y. S. Huang and J. L. Koenig, Appl. Spectr., 25, 620 (1971).
47. M. J. Gall, P. J. Hendra, C. J. Peacock, M. E. A. Cudby and H. A. Willis, Polymer, 13, 104 (1972).
48. M. J. Gall, P. J. Hendra, C. J. Peacock, M. E. A. Cudby and H. A. Willis, Spectrochim. Acta., 28A, 1485 (1972).
49. G. V. Fraser, P. J. Hendra, M. E. A. Cudby and H. A. Willis, J. Chem. Soc. - Chem. Comm. (1973), p. 16.
50. S. I. Mizushima and T. Shimanouchi, J. Am. Chem. Soc., 71, 1320 (1949).
51. W. L. Peticolas, G. W. Hibler, J. L. Lippert, A. Peterlin and H. Olf, Appl. Phys. Lett., 18, 87 (1971).
52. A. Peterlin, H. G. Olf, W. L. Peticolas, G. W. Hibler and J. L. Lippert, J. Polymer Sci., B9, 583 (1971).
53. H. G. Olf, A. Peterlin and W. L. Peticolas, J. Polymer Sci. - Phys., 12, 359 (1974).
54. J. L. Koenig and D. L. Tabb, J. Macromol. Sci. - Phys., B9, 141 (1974).
55. M. J. Folkes, A. Keller, J. Stejny, P. L. Goggin, G. V. Fraser and P. J. Hendra, Colloid and Polymer Sci., 253, 354 (1975).
56. R. F. Schaufele and T. Shimanouchi, J. Chem. Phys., 47, 3605 (1968).
56a. A. Hartley, Y. K. Leung, C. Booth and I. W. Shepherd, Polymer, 17, 354 (1976).
57. H. Okabayashi, M. Okuyama, T. Kitagawa and T. Miyazawa, Bull. Chem. Soc. Japan, 47, 1075 (1974).
58. M. Kobayashi, T. Uesaka and H. Tadokoro, Chem. Phys. Lett., 37, 577 (1976).
59. I. Modric, K. Holland-Moritz and D. O. Hummel, Colloid and Polymer Sci., 254, 342 (1976).
60. A. J. Melveger, J. Polymer Sci., Part A-2, 10, 317 (1972).
61. G. E. McGraw, A.C.S. Polymer Preprints, 11, 1122 (1970).

62. M. Kobayashi, K. Tashiro and H. Tadokoro, Macromolecules, $\underline{8}$, 158 (1975).
63. H. Evans and A. E. Woodward, Macromolecules, $\underline{9}$, 88 (1976).
64. J. L. Koenig and P. T. K. Shih, J. Polymer Sci., Part A-2, $\underline{10}$, 721 (1972).
65. S. Chow and Y. L. Chow, J. Appl. Polymer Sci., $\underline{18}$, 735 (1974).
66. D. L. Gerrard and W. F. Maddams, Macromolecules, $\underline{8}$, 54 (1975).
67. S. A. Liebman, C. R. Foltz, J. F. Reuwer and R. J. Obremski, Macromolecules, $\underline{4}$, 134 (1971).
68. S. A. Liebman, D. H. Ahlstrom, E. J. Quinn, A. G. Geigley and J. T. Meluskey, J. Polymer Sci., Part A-1, $\underline{9}$, 1921 (1971).
69. F. Inagaki, I. Harada, T. Shimanouchi and M. Tasumi, Bull. Chem. Soc. Japan, $\underline{45}$, 3384 (1972).
70. M. Tasumi, Chem. Lett. (1972), p. 75.
71. T. C. Strekas, D. H. Adams, A. Packer and T. G. Spiro, Appl. Spectr., $\underline{28}$, 324 (1974).
72. M. E. Heyde, L. Rimai, R. G. Kilponen and D. Gill, J. Am. Chem. Soc., $\underline{94}$, 5222 (1972).
73. P. R. Carey, A. Froese and H. Schneider, Biochemistry, $\underline{12}$, 2198 (1973).
74. P. R. Carey, H. Schneider and H. J. Bernstein, Biochem. Biophys. Res. Comm., $\underline{47}$, 588 (1972).
75. M. L. Howe, K. L. Watters and R. G. Greenler, J. Phys. Chem., $\underline{80}$, 382 (1976).
76. T. A. Egerton and A. H. Hardin, Catal. Rev. - Sci. and Eng., $\underline{11}$, 71 (1975).
77. R. P. Cooney, G. Curthoys and N. T. Tam, Adv. in Catalysis, $\underline{24}$, 293 (1975).
78. P. J. Hendra and E. J. Loader, Trans. Faraday Soc., $\underline{67}$, 828 (1971).
79. P. J. Hendra, I. D. M. Turner, E. J. Loader and M. Stacey, J. Phys. Chem., $\underline{78}$, 300 (1974).
80. C. L. Angell, J. Phys. Chem., $\underline{77}$, 222 (1973).
81. T. A. Egerton, A. H. Hardin and N. Sheppard, Canad. J. Chem., $\underline{54}$, 586 (1976).
82. R. G. Greenler and T. L. Slager, Spectrochim. Acta., $\underline{29A}$, 193 (1973).
83. M. Fleischmann, P. J. Hendra and A. J. McQuillan, Chem. Phys. Lett., $\underline{26}$, 163 (1974).
84. T. Takenaka and T. Nakanaga, J. Phys. Chem., $\underline{80}$, 475 (1976).

Discussion

On the Paper by N.J. Harrick

E. Kay (*IBM Corp.*): Please explain why total internal reflection is not an ideal approach for submonolayer vibrational spectra on metal substrates in the light of the following speakers remarked to the contrary.

N.J. Harrick (*Harrick Corp.*): For studying metal surfaces one must use caution in applying the internal reflection technique for two reasons. Firstly, it is difficult to obtain good contact between two solids (IRE and metal), and secondly, when "good" contact is obtained there is no longer total reflection; i.e., there is no longer a critical angle, no evanescent wave and the reflection is just that from a dielectric-metal interface[1] (see Fig. 4 page 201). The reflection curves for a dielectric-metal interface qualitatively resemble those from a free metal surface. Here, as for the free metal surface, maximum sensitivity is obtained by working with parallel polarization near the polarizing angle of incidence.

Dr. Jakobsen has overcome the two problems mentioned above by evaporating thin (∼20Å) metal films directly on the IRE - he thus obtained good contact with films of low conductivity. He has metal atoms present but not the high conductivity and, therefore, there is a (pseudo) critical angle with evanescent wave and the advantages of the internal reflection technique prevails. Such thin films of gold, platinum, tin oxide, indium oxide, etc., are frequently employed in electrochemical studies via Internal Reflection Spectroscopy.

1. *Internal Reflection Spectroscopy*, by N.J. Harrick, Interscience Wiley, 1967.

L.H. Lee (*Xerox Corp.*): I would like to thank Dr. Harrick for this excellent survey about the subject matter. Once again, infrared spectroscopy has demonstrated its lasting important role in the midst of new surface analytical tools discussed during this symposium.

On the Paper by D.L. Allara

D.T. Clark (*Durham University*):

1. Could you tell me how the films were produced. If by solvent casting (spinning), what solvents were used, were they degassed and was evaporation in air or an inert atmosphere?

2. Is it possible that migration of copper ions into the polymer occurred during evaporation?

3. Are you aware of the difficulties of producing unoxidized high density polyethylene samples from spinning or casting procedures? Our previous ESCA studies (cf. D.T. Clark, "Advances in Polymer Friction was Wear"5A", Ed L. H. Lee, Plenum Press (1974) have shown that hot pressing in an inert atmosphere under controlled conditions is a way of producing samples of a high degree of integrity. Such samples were pressed against aluminum foil, the external surface of which is coated with a fatty acid salt. In such samples we saw no evidence for oxidation or migration, the detection limit being a fraction of a monolayer. I note, however, that your time scales were extremely long compared with ours ($\sim$30 secs). What is the role of diffusion of oxygen into the polymer, (presumably diffusion coefficients, etc. are well known)?

D.L. Allara (*Bell Telephone Laboratories*):

1. The films were produced by spin coating from warm xylene or chlorobenzene solution. The solutions were not degassed and the final drying was done in a vacuum oven at $\sim$60° with air as the background atmosphere.

2. It is, of course, possible that trace amounts of copper ions could have migrated during evaporation but judging from our results at 60-90°C, where several hours in pure oxygen are required for measurable -CO_2- formation, it appears that the amounts of copper involved in the prereaction migration must be below our detection level (for the associated -CO_2^- ion). In the present experiments the minimum amount of carboxylate we could detect would be several monolayers. Future plans involve instrumentation capable of detecting significantly less than this amount.

3. There is no doubt that our "zero-time" samples are preoxidized. Examination of freshly prepared thick, (∿0.005") free-standing films by internal reflection spectroscopy shows traces of C=O type bonds. However, for the very thin films (<2000Å) on metals in our study, the carbonyl bonds are of too low an intensity to observe. In fact, it is my hypothesis that the rapid onset of $-CO_2-$ initial RCO_2H groups adsorbed at the Cu (oxide) surface. The optical absorption of a $-CO_2-$ group is stronger than a -COOOH group and this would enhance the early observation of the reaction product. In addition, the formation of a $-CO_2-$ group is presumably simultaneously accompanied by formation of Cu^{+2} ions which initiate oxidative catalyses of the polyolefin to more -COOH units. The latter react to give more $-CO_2-$ ions and an autocatalytic degration is thus established at the interface which leads to amounts of carboxylate greater than the equivalent of several monolayers, (perhaps up to 10-20). Accordingly, if one were to carefully exclude all traces of carboxylic acid in the starting polymer, one should observe onset of metal catalysis at the interface at drastically slower rates than observed in our work. We have carried out careful studies (Allara and Roberts, submitted for publication) on pure hexadecane ($C_{16}H_{34}$) over copper (oxide) surfaces and induction times at 100°C are ∿50 hours longer than for our poly(1-butene) films (with comparable surface areas of copper) at 90°. Addition of solutions (∿10^{-5}M) decreases the induction times. This work points out the importance of the acid chemisorption $-CH_2-$ groups over copper surfaces using the liquid-phase system rather than the polymer film system. Of course, in the "real" world, polyolefin films will always contain small amounts of oxidation products from processing and the present paper is thus relevant to such samples.

On the Paper by R.D. Andrews and T.R. Hart

L.H. Lee (*Xerox Corp.*): Dr. Andrews' excellent review about laser Raman spectroscopy has enabled us to appreciate its versatile applications to polymer chemistry. I would like to thank Dr. Andrews for his interesting presentation and thorough coverage of the subject matter.

S.Y. Tong (*University of Wisconsin - Milwaukee*): From the frequency of Raman spectroscopy, how good is the present theory for relating the frequencies to molecular structures? Whay type of model potentials are used? Empirical, pairwise potentials?

R.D. Andrews(*Stevens Institute of Technology*): The problem of assigning Raman frequencies is in about the same status as the assignment of IR frequencies, since they are really the same problem. Assignments are usually made by analogy with other compounds containing the same molecular groups, with supplementary help from normal mode analysis, using the force constants of the bonds. Although complete frequency assignments based on normal mode analysis have been made for small molecules, this usually cannot be done for polymers, where the assignments are usually deduced for one atomic group at a time. The analyses are usually not done from potential curves directly; but if this method were used, empirical pairwise potentials would probably be used.

W. Newby (*Aluminum Company of Canada Research Center*): The slide showing a peak for the lamellae of single crystals of polyethylene suggest that Raman spectra could be used for the detection of crystallinity of polymers. It this feasible?

R.D. Andrews: Yes, the Raman method can be used for the detection of polymer crystallinity in many cases; i.e., where crystallinity-sensitive bands are present. It should be noted, however, that where a lamellar frequency is used, the crystallinity detected will be only of the lamellar type and not of the micellar type, for example, which would exist together with lamellar crystals in drawn polyethylene. Overall crystallinity would presumably be measured in cases where the narrowing of a single line was used as the measure of degree of crystallinity, as in the work of Melveger on PET cited.

I.M. Stewart (*W.C. McCrone Associates*): One of the blind spots in microanalysis has been the organics. You gave "mini-sample" as one of the advantages of the Raman method. How "mini" is "mini"? Where does the best hope for "ultra-mini" lie------Raman or Fourier transform I.R.?

R.D. Andrews: Raman scattering measurements can generally be made on bulk samples with volume of the order of a cubic millimeter, which is the typical size of a laser beam. As mentioned in the paper, Raman work is also being done in reflection from surfaces, and spectra have been obtained from less than a monolayer of material adsorbed on a substrate surface. Ultra-mini samples (representing around 10^{-10} moles) can in very favorable cases be analyzed if it is possible to use resonance Raman scattering.

Fourier transform IR has been used for measurements by reflection from large surface areas (because of the small amount of energy available from black-body sources at long wavelengths), and therefore might be applied to adsorbed layers. The method has been used to study the surface structure of opaque solids (see, e.g., Barker, Phys. Rev. 132, 1474 (1963)), and can be extended into a range of very long wavelengths not normally accessible to IR measurements. A review of the method as applied to polymers has been given by Koenig (Appl. Spectr. 29, 293 (1975)), and as application of the method to the separation of the crystalline and amorphous contributions to the IR spectrum of polyethylene terephthalate has been described by D'Espositio and Koenig (J. Polymer Sci.-Phys. 14, 1731 (1976)). In the latter measurements, conventional thin-film samples were used in transmission, with a normal size IR beam (over an inch in diameter), but the improved signal-to-noise ratio provided the increased resolution needed for accurate subtraction of components of the spectrum. Unpublished work by Elliott at Exxon Research and Engineering (Analytical Div.) has shown that with the Fourier transform instrumentation, a usable spectrum from a small spot on a polyethylene film (a fraction of a mm in diameter) could be obtained in a normal period of time (20 min.) by passing the IR beam through a pin-hole in a metal foil at the point desired. Both Raman and Fourier transform IR therefore have definite potential in the area of mini-samples, but the superiority of one method or the other will depend on the specific problem or system involved.

PART III

Microscopy for Polymers

Introductory Remarks:

L.H. Princen

Northern Regional Research Center
Agricultural Research Service
U.S. Department of Agriculture
Peoria, Illinois 61604

Both optical and electron microscopy have always played important roles in characterization of surfaces. When the resolving power of the eye fails us, we tend to reach immediately for a hand lens or the nearest microscope for a closer look, rather than rely on some other form of inspection. Only when we have reached the limits of available equipment for surface image magnification do we sit back and search for other techniques to provide us with the desired information. This quest for always greater magnification has led us from the simple hand lens (10X) through the optical microscope (1,000X) to the transmission electron microscope (up to 1,000,000X).

The practical commercially available scanning electron microscope (SEM), based on an entirely different principle, has now been with us for 11 years, and is even more suited for surface study than the traditional forms of optical and electron microscopy, although the power of magnification (<100,000X) is not as high as that attainable in the transmission mode. The SEM, in its short history, has probably contributed more to the morphological studies of a greater variety of subjects than any other form of investigation.

When there is no way to visualize directly the surface structure of the object under study, there are indeed many other techniques available. However, in such forms of investigation the results become more susceptible to interpretation. Two good examples are provided in which such techniques can be applied successfully; i.e., light scattering and conductometric titration of latex particles in dispersion.

I deem it a great honor to have been given the opportunity by the Symposium chairman, Dr. Lieng-Huang Lee, to lead this session, and to be able to bring such a complete and varied program by recognized authorities.

Microscopical Analysis of Chemically Modified Textile Fibers

Wilton R. Goynes and Jarrell H. Carra

Southern Regional Research Center[1]
New Orleans, Louisiana 70179

Properties of textile fibers are altered to achieve specific qualities by treatment with various chemical finishes. These finishes produce changes that may be observed through direct or indirect microscopical procedures. In developing and evaluating finished fabrics, it is advantageous to determine sites of interaction of the finish and the fiber, since it is generally intended that finish chemicals be located in a particular area of the fabric structure. Microscopical techniques have been developed to show locations of these finishes. Scanning electron microscopy provided a means for showing deposition of finish on fabric and fiber surfaces, and energy dispersive X-ray analysis was used to show the presence of specific elements on or within fibers.

INTRODUCTION

Chemical finishes designed to improve or impart such properties as wrinkle resistance, water-, soil-, and oil-repellency, and flame resistance are commonly applied to textile fabrics. Location and distribution of finish chemicals in relation to structural elements of the fabric are important to the effectiveness and durability of the finish. Numerous chemical formulations are currently being developed to reduce fabric flammability. The actual effectiveness of these finishes is evaluated by standard flame tests; however, to better understand flame-retardant mechanisms and to evaluate finish procedures to determine if the flame-retardant chemical has interacted in its most effective site, the

---

[1] One of the facilities of the Southern Region, Agricultural Research Service, U.S. Department of Agriculture.

actual locations of chemicals in the fabric should be known. Thus, techniques that provide information on interaction of chemical and fabric subunits are important. Microscopical analysis is the only method available for providing such information at levels ranging from fabric to subfiber structure.

SCANNING ELECTRON MICROSCOPY

Changes that occur in fabric and fiber surfaces during chemical finishing affect characteristic properties such as hand, drape, and stiffness. Some finishes are designed to be surface adherent, and others are intended to penetrate into the interior areas of the fiber. Scanning electron microscopy (SEM) provides a means of directly studying sample surfaces to observe changes due to such finishes. Only simple preparative methods were required for SEM observations. A continuous conducting surface is necessary in SEM specimens in order to prevent charge buildup. If the specimen surface becomes electrically charged, image artifacts are produced in the form of bright patches or streaks, distortion, and shifting. Depending on the extent of charging, the resultant image may simply be of low resolution, or, in extreme cases, can be completely devoid of useful information. As textile fibers are not naturally conducting, a conductive surface was provided by attaching a small swatch of fabric to a sample stub and vacuum-coating it with up to 500Å of gold-palladium alloy. The fabric swatches were freehand cut with fresh razor blade edges so that undisturbed end views of yarn sections could be studied.

An image is formed in the scanning electron microscope when a specimen is bombarded with an electron beam. Interaction of the beam with the subatomic structure of the specimen generates several types of radiation. Information retrieved depends on the type of detector used. Secondary electrons emitted from the bombarded specimen are generally collected to form the SEM image, which may be readily identified since spatial detail is presented in a form similar to that seen with the naked eye.

An example of polymer deposition on fabric surfaces was shown in examination of a printcloth fabric made flame resistant by padding with a solution, containing 40% solids, of tetrakis(hydroxymethyl)phosphonium chloride (Thpc) that had been neutralized with sodium hydroxide, and curing in gaseous ammonia. This finish is referred to as tetrakis(hydroxymethyl)phosphonium hydroxide-ammonia (THPOH-NH_3).[1] Chemical reactions involved in fire retardants based on Thpc have been discussed by Drake and Reeves.[2] After finishing, the fabric contained 3.6% phosphorus and passed the standard vertical flame test.[3] SEM showed heavy deposits on both fabric and fiber surfaces. At low magnification (Fig. 1), the fabric shows little evidence of polymer buildup; however, at higher magnification

(Fig. 2) a heavy, uneven deposit can be seen. The amount of polymer that penetrates into the yarn can be determined by observing the yarn cross section (Fig. 3). Here, the polymer has bound together the fibers on the yarn surface but toward the center of the yarn the fibers are loose, with some polymer deposited between them. Physical appearance of surface-deposited polymer may change with finish variables such as solution concentration, curing time and temperature, and drying conditions. Figure 4 illustrates a THPOH-NH$_3$ finish similar to that in Figure 2 except for a shorter curing time. The surface polymer has a grainy appearance and was deposited deep into the yarn (Fig. 5). Penetration into fibers cannot be detected by this mode of SEM.

Fig. 1. Surface of cotton fabric treated with THPOH-NH$_3$.

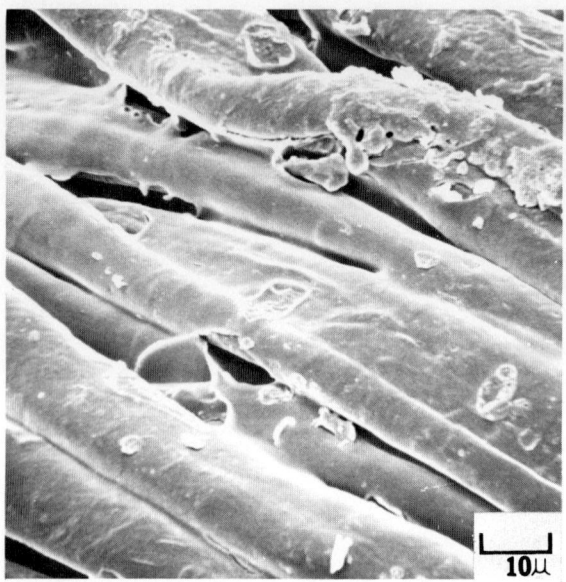

Fig. 2. Smooth polymer deposit on cotton fibers treated with THPOH-NH$_3$.

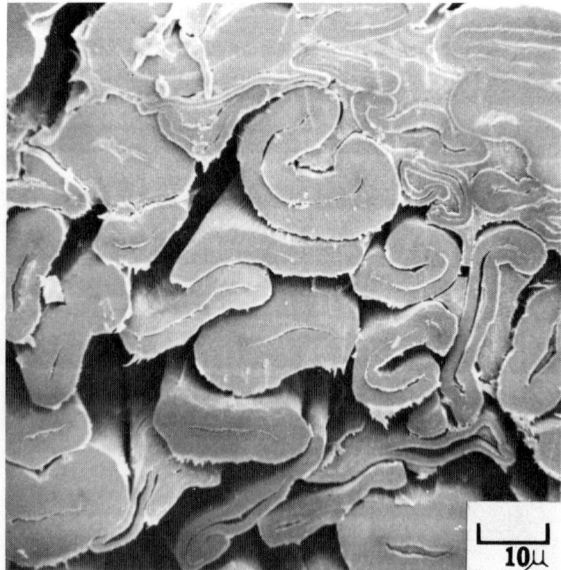

Fig. 3. Yarn cross section from cotton fabric treated with THPOH-NH$_3$.

Modified Textile Fibers   255

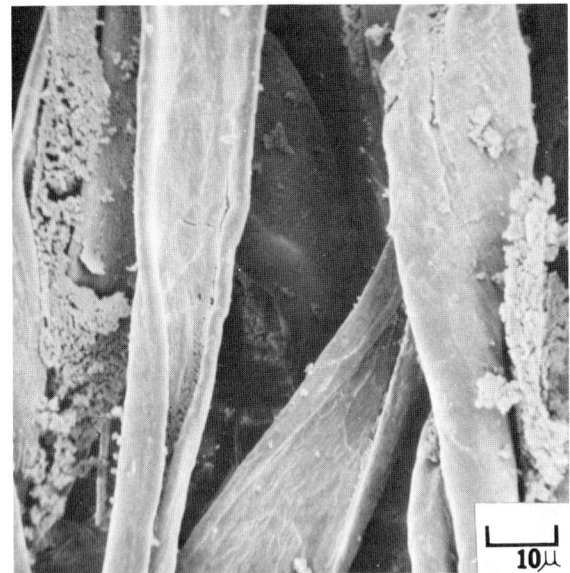

Fig. 4. Grainy polymer deposit on cotton fibers treated with THPOH-NH$_3$.

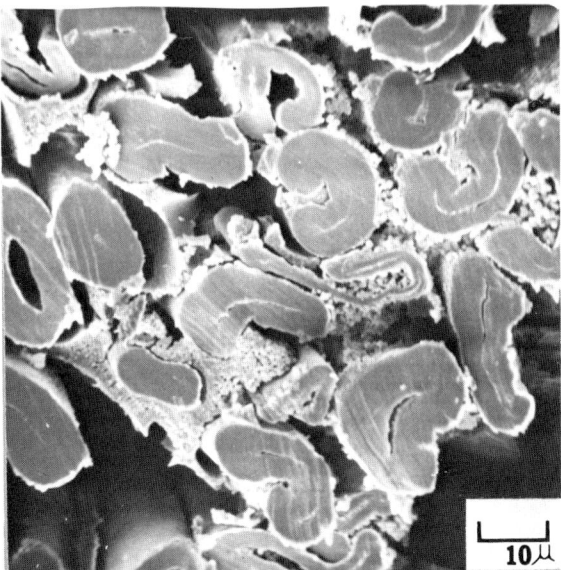

Fig. 5. Yarn cross section from cotton fabric treated with THPOH-NH$_3$.

Durability is of major concern in developing finishes. The finish must remain effective after repeated wear and laundering. With the THPOH-NH_3 finish illustrated in Figure 1, durability of surface polymer to laundering was studied through 1, 5, 20 and 50 cycles. After 1 laundry cycle, deposits between fibers had begun to break up (Fig. 6). After 5 cycles, much of the deposit had been removed (Fig. 7). After 20 and 50 laundry cycles, surface deposits were almost completely removed and abrasive fiber damage was evident, as illustrated in Figure 8. Even though only small amounts of polymer remained on fiber surfaces after extensive laundering, there was no significant decrease in flame resistance as measured by the standard vertical flame test.

The presence of chemical deposits on fiber surfaces affects fabric stiffness. In the fabric shown in the above laundering series, warp stiffness increased from 2.3×10^{-4} in.-lbs. in the original fabric to 28.7×10^{-4} in.-lbs. in the finished fabric, as measured by the Tinius-Olsen stiffness test.[4] Although the finished fabric was not excessively stiff, the increase in stiffness was evident in handling the samples. On laundering, warp stiffness was reduced to 6.8×10^{-4} in.-lbs. after the first laundry cycle, to 5.6×10^{-4} in.-lbs. after 5 cycles, and to 3.8×10^{-4} in.-lbs. after 20 cycles. This reduction in stiffness indicates that as the crusted surface polymer was cracked and loosened, as shown in Figures 6, 7, and 8, the fibers and yarns regained some of the freedom of movement lost in the original treatment.

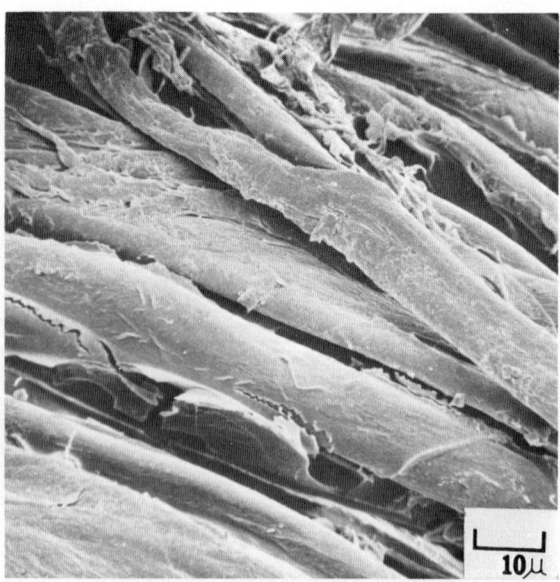

Fig. 6. Surface of cotton fabric treated with THPOH-NH_3, after 1 laundry cycle.

Modified Textile Fibers 257

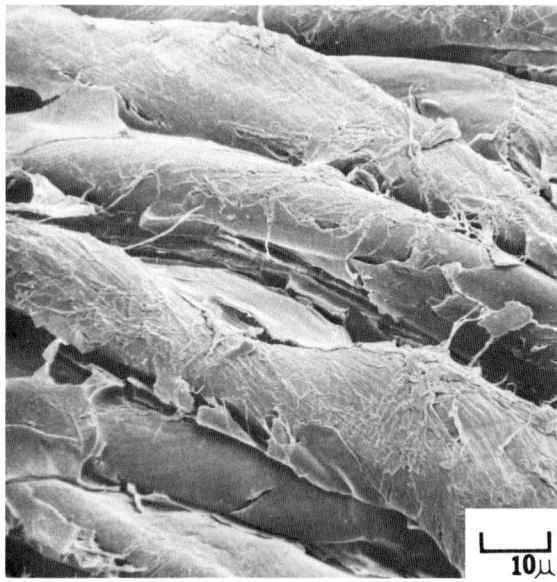

Fig. 7. Surface of cotton fabric treated with THPOH-NH$_3$, after 5 laundry cycles.

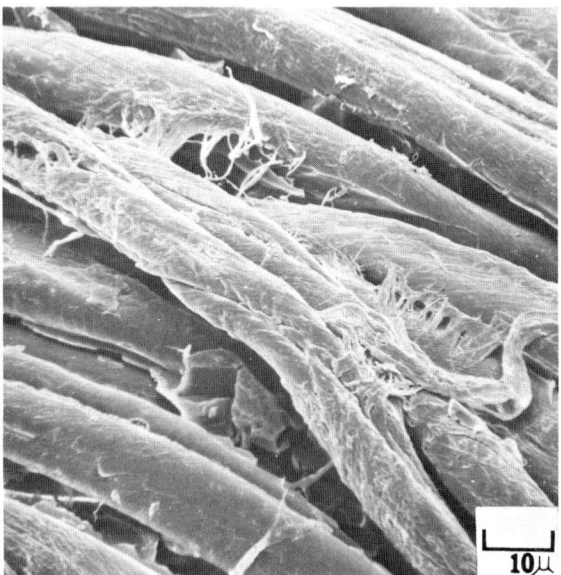

Fig. 8. Surface of cotton fabric treated with THPOH-NH$_3$, after 20 laundry cycles.

ENERGY-DISPERSIVE X-RAY ANALYSIS

In the analysis of finished textiles, it is important to distinguish between surface adhesion and surface penetration of the finish chemical and to determine depth of penetration into the fiber. Energy-dispersive X-ray (EDX) analysis provides a direct method for measuring such penetration by locating reacted or deposited elemental species in the specimen.[4] In combination with SEM, this technique permits mapping of elements deposited on fiber sections or penetrated into them.

After bombarding a specimen with an electron beam to produce SEM images, readjustment of electron energy levels after electron emission produces X-rays with energies characteristic of the particular elemental species. Collection of these X-rays and identification of their energies thus identifies the element from which they were emitted. When the X-ray information is fed through the cathode-ray display tube of the microscope, a distribution image is formed. Single elements are detected by adjusting the analyzer to the energy level of a particular element. When an X-ray of this energy is detected, a point is automatically brightened on the display tube. The resulting image corresponds spatially to the specimen observed, with the density of dots related to elemental abundance.

Since SEM provides images of surfaces only, to study internal fiber structure, sections must be cut and cut surfaces analyzed. Thick fiber sections were cut by modifying techniques for preparing ultrathin fiber sections for transmission electron microscopy.[5,6] To minimize chemical migration, either a hydrophobic (methacrylate) or a hydrophilic (polyvinyl alcohol) embedding medium was used, depending on the chemical nature of the finish.

For methacrylate embedding, a yarn was placed on a glass slide and coated with poly(butyl methacrylate) that was allowed to harden in an oven at 65°C. The embedment was thickened by addition of a 3/2 mixture of prepolymerized methyl and butyl methacrylates. After hardening, the embedded specimen was removed from the glass slide and trimmed for sectioning in a flat vise on a Porter-Blum[2/] MT-1 microtome. Thick sections were cut with a diamond knife by setting the thickness dial on the microtome to a maximum and, using the bypass arm, rotating the embedment through 6 to 8 turns before allowing it to pass over the knife. Variation in microtomes makes this setting arbitrary; experimentation with settings for

---

[2/]Use of a company of product name by the U.S. Department of Agriculture does not imply approval or recommendation of the product to the exclusion of others which may also be suitable.

desired section thickness is necessary with each microtome used.
After collection in a water-ethyl alcohol mixture in the boat,
sections were picked up with an eyelash mounted on a toothpick,
and placed on a carbon disk. The methacrylate embedding medium
was removed by placing the disk in a petri dish on a filter paper
saturated with methyl ethyl ketone. The dish was closed, placed
in a desiccator, and allowed to remain approximately 24 hrs.

Poly(vinyl alcohol) (PVOH) embedments were similarly prepared
by coating a yarn with an approximately 10% water solution of
Gelvatol 20-60. After the embedment had dried to a soft solid it
was freed from the glass slide, loosely sandwiched between two
glass slides and allowed to harden in an oven at 70°C. The embedment was trimmed and sections were cut as with methacrylate embedments, except that no liquid was used in the microtome boat
and after each section was cut it was removed from the knife edge
with an eyelash and placed on a carbon disk. When several sections had been collected, the disk was placed on water-saturated
cheesecloth in a covered petri dish. After approximately 12 hours,
deionized water was pipetted onto the carbon disk surface, allowed
to stand 1 hour, then carefully suctioned from the disk leaving
free sections. Since it is normally advantageous to provide an
SEM image of the specimen in order to correlate X-ray data with
fiber structure, the specimen was prepared for secondary image
production. This may be accomplished by evaporation of gold or
gold-palladium to form a continuous surface film. However, the
presence of these metals interfered with desired X-ray analysis;
consequently, development of an alternate procedure was necessary.
Carbon has been used as an aid in charge suppresion, and since
it does not interfere spectroscopically, the carbon disks containing the sections free of embedding medium were attached to
specimen stubs with silver paste, placed in a vacuum bell jar,
and coated with carbon. For these experiments, a Cambridge Mk IIA
scanning electron microscope with an EDAX X-ray system was used.

The carbon disks were scanned in the microscope to find
groups of sections of interest to be photographed prior to X-ray
examination. Low accelerating voltage is normally used in SEM of
textiles to further reduce the possibility of charging. Higher
voltages may be required for X-ray analysis of certain elements.
Long exposure of the specimen to the electron beam for X-ray
analysis may also cause damage, making the specimen unsuitable
for secondary image photography after X-ray mapping.

Samples analyzed included cotton printcloth treated with a
THPOH-NH_3 finish, a cotton-polyester blended fabric treated with
Thpc-urea-poly(vinyl bromide) (PVB), and a polyester fabric
treated with <u>tris</u>-2,3-dibromopropyl phosphate (TBPP).

Figure 9a shows thick cotton fiber cross sections from the fabric finished with THPOH-NH$_3$ and having a 3.6% phosphorus. Figure 9b is the phosphorus Kα X-ray map, showing that phosphorus is not only present in the polymer adhering to the fiber surface but also penetrated the entire fiber. Figure 10a shows cotton and polyester sections from the fabric treated with Thpc-urea-PVB, overlaid with the phosphorus map. Figure 10b shows the same sections overlaid with the bromine map. It is obvious that phosphorus is heavily deposited in the polymer adhering to the cotton fiber surfaces and has penetrated the fibers as well. The 1.5% bromine in the sample is concentrated only in the externally deposited polymer.

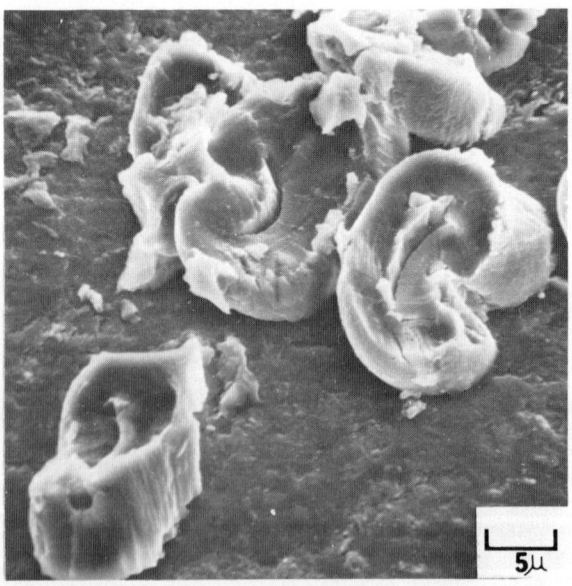

Fig. 9a. Cotton fiber cross sections from fabric treated with THPOH-NH$_3$.

Modified Textile Fibers 261

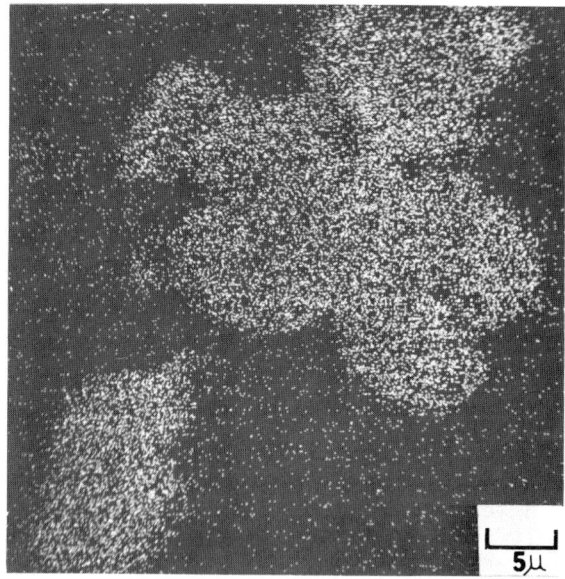

Fig. 9b. EDX Map showing location of phosphorus in sections shown in 9a.

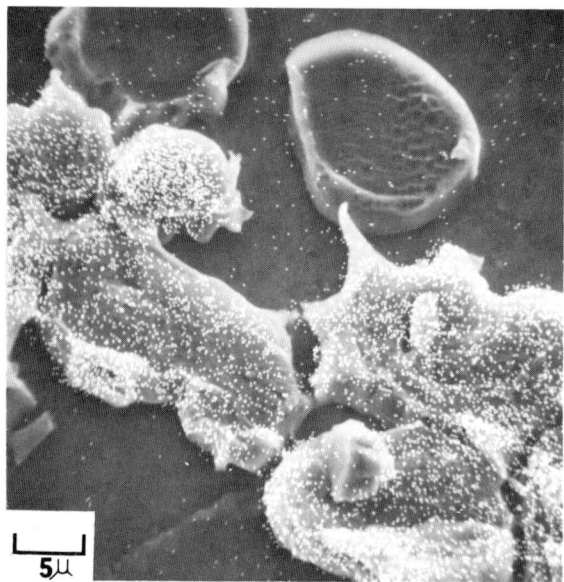

Fig. 10a. Cotton and polyester cross sections from fabric treated with Thpc-urea-PVB, with phosphorus EDX overlay.

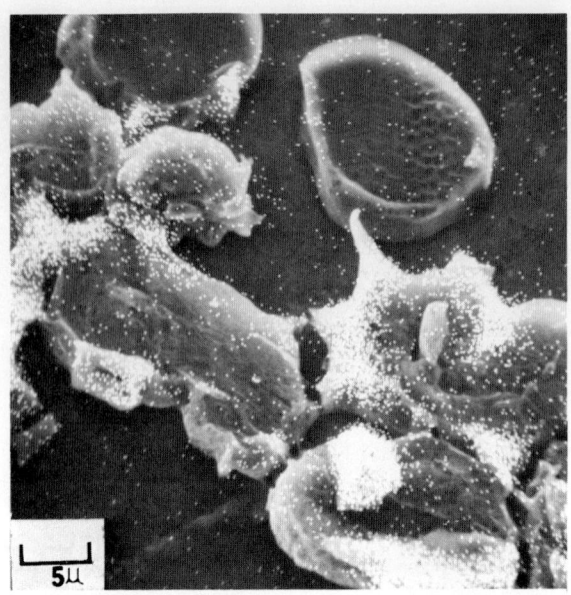

Fig. 10b. Cotton and polyester cross sections from fabric treated with Thpc-urea-PVB, with bromine EDX overlay.

Figure 11 is a polyester fiber section and bromine map from a fabric treated with TBPP and exhausted at 265°F. Bromine content of the fabric was 1.93%. There is no adhesive buildup on the fiber surface, and bromine has penetrated the fiber and is distributed throughout it. Figure 12 is a polyester section and bromine map from a fabric treated with TBPP in which no exhaustion was used; it contained 0.62% bromine, concentrated in the surface of the fiber.

CONCLUSIONS

Scanning electron microscopy has been used successfully to show changes in fabric and fiber surfaces brought about by chemical finishing. Surface deposits, especially those built up at yarn crossover points, increase fabric stiffness. SEM photographs of samples after laundering indicate that excess surface deposits may be removed, thereby decreasing stiffness. It appears that the surface polymers removed in laundering did not contribute significantly to the flame resistance of the fabric since flame resistance did not decrease greatly after 20 laundry cycles.

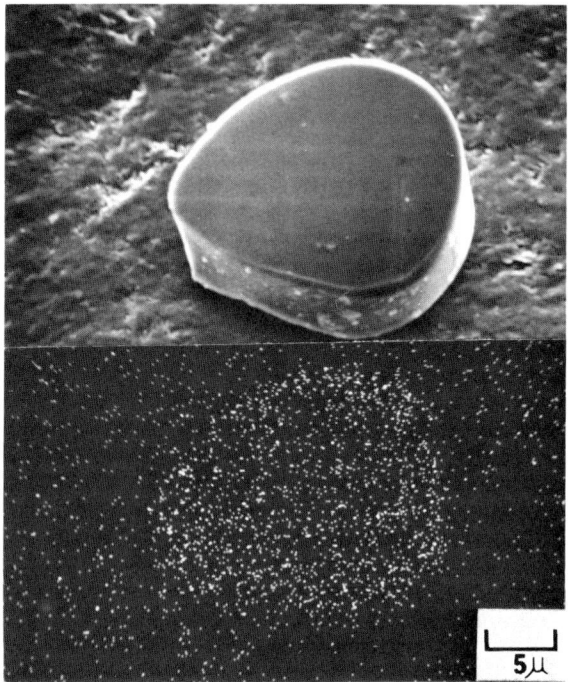

Fig. 11. Polyester cross section from fabric treated with TBPP at 265°F, with corresponding bromine EDX map.

Energy-dispersive X-ray analysis provided additional information on penetration of chemicals through fiber surfaces into internal areas. Elemental X-ray maps showed that in some finishes depth of chemical penetration was dependent on treatment conditions. Phosphorus was found to be distributed throughout the cotton fibers from the fabric treated with THPOH-NH_3. Bromine in polyester fiber samples finished with poly(vinyl bromide) was concentrated only in heavy polymer deposits on the fiber surfaces. With proper treatment conditions, the TBPP finish was found to have penetrated throughout the polyester fibers, producing a more durable finish.

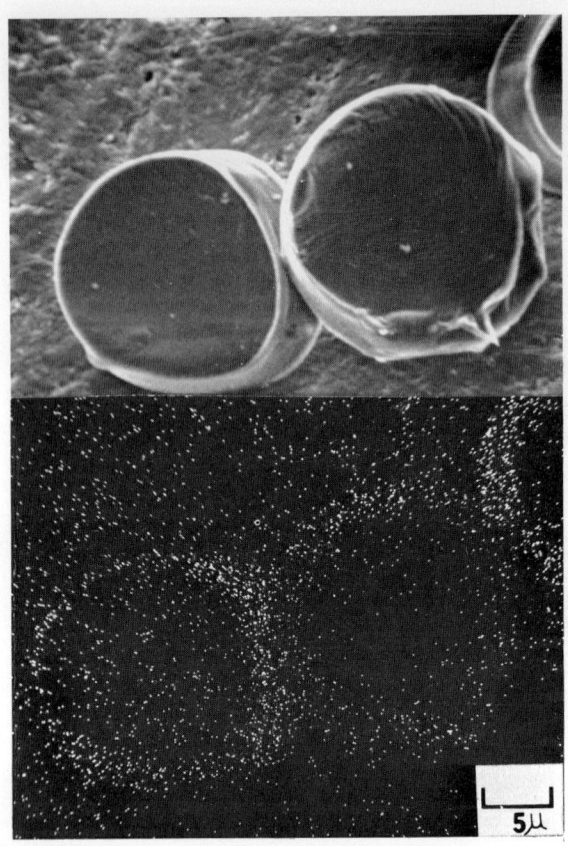

Fig. 12. Polyester cross sections from fabric treated with TBPP without exhaustion, with corresponding bromine EDX map.

Microscopical analysis of chemically finished fibers presents many problems. Fibers are generally nonconductive and require deposition of conductive materials on their surfaces to prevent electrical charging, which interferes with production of a true image. Elements analyzed are usually present in low concentrations and are often of low atomic number. Beam currents and accelerating voltages that are required to produce detectable X-rays may damage sensitive fiber structures or cause unacceptable amounts of surface charging. In spite of these problems, microscopical analysis provides visual information that is not available through any other source.

REFERENCES

1. L.W. Mazzeno, B.J. Trask, D.A. Yeadon, and G.F. Danna, J. Coated Fibrous Mater. $\underline{2}$, 174 (1973).
2. G.W. Drake, Jr., and W.A. Reeves, *Cellulose and Cellulose Derivatives*, Vol. 5, Part 5, N. Bikales and L. Segal, ed., Wiley-Interscience, New York (1971).
3. R.B. LeBlanc, American Association of Textile Chemists and Colorists, Committee RA46, AATCC Fire Resistance Test Revised. Textile Chem. and Col., 2(3), 47-8 (1970); Fire Resistance of Textile Fabrics---AATCC Test Materials 34----1969, ibid, 49-50.
4. J.C. Russ, *Elemental X-ray Analysis of Materials*, EDAX International, Inc., Prairie View, Ill. (1972).
5. M.L. Rollins, A.M. Cannizzaro, and W.R. Goynes, "Electron Microscopy of Cellulose and Cellulose Derivatives", in *Instrumental Analysis of Cotton Cellulose and Modified Cotton Cellulose*, R.T. O'Connor, ed., Marcel Dekker, Inc., New York (1972).
6. A.M. Cannizzaro, W.R. Goynes, and M.L. Rollins, Amer. Dyest. Rep. $\underline{57}$, 23 (1968).

Laboratory Study of Fiber Fracture Using the Scanning Electron Microscope

Alfredo G. Causa

Tire Textile Development Department
The Goodyear Tire and Rubber Company
Akron, Ohio 44316

Fibers in the form of cords are used as reinforcing materials for rubber in applications such as tires, conveyor belts and hoses. Consequently, a large number of tests are used in the tire and rubber industry for testing cords and cord-rubber composites. Each test is designed to produce a certain stress or combination of stresses in the test specimen.

This investigation deals with the study by scanning electron microscopy of the fracture surface topography of various tire textile cords, including aramid tire cord, failed under tensile and cyclic stresses.

INTRODUCTION

Scanning electron microscopy is an important tool in the study of fiber fracture and in the characterization of fiber surface topography.[1-6]

Fibers in the form of cords (cabled yarns)[7] are used as reinforcing materials for rubber in applications such as tires, conveyor belts and hoses. Consequently, a large number of tests are used in the tire and rubber industry for testing cords and cord-rubber composites. Each test is designed to produce a certain stress or combination of stresses in the test specimen.

It is the objective of this study to analyze fracture surfaces of various tire textile cords by scanning electron microscopy, and identify fractographic features that are characteristic of certain load history, testing mode or fiber microstructure.

EXPERIMENTAL

The same basic technique was used in the preparation and analysis of all the selected cord failures described in this study. The broken cord tips were mounted using customary procedures and were vacuum coated with carbon as a first layer, followed by a layer of gold.[8] In this work, the double vacuum coating technique was used to ensure uniformity of the layer and therefore effective suppression of charging, particularly on the irregular fracture surfaces and filament entanglements.

Oftentimes, low magnification photographic montages were made of the broken cord tips. These montages show the alteration of the geometric integrity of the cord and are helpful to select the particular areas, fibers or filaments for the higher magnification work. Individual fibers were also mounted by holding them with tweezers and cutting them near the fracture with a sharp razor blade with the aid of a stereozoom microscope. The scanning electron micrographs showing the fracture surfaces were obtained on a JEOL Model JSM-U3 scanning electron microscope operating in the secondary emissive mode.

This investigation covers the study of the fracture morphology of various tire textile cords, including aramid tire cord, subjected to the two types of stresses which are of major importance in cord-rubber composites, namely tensile and cyclic stresses.

Tensile Stresses

Tensile breaks were obtained on an Instron tester using a specimen gauge length of 25 cm, at a crosshead speed of 30 cm/min. The cords were conditioned before testing for a minimum of 6 hours in a controlled atmosphere of 24°C, 55% RH. Tensile breaks were also obtained on a Plastechon pneumatic impact tester[9] at a loading rate of 25,000 cm/min.

Cyclic Stresses

Various test concepts are used in the tire and rubber industry to generate cyclic stresses that lead to fatigue failure of the reinforcing cords.

Mallory tube test. In the Mallory tube test (Figure 1), the cords are embedded in a hollow rubber tube parallel to the longitudinal axis of the tube. The cords are located on a circle of specified radius.

The test consists of rotating the tube around its longitudinal axis after bending it at an angle. The tube is driven from one end

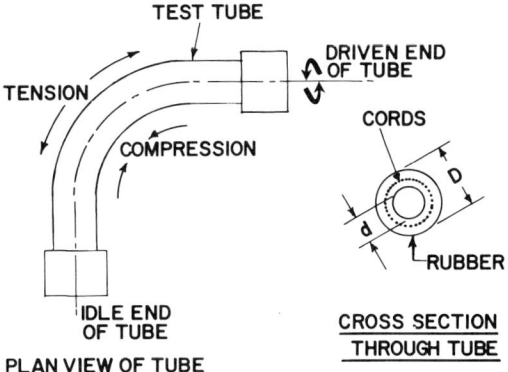

Fig. 1. Schematic diagram of Mallory tube test.

and has an idler bearing in the other. An internal air pressure is maintained in the tube during testing. The test applies an axial load, alternating sinusoidally from tension to compression in each of the cords. A bending stress is also applied to the cord as a result of its bending stiffness.

In the experiments described in this paper, the hollow rubber tube had a length of 23.5 cm, an outside diameter D=2.54 cm and an inside diameter d=1.27 cm. The cords were placed in the middle of the rubber annulus and there were eighteen cords per inch (2.54 cm) of the 1300/3 (1300 denier, 3 ply) experimental polyester-type cord and sixteen cords per inch of the 1500/3 aramid cord. The tubes rotated at 600 rpm, at a 90° flex angle while maintaining 50 psi (3.52 Kg/cm^2) air pressure inside the tube. A timing mechanism reversed the direction of rotation every thirty minutes, and the test was run until the onset of failure. In this test, kilocycles to failure as well as the temperature of the tube due to hysteresis can be measured.

Compression flex test. In the compression flex test, a belt made up of two cord bands embedded in rubber is bent over a spindle of a given diameter (Figure 2). The ends of the belt are gripped in clamps, and a vertical load is applied upward on the spindle, thus applying a tension to the belt. A crank mechanism is used to apply a sinusoidal motion to the ends of the belt, with a double amplitude of 13.34 cm at a frequency of 250 cycles per minute. Dynamic flexing of the belt over the spindle results.

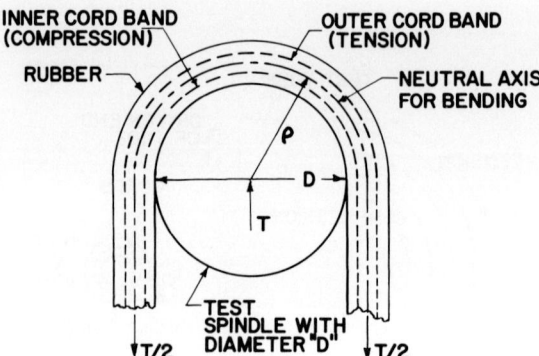

Fig. 2. Schematic diagram of compression flex test.

The test samples are contained in an environmental chamber at 80°C. As a result of bending of the belt, the outer band is in axial tension and the inner band is in axial compression. An additional tension is exerted on each band.

In this investigation, the belt was 25.4 cm long, 2.54 cm wide and 0.32 cm thick. Each band contained sixteen 1500/3 aramid cords per 2.54 cm and eighteen cords per 2.54 cm of the 1300/3 experimental polyester cord, lying parallel to the length of the belt. The spindle had a diameter of 2.54 cm, and the test was run until the specimen failed.

<u>MTS testing machine</u>.[10] The MTS system is a closed-loop electrohydraulic test machine. Periodic functions of stroke or load can be programmed to the hydraulic actuator over a wide range of frequencies. The load in the test specimen is accurately measured with a strain gage load cell and the actuator stroke with a LVDT (Linear Variable Differential Transformer). Both transducers are a part of the test machine. Readout of parameters is accomplished with a high frequency response pen recorder and an oscilloscope. An environmental chamber is used to control the elevated temperature of the test specimen. Either bare cords or cord-rubber composites can be tested.

<u>SCEF fatigue tester</u>.[11] In the SCEF (shear, compression, elongation, flex) fatigue tester, compression-tension cycles are applied to the cords through a shearing action at elevated temperatures. A cord-rubber composite is used as test specimen.

In this study, the test specimen was a rubber belt 20.32 cm long, 1.91 cm wide and 0.32 cm thick, containing in the test zone three 1500/3 aramid tire cords cured in the middle and lengthwise of the rubber belt. The belt was positioned in the apparatus at a 45° angle to the direction of oscillation of a central floating clamp which operated at a frequency of 380 cycles per minute with a stroke of 1.27 cm. The environmental chamber of the tester was maintained at 100°C and the test was run until failure of the specimen occurred. When the test cord fails, the apparatus shuts off automatically and the time to failure is recorded on a counter.

RESULTS AND DISCUSSION

It is important to recognize that fractography of textile cords has some limitations due to the complex stress patterns developed in a cord and in cord-rubber composites. Furthermore, fracture surfaces are sometimes damaged in laboratory testing by smearing or rubbing and are of no fractographic value. The fracture surfaces analyzed in this investigation were carefully selected so that they are representative of each testing mode, and free from any experimental artifact.

Fracture under Tensile Stresses

Figure 3 shows the typical topography observed at practically all the fiber fractured ends in an Instron tensile break of 1000/3 (1000 denier, 3 ply) polyester cord. A Plastechon tensile break of the same cord is illustrated in Figure 4, both micrographs at the same magnification. It can be noticed that a change from a globular shape to a flatter surface configuration at the fractured fiber tips was brought about by the approximately 800-fold increase in tensile loading rate. The same observation was recorded upon comparing an Instron and a Plastechon tensile break of 1260/2 nylon 66 cord (Figures 5 and 6).

Aramid tire fiber is the strongest fiber ever synthesized[12]; in fact, a typical aramid industrial cord for tire reinforcement is approximately five times stronger, on a weight basis, than a typical steel cord: 350 m N/tex for the tenacity of steel cord versus 1800 mN/tex for aramid cord. This high tenacity is undoubtedly due to the high degree of orientation of non-folded chains in the fiber spun from an anisotropic solution.[13-16]

Under the scanning electron microscope, a filament in untested aramid tire yarn has the appearance of a smooth cylinder of 10-12μm diameter. There are no distinctive topographical features; nicks, small ridges and some debris that are commonly found on fiber surfaces are also observed on the surface of aramid fiber (Figures 7 and 8).

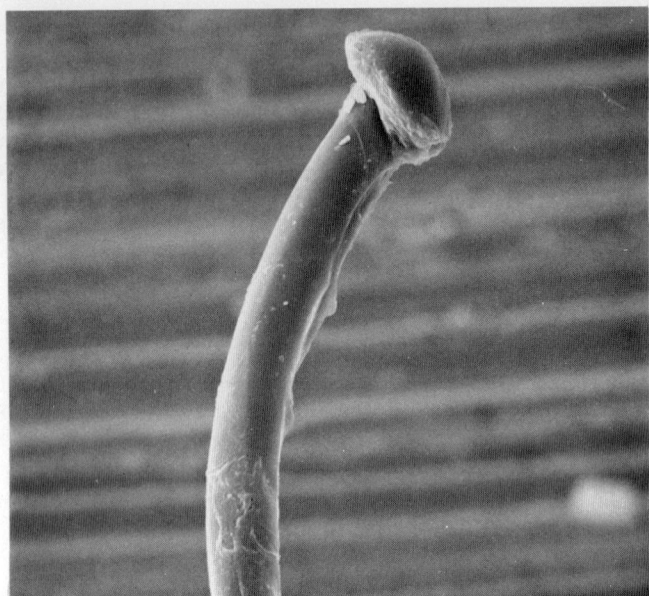

Fig. 3. Typical Instron tensile break of a polyester fiber(500X).

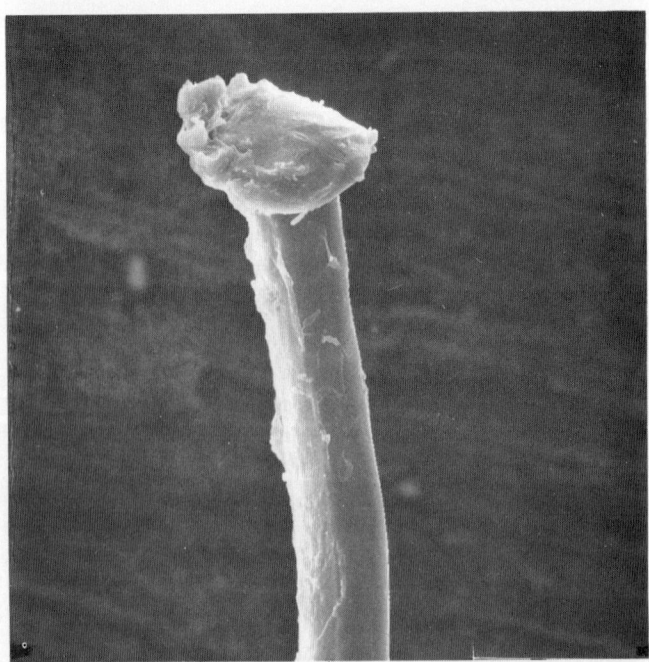

Fig. 4. Polyester fiber broken in tension on the Plastechon impact tester (500X).

Study of Fiber Fracture 273

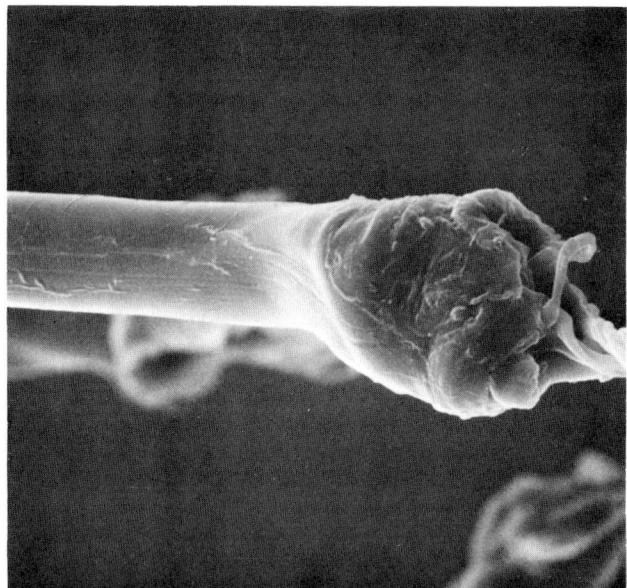

Fig. 5. Instron tensile break of a nylon 66 fiber(500X).

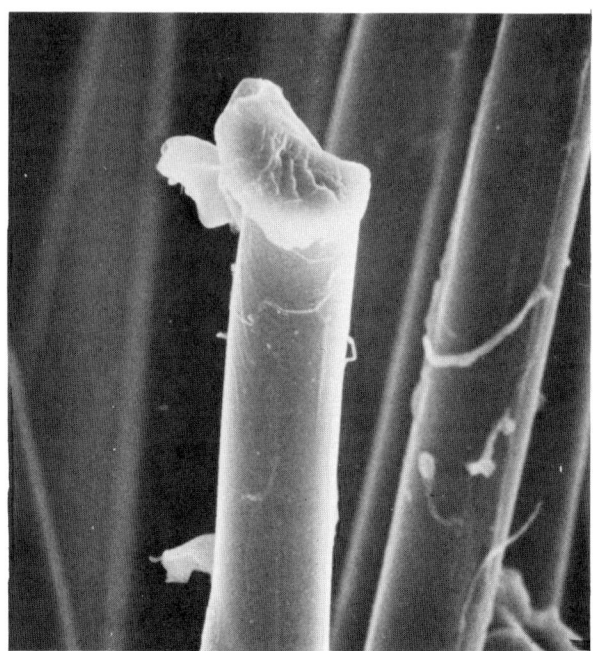

Fig. 6. Nylon 66 fiber broken in tension on the Plastechon impact tester(500X).

Fig. 7. Surface topography of untested aramid tire yarn(3,000X).

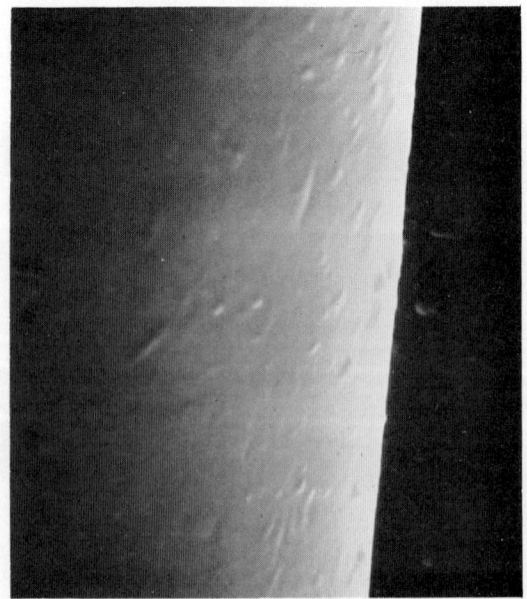

Fig. 8. Surface topography of untested aramid tire yarn(10,000X).

Characteristic aramid filament tensile breaks are shown on the micrographs in Figures 9 and 10. Axial splitting, with greater or lesser degree of fibrillation, is the main surface topographical feature which does not change substantially on changing the loading rate.

Longitudinal splitting and fibrillar type of fracture under tensile deformation have been well documented in the literature. Laible, Figucia and Kirkwood[17] described this failure mode in aromatic polyamides both under impact and slow speed tensile test conditions. More recently, a similar observation has been recorded by Bunsell[5] and by Carter and Schenk.[18] In a particularly interesting and revealing experiment, Goswami and Hearle[6] have shown that axial splitting and fibrillation are the predominant morphological features in the tensile fracture of a 15-denier nylon 66 filament at liquid nitrogen temperature. Mechanical anisotropy (axial strength larger than transverse strength) brought about by molecular orientation is the main microstructural factor responsible for this tensile fracture morphology, as documented in the experiments of Curtis and Treloar.[19]

Fracture under Cyclic Stresses

Figures 11 and 12 show the fatigue fracture surface of an experimental polyester fiber failed in the Mallory tube test. Microstructural changes can be induced in this fiber under the action of cyclic loadings at high temperature on the MTS apparatus. When subjected to zero-tension cycling to 75% of its tensile breaking load, at 15 Hz frequency in a haversine function, at 150°C, the fiber underwent a reduction in molecular weight with a concomitant increase in crystallinity, and presented the unusual conchoidal fracture shown in Figure 13. In this experiment, the cord was encapsulated in rubber, and the test was conducted until the specimen failed.

The failure mode of aramid tire cord under Mallory tube testing conditions is dominated by axial splitting and fibrillation, as illustrated in Figure 14. Cross striations or markings most probably due to compressive stresses, are also noted (Figure 15). A head-on view of the fracture surface morphology of an aramid filament ruptured in the Mallory tube test is shown in Figure 16.

The fracture surface topography of aramid tire cord, both bare and embedded in rubber, after cyclic loading to failure on the MTS apparatus was also investigated. The experiments were conducted under conditions of zero-tension cycling to 85% of the fiber's tensile breaking load, at 15 Hz frequency in a haversine mode, at 150°C. Fibrillation at the fiber broken tips (Figure 17) and longitudinal splitting (Figure 18) are again the predominant

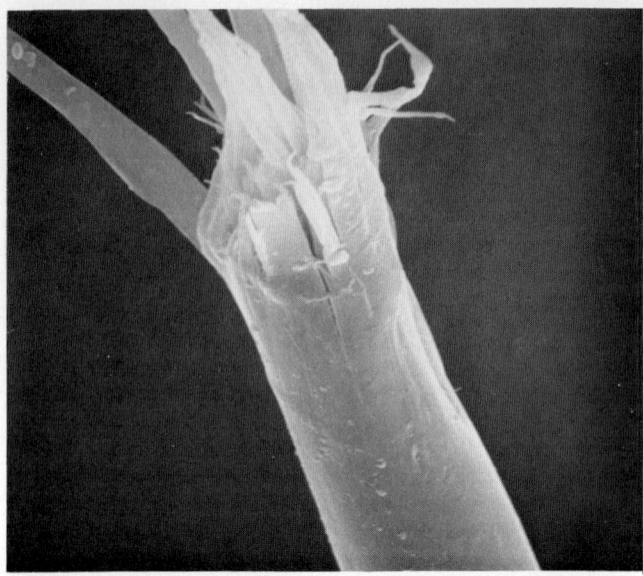

Fig. 9. Typical tensile fracture of aramid tire fiber showing axial splitting and fibrillation(3,000X).

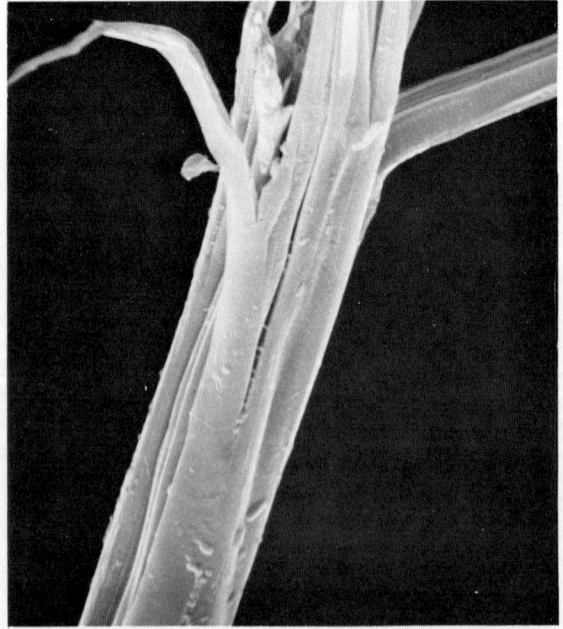

Fig. 10. Another view of an aramid tire fiber broken in tension on the Instron tester(3,000X).

Study of Fiber Fracture 277

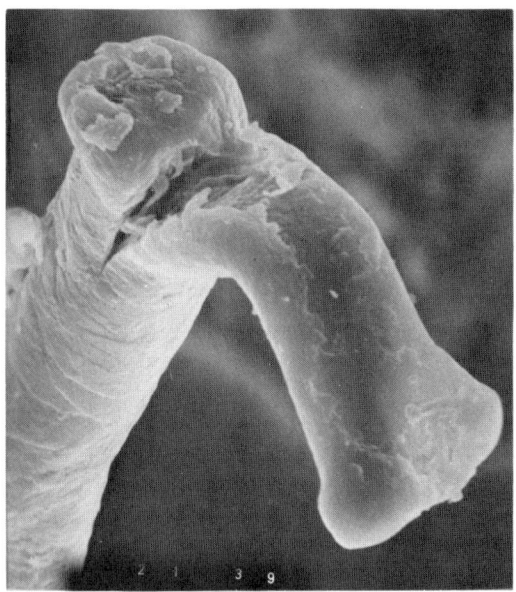

Fig. 11. Mallory tube fatigue fracture of an experimental polyester fiber(1,400X)

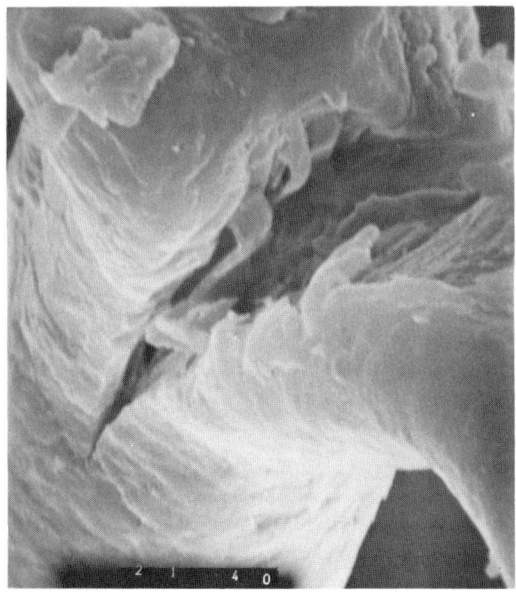

Fig. 12. Detailed view of Mallory tube fatigue fracture shown on Figure 11(4,000X).

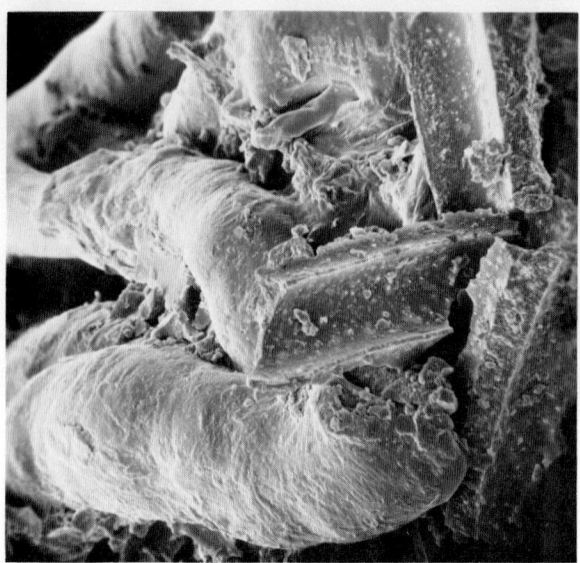

Fig. 13. Fracture of experimental polyester fiber (Figures 11 and 12) after cyclic loading on MTS apparatus (500X).

Fig. 14. Aramid tire cord failed under Mallory tube testing conditions (1,000X).

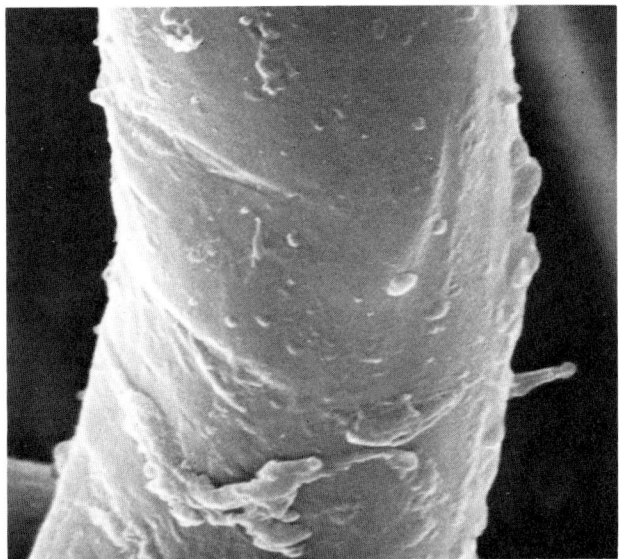

Fig. 15. Surface topography of aramid tire fiber failed in Mallory tube test (5,000X).

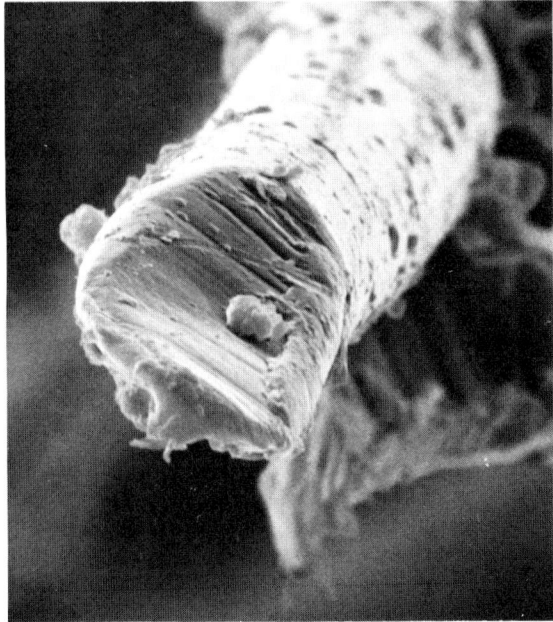

Fig. 16. Fracture surface morphology of aramid tire fiber ruptured in the Mallory tube test (3,000X).

Fig. 17. Fibrillated broken tip, aramid tire fiber, MTS cyclic loading(500X).

Fig. 18. Longitudinal splitting, aramid tire fiber, MTS cyclic loading(1,000X).

fractographic features. A close-up view of a longitudinal split and surface cavitation (gouging) is illustrated in Figure 19. Longitudinal splitting and formation of kink bands[2,20,21,22,23] at the intrados of the bent filament-symptomatic of compressive buckling-are shown in Figures 20 and 21.

The interior of the cavity formed upon filament splintering offers a glimpse of the fiber microstructure as illustrated in Figure 22. An excellent view of the internal microfibrillar structure of an aramid tire fiber can be obtained by the Scott peeling technique[24,25] (Figure 23).

In another experiment, bare 1260/2 nylon 66 cord was ruptured on the MTS apparatus by zero-tension cyclic loading to 75% of its tensile breaking load, at 15 Hz frequency in a haversine function, at 150°C. Typical filament fatigue fracture is shown in Figure 24.

Fractographic analyses were made of aramid tire cord failed under the deformation mode known as compression flex fatigue test. Fibrillation and splintering are readily observed, but another interesting fracture mode is also present in the broken filaments of the failed specimen. Figure 25 shows a V-shaped fracture in an aramid filament taken from either the tension or the compression ply of the compression flex test belt. The shape of the fracture suggests that it started from the convex side of the bend and propagated towards the concave side which still remained intact.[26]

An ill-defined fracture at the convex side of the bent filament was also noted when analyzing the failed area of the compression flex test specimen reinforced with the experimental polyester fiber previously used in this work (Figure 26).

Considerable amount of rubbing and smearing of the fracture surfaces is noticed in this test thereby making any further fractographic interpretation a matter of dubious value.

When aramid tire cord is tested to failure in the SCEF fatigue tester, the characteristic splintering is observed at the fractured fiber tips. Micrographs taken at the fiber-rubber interface show that fracture has occurred between the rubber-covered cord and the rubber matrix (Figures 27 and 28). Investigation with other fibers proceeds in our laboratory.

CONCLUSION

The experiments described in this paper provide evidence of the usefulness of scanning electron microscopy to analyze failure modes in textile cords, either bare or embedded in rubber.

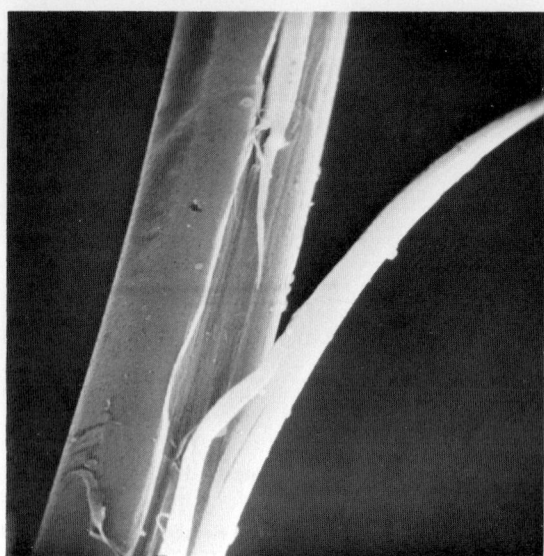

Fig. 19. Longitudinal split and cavitation, aramid tire fiber, MTS cyclic loading (2,500X).

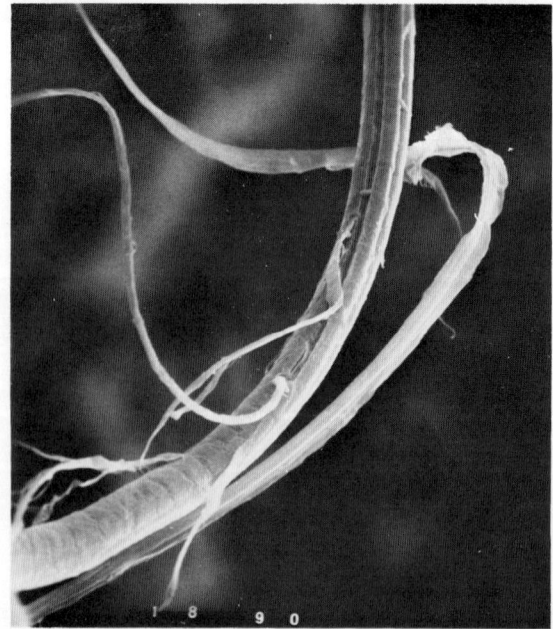

Fig. 20. Aramid tire fiber failed under MTS cyclic loading conditions, showing axial splitting and kink band formation (1,000X).

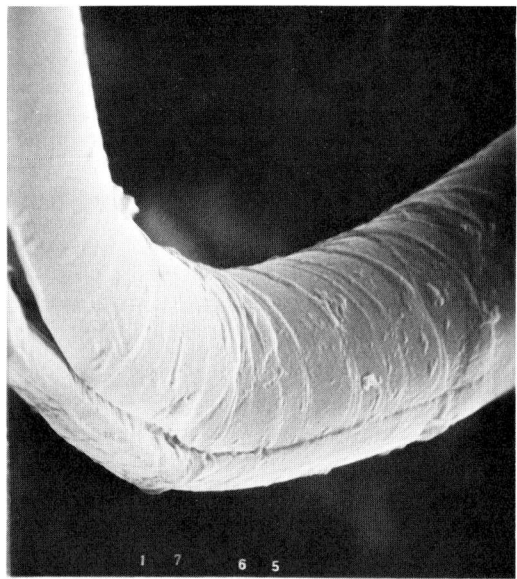

Fig. 21. Aramid tire fiber failed under MTS cyclic loading conditions, showing longitudinal crack and kink band formation (3,000X).

Fig. 22. Splintered aramid tire fiber, MTS cyclic loading, showing microstructure inside cavity (3,000X).

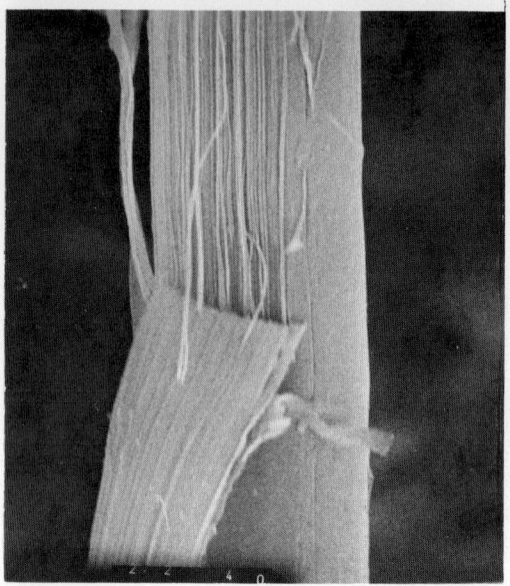

Fig. 23. Peeled aramid tire fiber (Scott peeling technique) showing microfibrillar structure(3,000X).

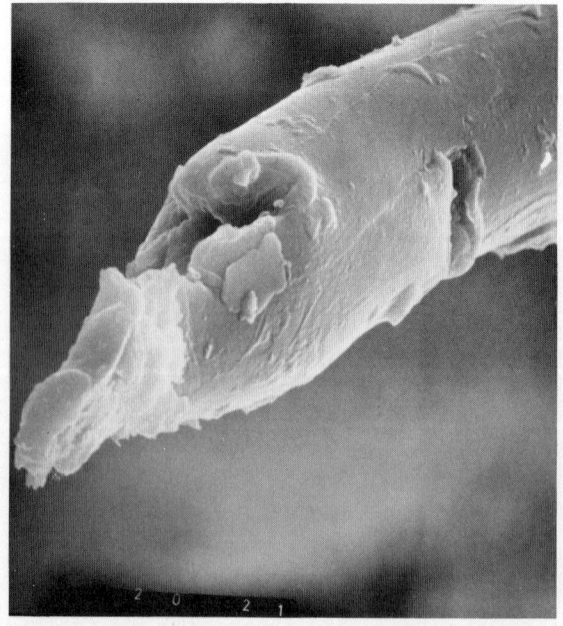

Fig. 24. Nylon 66 tire fiber failed under MTS cyclic loading conditions(1,300X).

Study of Fiber Fracture 285

Fig. 25. V-shaped fracture formed in aramid tire fiber failed in compression flex fatigue test(3,000X).

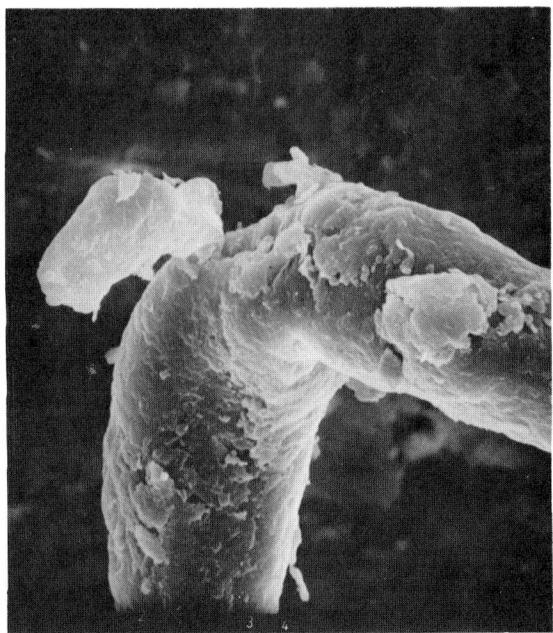

Fig. 26. Experimental polyester failed under compression flex testing mode(1,000X).

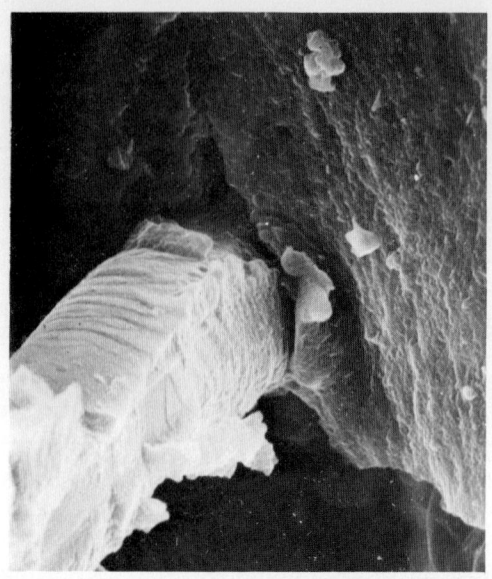

Fig. 27. Aramid tire cord failed under SCEF testing mode showing rubber-to-rubber fracture(1,000X).

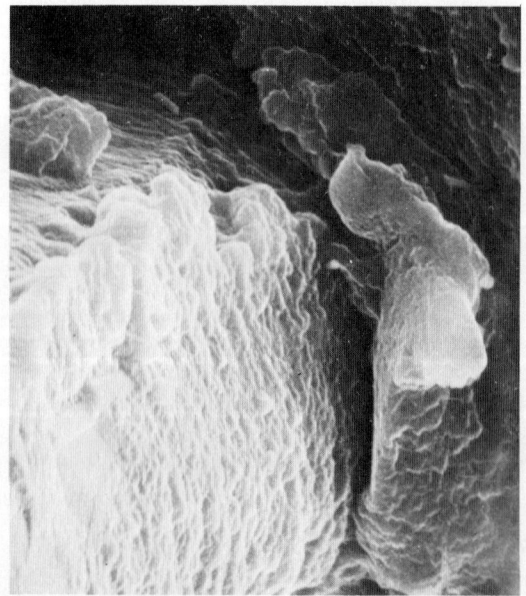

Fig. 28. Close-up view of rubber fracture shown on Figure 27 (Aramid tire cord, SCEF test) (3,000X).

ACKNOWLEDGEMENTS

The author wishes to express his appreciation to The Goodyear Tire and Rubber Company for permission to publish this paper. Thanks are also due to Mr. Robert McMillan, Goodyear Research, for his help in obtaining some of the scanning electron micrographs used in this study.

REFERENCES

1. A.R. Bunsell and J.W.S. Hearle, J. Appl. Polym. Sci., 18, 267 (1974).
2. A.R. Bunsell and J.W.S. Hearle, Rheol. Acta., 13, 711 (1974).
3. J.W.S. Hearle, B. Lomas and A.R. Bunsell, Applied Polymer Symposium 23, 147 (1974).
4. A.R. Bunsell, J.W.S. Hearle, L. Konopasek and B. Lomas, J. Appl. Polym. Sci., 18, 2229 (1974).
5. A.R. Bunsell, J. Mat. Sci., 10, 1300 (1975).
6. B.C. Goswami and J.W.S. Hearle, Text. Res. J., 46, 55 (1976).
7. J.W.S. Hearle, P. Grosberg and S. Backer, *Structural Mechanics of Fibers, Yarns, and Fabrics*, Vol. 1, Wiley, 1969, p. 63.
8. J.W.S. Hearle, J.T. Sparrow, and P.M. Cross, *The Use of the Scanning Electron Microscope*, Pergamon Press, 1973, p. 76, 77 and 144.
9. W.J. Lyons, *Impact Phenomena in Textiles*, MIT Press, 1963, p. 67.
10. K.D. Zell, *A Review of Advanced Electrohydraulic Systems for Fatigue Testing*, Publication 160.02-01, December 1971, MTS Systems Corporation, Minneapolis, Minnesota.
11. M.J. Forster and I.B. Prettyman, Rubber Chem. Technol., 42, 1000 (1969).
12. R.E. Wilfong and J. Zimmerman, J. Appl. Polym. Sci., 17, 2039 (1973).
13. S.L. Kwolek, U.S. Patent 3,671,542, June 20, 1972, E.I. duPont de Nemours and Company.
14. H. Blades, U.S. Patent 3,767,756, October 23, 1973, E.I. duPont de Nemours and Company.
15. H. Blades, U.S. Patent 3,869,429, March 4, 1975, E.I. duPont de Nemours and Company.
16. H. Blades, U.S. Patent 3,869,430, March 4, 1975, E.I. duPont de Nemours and Company.
17. R.C. Laible, F. Figucia and B.H. Kirkwood, Applied Polymer Symposium, 23, 181 (1974).
18. G.B. Carter and V.T.J. Schenk, in *Structure and Properties of Oriented Polymers* (I.M. Ward, Ed.), Wiley, 1975, p. 478.
19. J.W. Curtis and L.R.G. Treloar, Nature, 220, 60 (1968).
20. D.A. Zaukelies, J. Appl. Phys., 33, 2797 (1962).
21. D.E. Bosley, Text. Res. J., 38, 141 (1968).
22. B.M. Chapman, J. Text. Inst., 64, 323 (1973).

23. C. Varela, S.K. Batra, and S. Backer, Text. Res. J., 45, 307 (1975).
24. R.G. Scott, ASTM Bull., No. 257, 121 (1959).
25. R.D. VanVeld, G. Morris and H.R. Billica, J. Appl. Polym. Sci., 12, 2709 (1968).
26. M.M. Schoppee and J. Skelton, Text. Res. J., 44, 971 (1974).

The Investigation of Poly(tetrafluoroethylene) Wetting Behavior by Scanning Electron Microscopy

N. E. Weeks, G. M. Kohlmayr and E. P. Otocka

*Materials Engineering and Research Laboratory
Pratt & Whitney Aircraft
Middletown, Conn. 06457*

The relationship between PTFE disordering kinetics and wetting kinetics of individual PTFE resin particles was studied. Resin particles were heated on pyrolytic graphite (HOPG) basal surfaces, quenched, and examined by scanning electron microscopy (SEM). Particles heated for extended times above the equilibrium melting point do not assume the classical spreading drop configuration. Anisotropic melting and wetting behavior is evident. The retained order in the melt is manifested by the continuing existence of the original particle shape. Resin flow during wetting occurs as (1) a highly mobile thin film covering distances of $\geq 1\mu$ away from the particles, and (2) a quasi-spreading drop with finite contact angle. Na-napthalene etching of the PTFE facilitated SEM observation of the spreading film and its morphology. Evidence of epitaxial crystallization by PTFE on the HOPG substrate is presented. Similar control of PTFE morphology was not observed in other PTFE/substrate systems under study.

INTRODUCTION

As described in recent studies,[1-7] the kinetics of substrate wetting by polymer melts are controlled by a combination of surface energy forces at the gas-liquid-solid interface and viscous properties of the polymer melt. A dominant surface force component leads to droplet spreading on the surface until an equilibrium contact angle is attained. Viscous control of spreading behavior results in the formation of an advancing "foot" on the spreading drop.[3,7]

A special aspect of the general wetting problem is the sintering behavior of polymer particles. The fundamental parameter in sintering is the growth of the interface between contacting resin particles. In either wetting or sintering, interface growth is a function of the ratio, γ/η, where γ is the droplet surface tension and η is the bulk viscosity. The interface growth equation must also include terms accounting for the viscoelastic properties of the polymer when sintering within the viscoelastic (VE) region of the bulk flow spectrum.[8] Poly(tetrafluoroethylene) (PTFE), with its high molecular weight and chain stiffness, is a polymer where the VE flow properties could be expected to play a dominant role in sintering and wetting behavior.

The substrate wetting kinetics of PTFE resins have not been investigated, however. With model sintering experiments using extruded PTFE rods, Lontz[9] has shown that (a) interface growth is limited, and (b) a time dependent viscosity term must be included in the interface growth equation to explain the experimental data. The time dependent term is a manifestation of the VE forces present in the high viscosity ($\sim 10^{11}$ poise)[10] PTFE melt. The Lontz equation is specific for PTFE, and Steiner's general growth equation does not describe the observed PTFE behavior.[8]

An analysis of PTFE wetting (and sintering) behavior is complicated by the morphology of the virgin resin particles. Prior to their initial fusion, PTFE resin particles are highly crystalline ($\geq 90\%$), with a morphology which is irreversibly destroyed during the first fusion. Bassett[11] has shown that the virgin resin morphology contains a superheatable component with an equilibrium melting point of $\sim 331°C$. Evidence also indicates order is retained in the melt up to five hours (at 375°C)[10] after fusion of the virgin resin morphology.

Possible interaction between melting and wetting kinetics has not been considered in previous studies of polymer wetting behavior. More specifically, the effect of substrate-polymer interactions upon either the fusion kinetics of the PTFE morphology or the melt order retention has not been investigated. Literature data[12] for low molecular weight fluorocarbons suggest PTFE molecules may be strongly adsorbed on a highly oriented pyrolytic graphite basal surface (HOPG). Such an effect would provide an additional driving force for substrate wetting without affecting the growth of the interface between contacting particles (i.e., sintering behavior).

This work was carried out to determine the wetting behavior of PTFE dispersion resin particles on various substrates, with the emphasis on highly oriented pyrolytic graphite basal plane surfaces. The depth of field and resolution capabilities of the scanning electron microscope (SEM), unattainable with other microscopic techniques, provides a unique opportunity to investigate the wetting behavior of the ultimate PTFE resin particles (d = < 0.5μ). Previous polymer wetting studies have been limited to resin droplets ≥10^2 times larger.

EXPERIMENTAL

Four PTFE dispersion resins with different molecular weights and fusion properties were chosen for the wetting studies. Resin molecular weights were determined from recrystallization behavior using the method of Suwa, et al.[13] Number average molecular weight values, M_n, are listed in Table 1. Initial crystallinities determined by x-ray and heats of fusion were ≥90% in all cases.

For SEM study, PTFE resin particles were removed from dilute solutions of the original latices by filtration, washed with isopropanol to remove surfactant, and ultrasonically redispersed in Freon. Significant agglomeration of the resin was prevented by carrying out all steps rapidly at temperatures below 19°C (PTFE phase transition temperature). X-ray data show no disruption of the morphology by the ultrasonic agitation.

Resin particles were deposited upon the substrates by evaporation of Freon. HOPG substrates were provided by Dr. A. W. Moore of Union Carbide Corp., Parma Technical Center. Other substrates were freshly cleaved mica and Pt foil.

Table 1

Molecular Weights of PTFE Dispersion Resins

Resin	M_n (x10^{-6})
D-1	1.5
I-1	2.0
D-2	4.6
D-3	11.0

Virgin resin fusion behavior was determined with a DSC-1B differential scanning calorimeter (DSC) (Perkin Elmer Corp.) at heating rates of 5, 10, 20 and 40°C/min. Heating of PTFE-substrate systems was also done in the DSC at temperatures of 335, 341 and 355°C for various lengths of time. Thin substrates ($\sim$.03" thick) and small resin particle size combine to eliminate temperature gradients across the sample.[3] The major portion of this study utilized the SEM (JEOL JXA 50A) with viewing angles of 45° and <15° from the substrate plane. SEM samples were sputter coated with 100Å of platinum to eliminate charge buildup. The necessity of Pt coating samples before SEM examination precluded use of a heating stage for direct monitoring of PTFE wetting within the microscope. Instead, samples heated in the DSC were cooled to room temperature at 80°C/min., minimizing distortion of the molten particle shape by limiting the extent of recrystallization.

To develop crystal structure in the resin, various slow cooling and annealing programs were used. Also to develop structure, etching of PTFE resin was carried out using a Na-naphthalene solution.[14] Substrates containing fused PTFE resin particles were immersed for 10 seconds, followed by rinsing successively in THF/methanol and pure THF.

RESULTS AND DISCUSSION

1. Fusion Behavior of PTFE Resins

The fusion endotherms of the four virgin resins contained from one to three peaks. Similar PTFE fusion behavior has been found by Bassett[11] and Suwa.[15] All of the resins studied show evidence of superheating. Extrapolation of the (highest) peak melting temperatures to zero heating rate gives apparent equilibrium melting points of $\sim$344 to $\sim$335°C for the various resins.

A more precise study of the effect of time and temperature on fusion of the virgin resin morphology was carried out through isothermal melting (annealing) experiments. Temperatures were 327, 335 and 338°C. After holding samples at these temperatures for various times, the samples were cooled rapidly (80°C/min.) to room temperature. Complete fusion cycles were then run at 10°C/min.

All four resins retain their original melting endotherm after annealing at 327°C up to 30 minutes. Upon annealing at 335°C, the original endotherms split into two distinct peaks. As shown for D-3 resin, in Figure 1, a lower melting peak appears at $\sim$327°C, the melting point of once melted PTFE. The magnitude of this peak increases with time at 335°C. The highest original melting peak ($\sim$341°C) is retained but decreases in magnitude with time. The rapid conversion of the mass fraction melting in the range 335-340°C

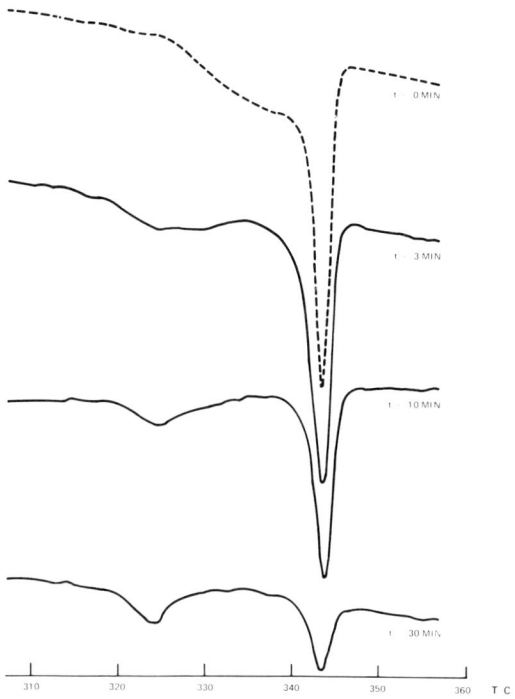

Fig. 1. Fusion endotherms after annealing for indicated time.

into the once melted form and the slow conversion of the (superheating) fraction melting at ∼341°C are consistent with the presence of two levels of order in the resin particles. With the exception of D-1, similar comments apply to the other resins in this study. D-1 resin is the only resin exhibiting a single peak in the initial fusion endotherm. Thus, D-1 resin particles contain a single, slightly superheatable structure.

Stability of the superheatable morphology decreases as follows: D-3>D-2>I-1>D-1. The variation is evident in Figure 2, where the retained fraction of the original peak height is plotted as a function of annealing time. Similar effects are observed at the 338°C annealing temperature with the melting rate approximately doubled. Consistent with Bassett's work[11], a significant fraction (>50% for D-2 and D-3) of the original morphology is superheatable with an equilibrium melting point between 327 and 335°C. More precise values for the equilibrium melting points of the various

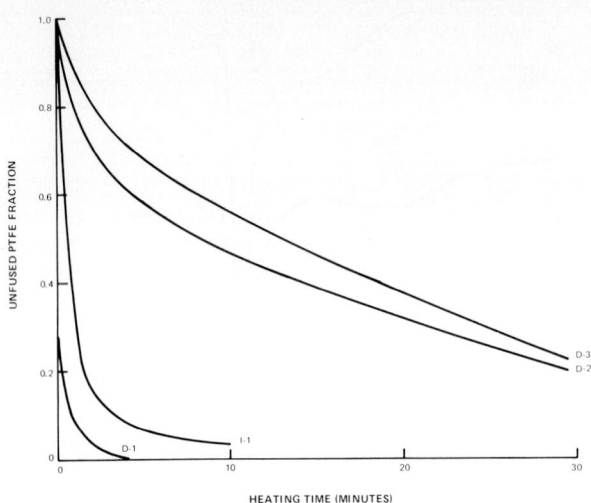

Fig. 2. Fraction of initial fusion peak height remaining as a function of annealing time.

resin morphologies require additional study.

Heating rate data do not provide an accurate indication of the slow virgin resin melting rates, particularly in the case of the higher molecular weight D-3 resin. Consequently, equilibrium melting points derived from heating rate data are too high.

The correlation between melting rate and molecular weight of the various resins, evident in Figure 2, is consistent with earlier observations on PTFE superheating and molecular weight.[15,16] Based upon recent investigations[15,17] of PTFE resin morphology, all resin particle morphologies under study consist of extended chain PTFE crystals in the shape of folded ribbons. These ribbons are composed of PTFE chains extended along the long axis of the ribbon. Suwa, et al.[15] have proposed that the lower melting peak(s) of the virgin resin fusion endotherms are produced by the less perfect ribbon folds. The wetting experiments discussed in the next section indicate an independent, less stable component of the resin particle morphology is a more likely source of the rapid melting resin fraction. The relation between melting rate and molecular weight may not be universal, since variations in resin morphology may be introduced by other factors, i.e., polymerization reactor conditions.

At temperatures above 338°C, the rate of melting as measured by changes in fusion endotherms is rapid for all the resins under investigation. However, the previously noted evidence of order retention in the melt is an indication that the factors responsible for the slow melting behavior remain active at these higher temperatures.

2. PTFE Wetting

The HOPG basal plane substrate provides an extremely smooth, low energy surface for the initial PTFE wetting studies. An SEM micrograph of unsintered resin particles on an HOPG substrate is presented in Figure 3. No substrate wetting or particle coalescence is evident. A similar granular appearance of the particle surface has been observed previously.[18] Whether this is due to nucleation of the Pt coating or a manifestation of surface structure is uncertain.

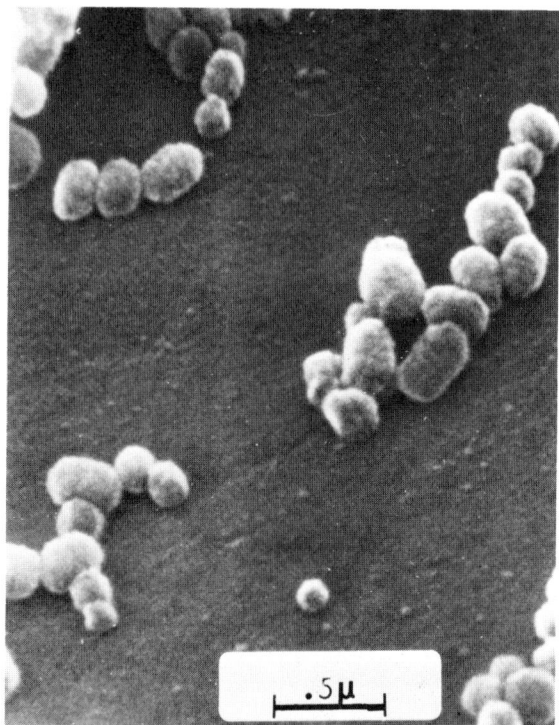

Fig. 3. PTFE dispersion resin particles on HOPG substrate.

The complex wetting behavior of PTFE is evident in the Figure 4 SEM micrograph of D-1 heated to 341°C for 100 minutes. The virgin resin superheatable morphology is eliminated in <1 min. for all resins at this temperature. Three distinct features are present. First, a fraction of the original particle resists flow and retains its shape. Second, a fraction of flowing PTFE assumes a conformation, resembling a spreading drop with a finite contact angle (<30°C), around the remains of the original particle, and finally, thin film flow with a zero contact angle occurs over distances of >1µ away from the particles. Film thickness is >200Å. A variety of treatment conditions were investigated to determine the sequence in which these features occurred.

In Figures 5 and 6, resin D-3 was heated on HOPG for 50 and 100 minutes, respectively, at 341°C. After 50 minutes a well developed "spreading drop" component is evident. Little, if any thin film flow can be seen (note the exposed surface debris).

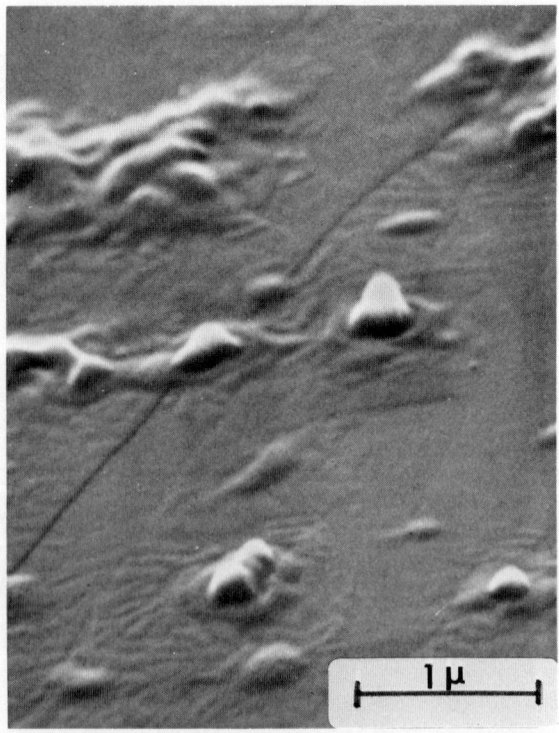

Fig. 4. D-1 resin on HOPG substrate after 100 minutes at 341°C.

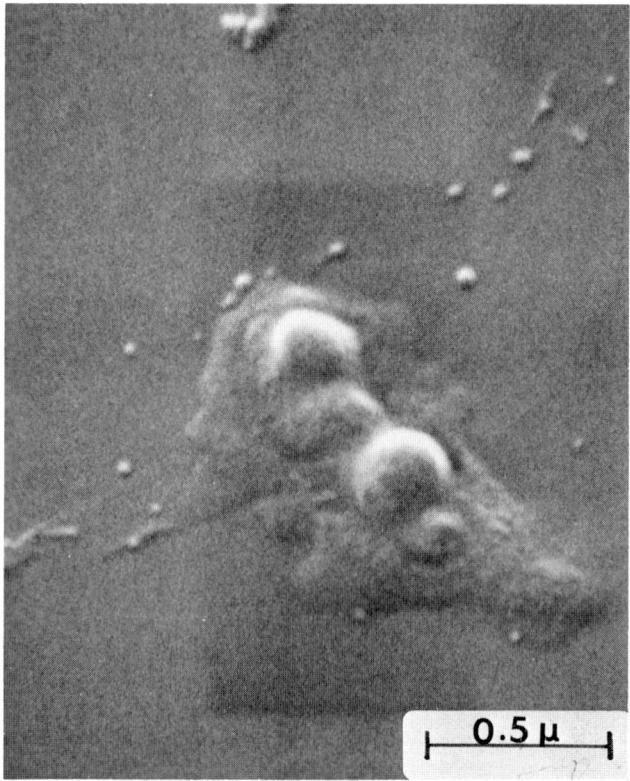

Fig. 5. D-3 resin on HOPG substrate after 50 minutes at 341°C.

After 100 minutes, resin flow onto the surface is more extensive but the original particle shape is retained. The flow anisotropy observed here is a characteristic of PTFE substrate wetting. To a first approximation, the residual resin particle structure has not contributed to the additional flow seen in Figure 6. Rather, the resin mass in the "spreading drop" of Figure 5 has continued to spread. This emphasizes the two component nature of the particle morphology.

The decrease in the magnitude of thin film flow with reduced resin fusion rate is evident at temperatures even further into the PTFE melt region. D-1 and D-3 substrate wetting after 10 minutes at 355°C is shown in Figures 7 and 8, respectively. The less pronounced ripple topography on the substrate and the high contact angles retained by several D-3 resin particles are evidence of a more limited spreading by this resin system. In both resin

Fig. 6. D-3 resin on HOPG substrate after 100 minutes at 341°C.

systems, the original particle shape is maintained, and at the same time significant resin flow occurs.

The poor SEM visibility of the thin film flow occurring during PTFE substrate wetting necessitated the development of techniques to improve observation of these films. SEM resolution is limited to surface features >200Å in size. Structural detail developed by the selective removal of non-crystalline PTFE during Na-napthalene etching was utilized to enhance the visibility of thin resin films and to provide information on resin morphology. The contrast between resin and substrate in SEM micrographs was also increased by etching. The improved resin visibility after etching is obvious in the before-and-after sequence of Figure 9. The absence of surface detail after similar treatment of unheated samples indicates artifacts are not introduced by the etching process.

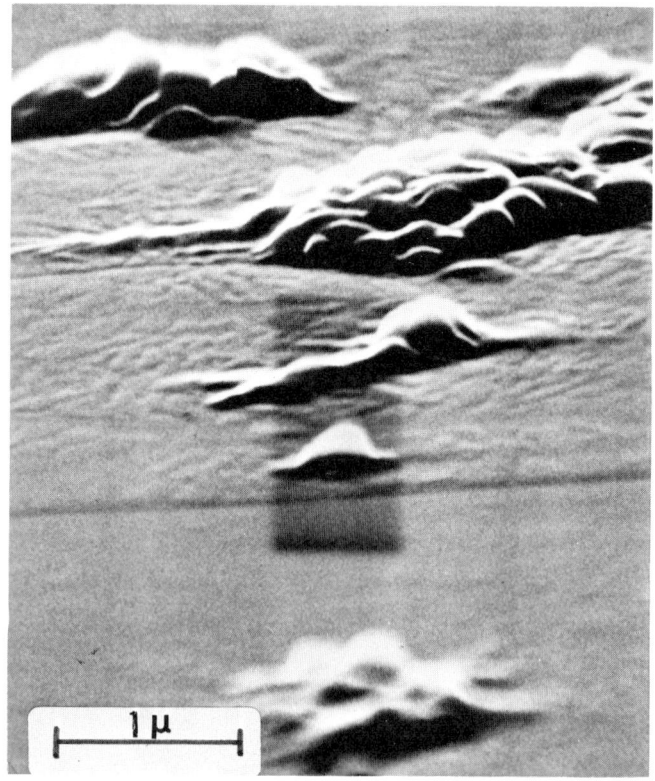

Fig. 7. D-1 resin on HOPG substrate after 10 minutes at 355°C.

In Figure 10, the extreme mobility but limited volume of the D-3 spreading component becomes more evident after etching following a 355°C, 10 min. treatment. In contrast, in Figure 11, etching illustrates the greater volume of I-1 resin spreading after similar thermal treatment. Almost total surface coverage by the rapid melting I-1 resin held at 341°C for 10 minutes is observed in Figure 12. The contrasting film thickness and surface structural detail in Figures 11 and 12 are likely due to a temperature related variation in the total volume of spreading resin.

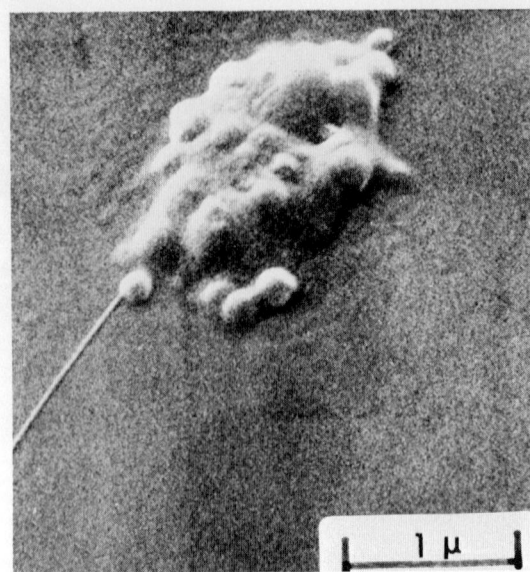

Fig. 8. D-3 resin on HOPG substrate after 10 minutes at 355°C.

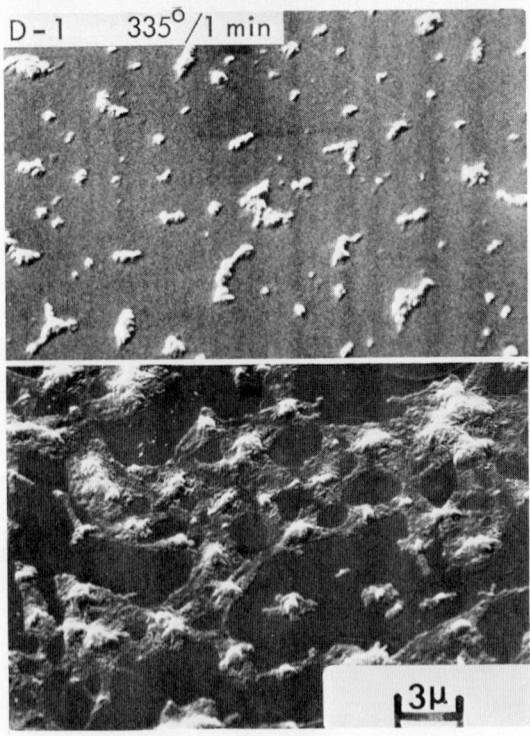

Fig. 9. The effect of etching on resin visibility.
Upper: Before etching; Lower: After etching.

Poly (tetrafluoroethylene) Wetting Behavior 301

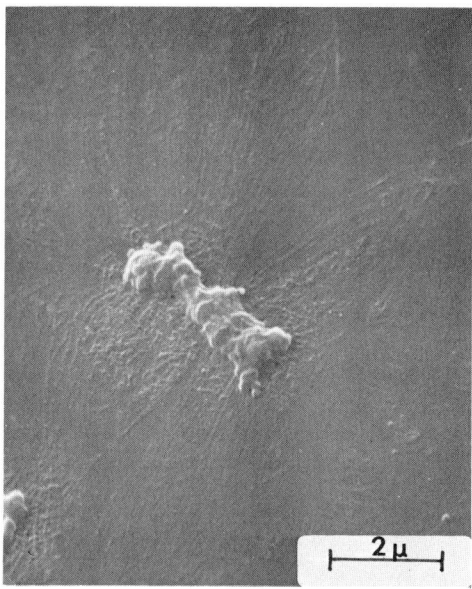

Fig. 10. D-3 resin on HOPG substrate etched after 10 minutes at 355°C.

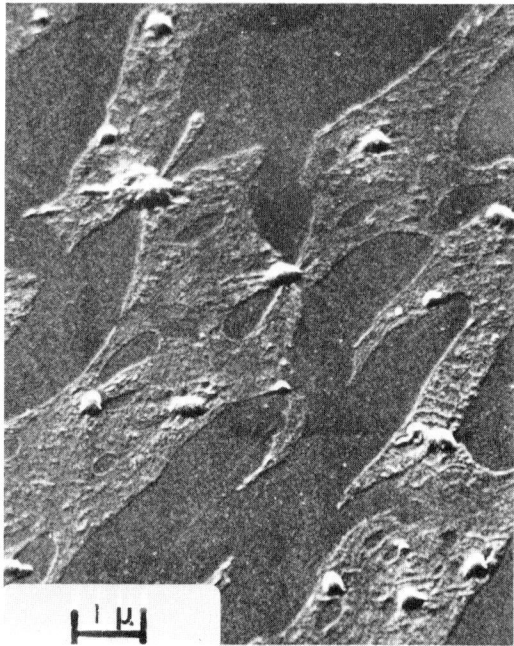

Fig. 11. I-1 resin on HOPG substrate etched after 10 minutes at 355°C.

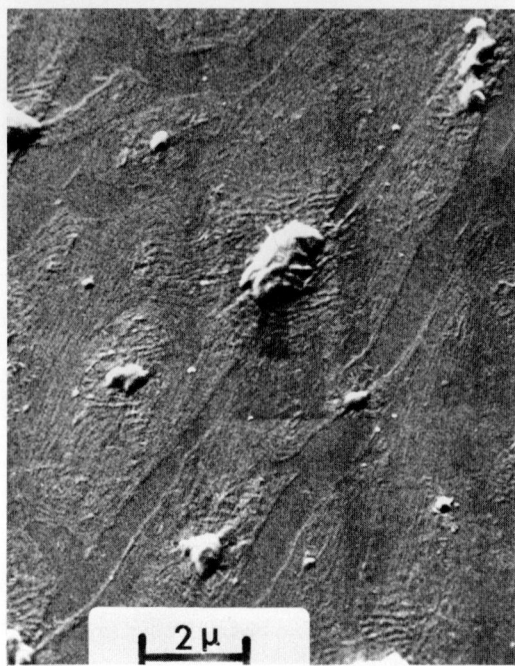

Fig. 12. I-1 resin on HOPG substrate etched after 10 minutes at 341°C.

The mobility of the spreading resin film, as well as the relation between fusion and wetting rates, is readily apparent upon examination of etched D-1 and D-3 resins in Figures 13 (D-1) and 14 (D-3). Both resin samples were heated to 335°C for one minute, a condition where limited melting of D-3 and extensive fusion of D-1 would be expected (see Figure 2). The original D-3 resin particle shape is retained with little, if any, reduction in the initial contact angle. Concurrently, a thin resin film has spread over distances of >2μ. In contrast, substantial bulk flow and contact angle reduction of D-1 resin has occurred.

The anisotropy of the film spreading and the resemblance of crystallite orientation on the HOPG surface were strongly suggestive of epitaxy (see Figures 4, 7, 10 and 12). Similar interactions between graphite basal planes and a variety of polymers have been reported by others.[19,20,21] Various thermal histories were employed in the cooling phase to increase the degree of crystallinity of the PTFE and enhance epitaxy. Both resin molecular weight and cooling rate affect the degree of epitaxial perfection, as shown in Figures 15 and 16. Figures 17 and 18 show two views of I-1 resin with the same heating history as shown in Figure

Poly (tetrafluoroethylene) Wetting Behavior 303

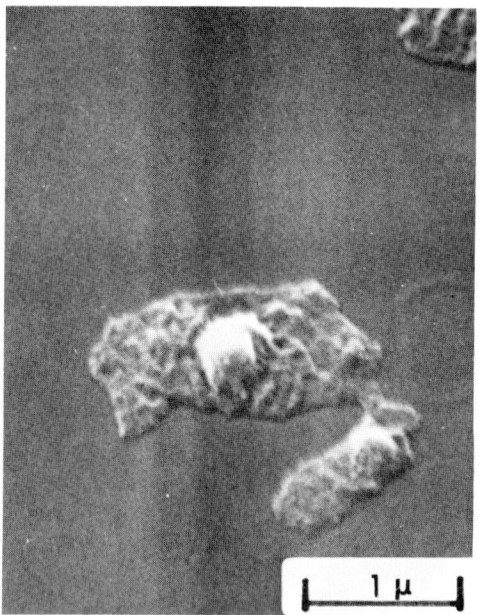

Fig. 13. D-1 resin on HOPG substrate etched after 1 minute at 335°C.

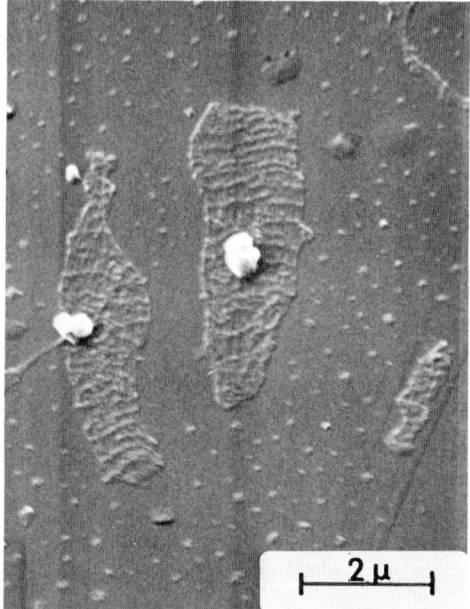

Fig. 14. D-3 resin on HOPG substrate etched after 1 minute at 335°C.

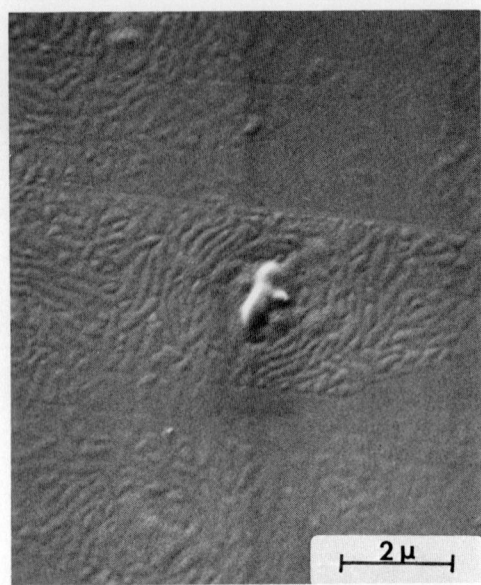

Fig. 15. D-1 resin on HOPG substrate annealed after 1 minute at 341°C.

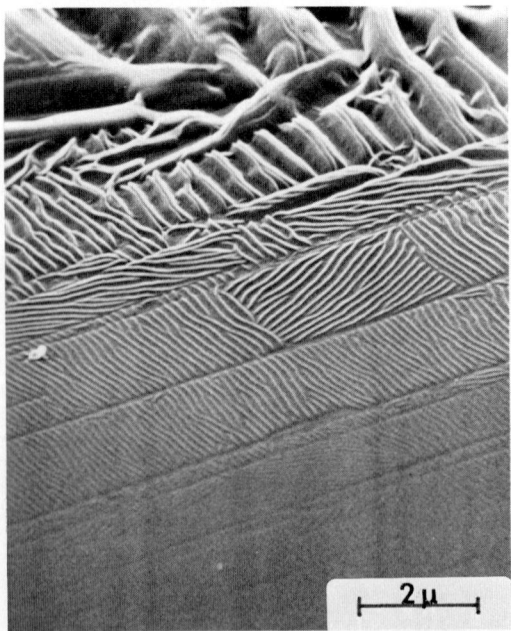

Fig. 16. Low molecular weight PTFE resin annealed after 10 minutes at 341°C.

Poly (tetrafluoroethylene) Wetting Behavior 305

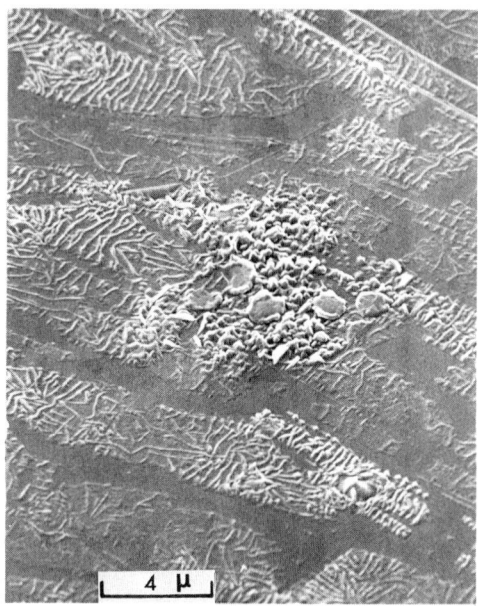

Fig. 17. I-1 resin on HOPG substrate annealed and etched after 10 minutes at 341°C.

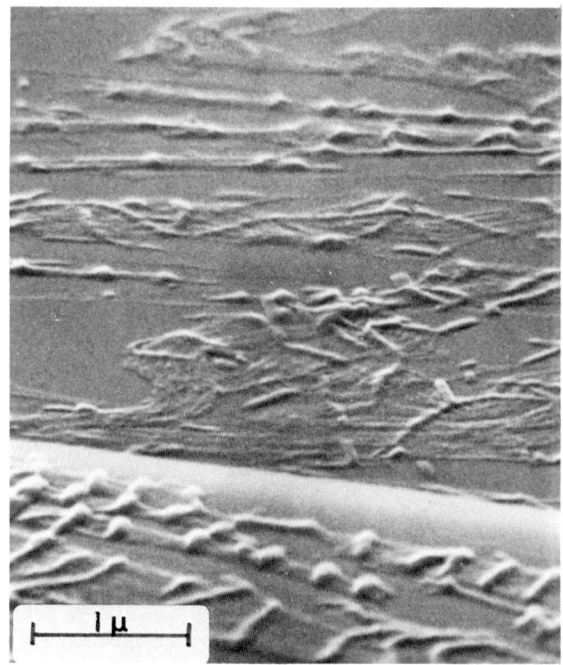

Fig. 18. I-1 resin, higher magnification of Figure 17.

12, but with slow cooling instead of quenching.

PTFE wetting of substrates other than HOPG does not show film structural development. Resins heated on Pt foil wet the substrate and show evidence of both the "advancing foot" and "spreading drop" behavior. The extent of spreading (if any) as a thin film is difficult to ascertain. A unique feature of the wetting of Pt is shown in Figure 19. After heating I-1 resin for 10 minutes at 341°C, "worm-like" structures are distributed over the surface. Similar results were obtained with the other resins.

Examination of a heating time sequence shows that the "worms" are a fundamental part of the resin spreading mechanism on Pt. In Figures 20-22, "worms" are found extending from the etched resin particles, increasing in length to $\sim$1-2μ, after which they break apart and become arrayed on the Pt surface. The contact angles between resin particles or "worms" and the substrate are large. The maintenance of the structures over such a long time period indicates the "worms" do not wet Pt in the classical sense. The crystallinities of these structures must be quite high to survive the etch treatment. There also must be a substrate controlled potential energy map which governs their motion because large areas of preferred "worm" orientation can be found, as shown in Figure 23. Although substrate wetting is apparent in Figure 19,

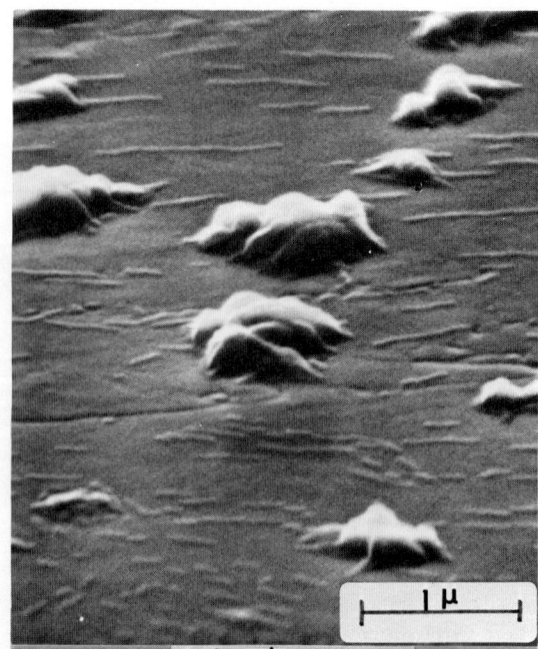

Fig. 19. I-1 resin on Pt substrate after 10 minutes at 341°C.

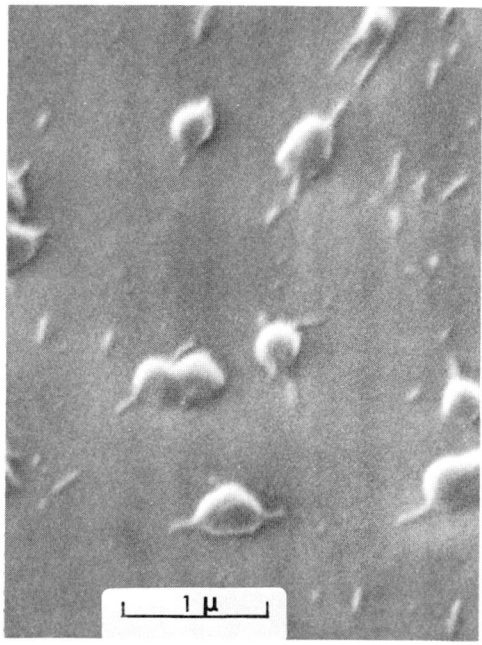

Fig. 20. I-1 resin on Pt substrate etched after 1 minute at 341°C.

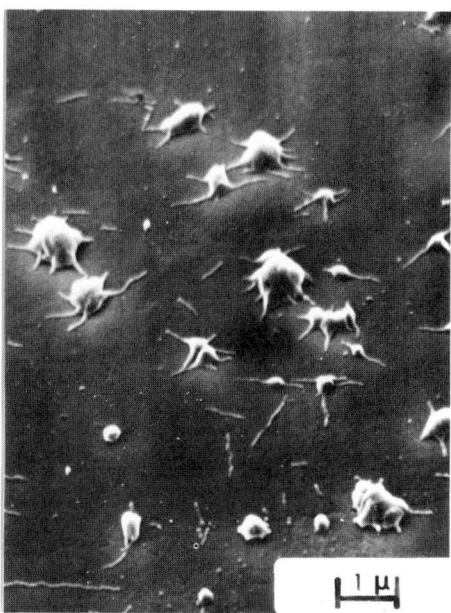

Fig. 21. I-1 resin on Pt substrate etched after 10 minutes at 341°C.

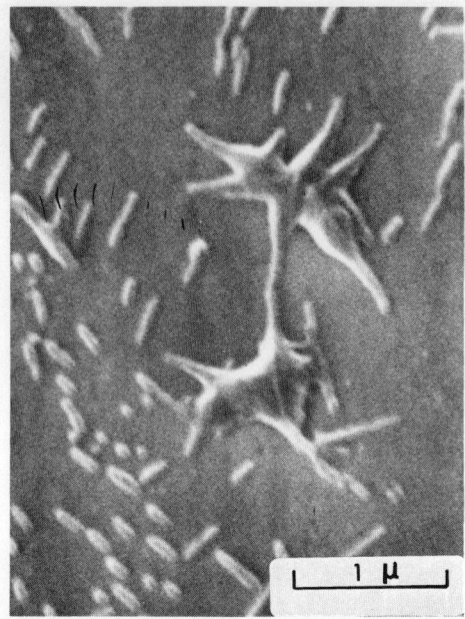

Fig. 22. I-1 resin on Pt substrate etched after 100 minutes at 341°C.

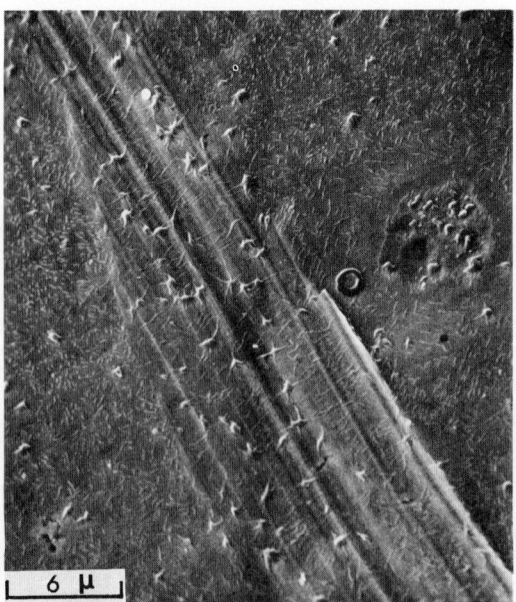

Fig. 23. I-1 resin, same conditions as Figure 22. Note the aligned PTFE structure in the scratch.

etching and/or annealing procedures provide no evidence of thin film flow onto the Pt surface.

The mechanism of "worm" formation is unknown. Under shear, PTFE dispersion resin particles develop fibrils. It is possible that the "worm" structures are similar to these fibrils and are related to the extended chain ribbons which comprise the resin particles.

PTFE wetting behavior on mica and evaporated carbon films was quite simple and unspectacular. Wetting occurred with the development of an advancing foot and the retention of the original particle shape. Neither etching nor annealing developed surface structure attributable to a spreading resin film.

Combining the fusion and wetting data, the most consistent features of the preceding SEM micrographs are the retention of the original particle shape long after the original fusion endotherm has been eliminated. Such behavior is a manifestation of the previously noted order retention in the melt. Additional evidence of melt order is the insensitivity of the retained original particle volume fraction to heat treatment time and temperature.

Resins with the least stable morphology, such as I-1 and D-1, develop the quasi-spreading drop flow most rapidly. On the other hand, the resin with the most stable morphology, D-3, has a component which still participates in thin film flow. I-1 resin also has more extensive film flow than D-1 resin.

The common factor linking resins exhibiting rapid "spreading drop" flow is the limited stability of the superheatable morphology At the same time, resins with multiple peaks in their virgin fusion endotherms (D-2, D-3, I-1) exhibit more extensive thin film flow than the single peaked D-1 resin. Formation of the spreading resin film, particularly for D-3 resin, occurs within the same time scale as the disappearance of the lower melting peaks on the virgin resin fusion endotherm. Contrary to Suwa, et al.,[15] this would indicate that an independent component of the resin morphology rather than less perfect crystals located in the (extended chain) ribbon folds, was responsible for the lower melting peak(s).

The high mobility of the thin film flow may be a consequence of the potentially strong interaction between the PTFE chains and the HOPG surface. Mechanisms of the film formation include surface diffusion similar to that reported for organic liquids.[22] Data on surfaces which do not have a potentially strong interaction with the resin chains are ambiguous and additional data are required for clarification.

Even though the exact mechanisms of PTFE resin sintering and wetting are unresolved, it is clear that sufficient resin flow does occur on a microscopic scale for coalescence of resin particles. The rapid initial resin spreading and subsequent immobility of the resin fraction retained in the original particle mirrors the VE behavior of PTFE observed by Lontz.[9] If, after long sintering times, the order retention represented by the original particle persists, bulk resin flow will be restricted. These ordered regions would act to limit the bulk flow which Lontz was observing during his sintering experiment. At the same time, flow over distances <1μ occurs readily, resulting in coalescence of the original resin particles.

The complexity of the PTFE flow behavior prohibits the direct application of the usual wetting equations. However, an approximate value for the viscosity of the flowing resin can be obtained by ignoring the original particle region and considering only the spreading fronts. The assumption of isotropic flow is also necessary. As noted earlier, droplet spreading is a function of the ratio γ/η and time. Compared with the γ/η ratios of polyethylene and polymethyl methacrylate ($\sim 10^{-5}$), the ratio for PTFE ($\sim 2 \times 10^{-10}$, using bulk PTFE data) is too small. Wetting rates in all three cases are not sufficiently dissimilar to account for the 10^5 x difference in ratio. Such a difference in the ratio can only be corrected if the microscopic viscosity of flowing PTFE is significantly below the bulk value of $\sim 10^{11}$ poise.

CONCLUSIONS

1. Earlier evidence of an ordered component in the PTFE morphology that is metastable above an equilibrium melting point in the range 327-335°C is confirmed. The amount and stability of this component varied directly with the resin molecular weight. The presence of such a structure is consistent with previous reports of order retention in the melt over extended time periods.

2. PTFE flow on the microscopic level during wetting of HOPG was extensive. Classical spreading behavior was not observed due to the retention of order above the resin melting point. Three distinct types of flow behavior were observed. A mobile component flowed out into a thin film initially. Retention of order in the melt was manifested by the continuing presence of the original particle conformation at long heating times. This structure was surrounded by a flowing phase resembling the classical spreading drop. Evidence of increased flow in both thin film and "spreading drop" modes with decreasing morphological stability supports the hypothesis that wetting behavior is controlled by the morphological disordering kinetics.

3. An independent component of the virgin resin morphology was responsible for the lower melting peaks on the virgin resin fusion endotherms and was primarily responsible for the initial rapid thin film flow on HOPG.

4. PTFE epitaxially crystallized on HOPG. Similar behavior was not observed on Pt or mica.

5. PTFE resin flow onto Pt substrates was both extensive and unique. "Worm-like" structures moved onto the substrate from the resin particles, break off and array themselves on the substrate.

6. Mica, Pt and evaporated carbon surfaces were wet by PTFE. Annealing and/or etching procedures produced no evidence of the extensive thin film flow found on HOPG substrates.

7. Na-napthalene etching enhanced the observation of thin PTFE films spreading on HOPG substrates.

8. On the individual resin particle level, coalescence of PTFE resin was complete during sintering. The limited coalescence reported by Lontz is a result of significant differences between microscopic and bulk PTFE flow behavior.

ACKNOWLEDGEMENTS

The authors wish to thank Drs. W. Vogel, L. Christner and K. Klinedinst for many helpful discussions concerning their work on PTFE wetting phenomena. The contributions of Mr. K. Gumz (SEM data) and Mr. I. Mittleman (DSC data) are also gratefully acknowledged.

REFERENCES

1. H. Schonhorn, H.L. Frisch and T.K. Kwei, J. Appl. Physics, $\underline{37}$, 4967 (1966).
2. T.K. Kwei, H. Schonhorn and H.L. Frisch, J. Colloid Interface Sci., $\underline{28}$, 543 (1968).
3. R.H. Dettre and R.E. Johnson, Jr., J. Adhesion, $\underline{2}$, 61 (1970).
4. W.W.Y. Lau and C.M. Burns, J. Polym. Sci., Poly. Phys., $\underline{12}$, 431 (1974).
5. S. Newman, J. Colloid Interface Sci., $\underline{29}$, 174 (1969).
6. T.P. Yin, J. Phys. Chem., $\underline{73}$, 2413 (1969).
7. W. Radigan, H. Ghiradella, H.L. Frisch, H. Schonhorn and T.K. Kwei, J. Colloid Interface Sci., $\underline{49}$, 174 (1969).
8. G. Steiner, Ph.D. Thesis, Lehigh University (1969).
9. J.F. Lontz in *Fundamental Phenomena in Materials Sciences*, Vol. 1, ed. by L.J. Bonis and H.H. Hausner, Plenum, New York (1964).

10. S. Sherratt in *Encyclopedia of Chemical Technology*, Wiley, New York, 9, 805 (1966).
11. D.C. Bassett and R. Davitt, Polymer, 15, 721 (1974).
12. I.A. Dolova, A.V. Kiselev and Ya. I. Yashin, Zh. Strukt. Khim., 13, 162 (1972).
13. T. Suwa, M. Takehisa and S. Machi, J. Appl. Polym. Sci., 17, 3253 (1973).
14. A.A. Benderly, J. Appl. Polym. Sci., 6, 221 (1962).
15. T. Suwa, T. Seguchi, M. Takehisa and S. Machi, J. Polym. Sci., Polym. Phys., 13, 2183 (1975).
16. E. Hellmuth, B. Wunderlich and J.M. Rankin, Jr., Appl. Polym. Symp., 2, 101 (1966).
17. F.J. Rahl, M.A. Evanco, R.J. Fredricks and A.C. Reimschuessel, J. Polym. Sci., A-2, 10, 1337 (1972).
18. N.K.J. Symons, J. Polym. Sci., A, 1, 2843 (1963).
19. F. Tuinstra and E. Baer, Polymer Letters, 8, 861 (1970).
20. P. Bless, J. Semen and J.B. Lando, J. Appl. Polym. Sci., 19, 141 (1975).
21. J.L. Kardos, J. Adhesion, 5, 119 (1973).
22. W.D. Bascom, R.L. Cottington and C.R. Singleterry, Adv. Chem. Series, 43, 355 (1964).

Fluoropolymer Surface Studies. II.

David W. Dwight

Department of Chemistry
Virginia Polytechnic Institute and State University
Blacksburg, Virginia 24061

Fluoropolymers are known for their unique surface properties, but it is not generally recognized that these properties can vary appreciably, depending upon the specifics of preparation. This report highlights the surface characterization of films (1) cast from aqueous PTFE dispersion, (2) skived from a sintered PTFE billet, and (3) sprayed, using a "Teflon" FEP/ epoxy enamel. ESCA revealed changes in the fluorocarbon/ hydrocarbon ratio with variations in process conditions, and SEM showed effects on roughness. The receding contact angle was most sensitive to changes in high surface-energy fraction. A remarkable, new surface - long fibers composed of ∿5000 parallel PTFE molecules - was obtained by annealing the PTFE samples.

INTRODUCTION

Combinations of x-ray photoelectron spectroscopy (ESCA), contact-angle hysteresis and scanning electron microscopy (SEM) provide detailed analysis of the chemical and physical nature of solid surfaces. The fundamentals of these techniques, and their application to sodium-etched and gold-crystallized fluoropolymers were the subjects of our first report[1]. Briefly, all three techniques analyze only the outer tens of Angstrom units of the sample; ESCA provides elemental analysis, distinguishing between fluorocarbon and other types of carbon by distinct chemical shift, SEM provides a view of the surface at high magnification and depth-of-field, and contact angles reflect both chemical and physical effects. Correlations emerge when two or three techniques are used on the same sample, allowing a detailed perception of surface structures and processes.

Other reports of surface measurements on fluoropolymers, especially contact-angle data, are numerous, but the earlier results scatter significantly. The situation was clarified by Allan and Roberts[2] in a study of the wettability of "Teflon" PTFE and FEP resins, and the effects of several methods of preparation designed to vary roughness of the polymer surfaces. Similar results showing a relationship between roughness, stretching and anisotropic wettability, have been reported in later work[3,4] and SEM was used to determine the microtopography. Johnson and Dettre[5] show correlations between theory and experiment for the effects of both roughness and hydrocarbon surface fraction on contact-angle hysteresis of some fluorocarbons. ESCA data on fluoropolymers reported by Clark et al.[6,7] and Ginnard and Riggs[8], indicate the possibility of quantitative surface analysis. In a recent symposium, there were several reports of the use of ESCA and contact angles to characterize fluorocarbon films deposited by glow discharge or plasma polymerization[9,10]. Our earlier work[1] demonstrated that characterization of some treated "Teflon" surfaces by the combination of ESCA, SEM, and contact-angle hysteresis helped elucidate the behavior of these systems. This report completes our characterization of the chemical and physical nature of the most common fluoropolymer surfaces.

The most common fluoropolymer film is "Teflon" FEP (Type A), i.e. poly-(tetrafluoroethylene/hexafluoropropylene) extruded from the melt. This copolymer is also blended with thermosetting resins and solvents to give a hard, durable enamel with a lubricated, release surface. Poly-(tetrafluoroethylene) homopolymer cannot be extruded, thus films are formed either by skiving a billet of sintered PTFE molding powder or sintering a film of aqueous PTFE dispersion. A variety of chemical and physical structures arise from these diverse materials and processes.

EXPERIMENTAL

Materials and Procedures

Films of aqueous PTFE dispersion (DuPont "Teflon" T-30, stabilized with 6% surfactant) were prepared on fiber glass cloth, using a continuous process shown schematically in Figure 1. Two pairs of samples were studied, representing constant conditions in all but one process variable. For one pair of samples, an increase in dispersion viscosity was used to obtain twice the weight pick-up of dispersion per pass through the dip tank. Thus, the same weight of PTFE was applied in four passes on one sample and eight passes on the other. For the second pair of samples, eight passes were used (thinner coats), but the temperature of the sintering oven was lowered stepwise on the seventh pass until the coated fabric would no longer retain a uniform film of dispersion after the subsequent pass through the dip tank. (This is termed "poor re-wetting" in commercial practice.)

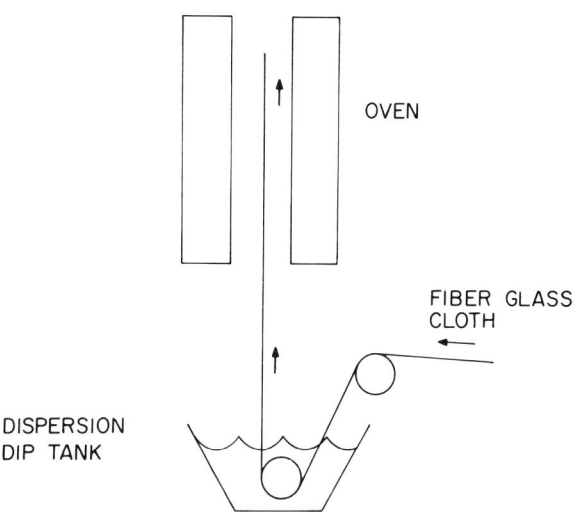

Fig. 1. Schematic diagram of the dip-coating process used to produce "Teflon"-coated fiber glass cloth by drying and sintering a film of aqueous PTFE dispersion.

The pair of samples was taken after the seventh pass through the oven, one above and one below the temperature at which the onset of poor rewetting occurred.

Also PTFE films were prepared by skiving (shaving with a sharp blade in a lathe) from a billet of free-sintered molding powder, a procedure known to give anisotropic and variable wettability[2]. A new surface on both skived and dispersion-cast films appeared after annealing by the following procedure (ASTM D-1457-69): samples were placed in a preheated sintering oven at 300°C, then heated at 2°/min to 380°C, maintained at that temperature for 30 min, and then cooled at 1°/min to 295°C. The temperature was maintained at 295°C for 25 min. after which the samples were removed from the oven and allowed to cool to room temperature.

One form of poly-(tetrafluoroethylene/hexafluoropropylene) film was prepared from a blend of epoxy and "Teflon" FEP resins, known to give a surface with fluoropolymer properties[11]. One set of samples utilized standard 954-line "Teflon-S" enamel (DuPont, Fabrics & Finishes Dep't), which was reduced approximately 20% by volume with methyl ethyl ketone and filtered through cheese cloth into a standard air-spray paint gun. Using 40 psi air pressure at the gun, the enamel was sprayed for a few seconds onto aluminum foil panels from a distance of about three feet, resulting in a dry film thickness of 0.5 to 1.0 mil. The panels were allowed to air dry for at least ten minutes. Identical panels were heated for 15 min. in an air oven at 350°, 450° or 500°F. A second set of samples was prepared in the same way, except a proprietary form of poly-(tetrafluoroethylene/hexafluoropropylene) having low-molecular weight and functional species extracted, was used.

Surface Analysis

Experimental details have been described for collecting the ESCA data and for water contact angles by the goniometer technique[1]. Scanning electron micrographs were obtained at 25 kV on a Jeolco JSM-3, and observation of small surface features was enhanced by tilting the sample at approximately 45° to the electron beam. To provide conductivity for fluoropolymer samples, a coating ($\sim 200 \mathring{A}$) of gold-palladium was applied in a vacuum evaporator.

The goniometer technique was used to obtain water and hexadecane contact angles on the PTFE and "Teflon-S" samples. The alkane was obtained from Burdick and Jackson Laboratories and percolated through silica (Fisher, 28-200 mesh) and alumina (Woelm, Neutral Grade). Its surface tension was 27.2 ± 0.2 dynes/cm at 25°C. Precision was reduced by the anisotropic roughness of the skived films: while liquid was being added, the drop would often move easily to one side and not at all to the other, or move in jumps. Therefore we also used the wetting balance technique described by Johnson and Dettre[5]. Film samples were fastened around a cylindrical sample holder, giving rigidity to the portion of the film protruding below the sample holder, and a beaker containing water or hexadecane was moved automatically, advancing and retracting the liquid at 0.1 inches per minute. During immersion the contact angle builds up to a steady state value, Θ_a, with a corresponding decrease in force. When Θ reaches Θ_a, the force-vs-depth curve becomes a straight line with a slope due to buoyancy. Extrapolation of the linear portion of the buoyancy slope to zero depth-of-immersion allows calculation of the advancing contact angle. Similarly, the receding contact angle was calculated from the buoyancy slope obtained while the film was being withdrawn from the probe liquid.

The reason for low precision of contact-angle data on rough surfaces is obvious from inspection of Figure 2, which compares plots of force vs depth-of-immersion for skived "Teflon" film parallel and perpendicular to the skiving direction. When the liquid moves perpendicular to the skive marks, the plot shows a pronounced saw-tooth pattern, corresponding to the "jumping forward" of the drop front observed during the goniometer measurements.

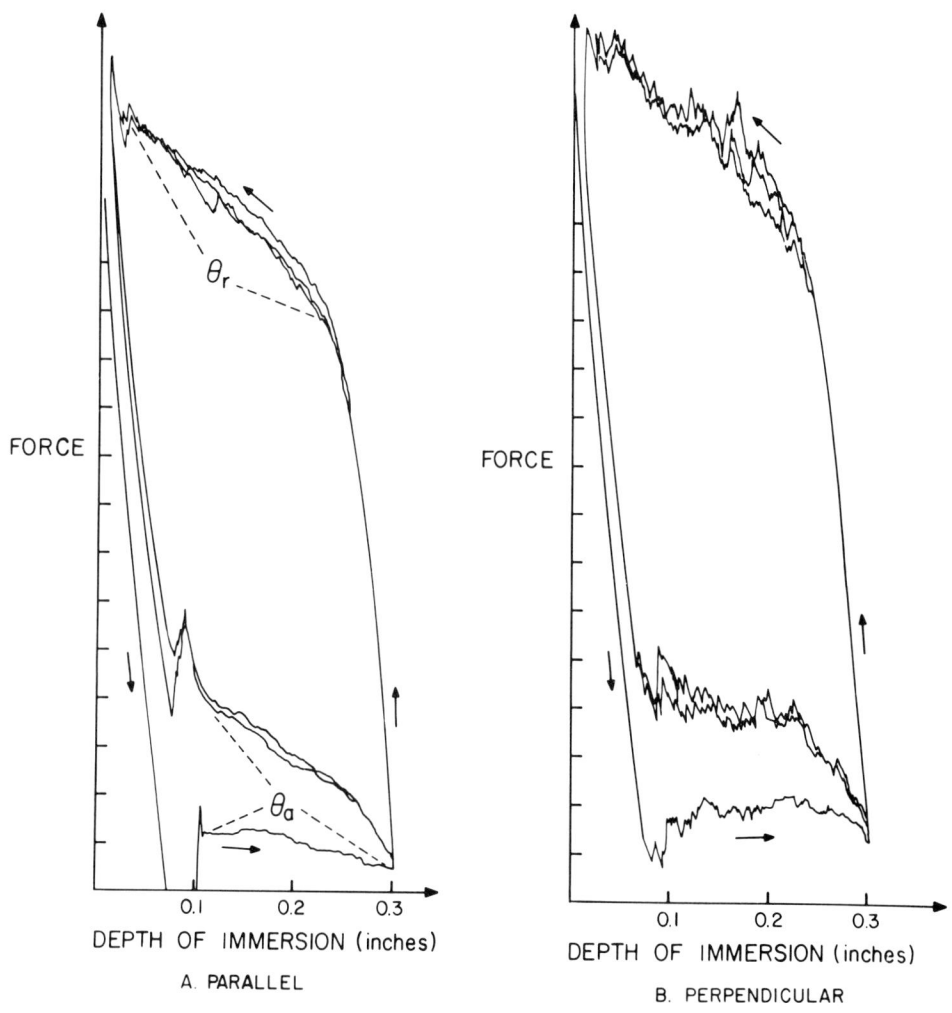

Fig. 2. Recorder traces of force (relative) vs depth-of-immersion obtained from the wetting balance with skived "Teflon". A. Water advanced parallel to skive marks. B. Water advanced perpendicular to skive marks.

These results corroborate the work of Johnson and Dettre[12], who used a model in which rugosities are energy barriers that must be surmounted by the moving liquid front.

Another feature of the wetting balance traces in Figure 2 deserves comment: The advancing buoyancy line on the first immersion shows a lower slope, while subsequent immersions have a slope roughly parallel to the receding slope. These results suggest that some of the test liquid is left behind on the fluoropolymer surface after immersion. If enough time elapses before re-immersion, the original, lower slope is obtained again. Advancing angles were calculated on the first immersion.

RESULTS AND DISCUSSION

Poly-(tetrafluoroethylene)

<u>Dispersion-Coated Fiber Glass Cloth</u>. Analysis by ESCA, SEM and contact-angle hysteresis gave essentially identical results from both the sample prepared with thicker coats per pass through the PTFE dispersion and the sample sintered at the lower oven temperature. Likewise, the other pair of samples (thinner coats or higher oven temperature) showed similar surface characteristics. The results indicate that the surfactant in the PTFE dispersion must diffuse to the surface of the film deposited on the fiber glass cloth and then volatilize or pyrolyze. Apparently, thicker coatings and lower oven temperatures do not allow the complete removal of surfactant residues.

The evidence for these conclusions is presented in Figures 3 and 4. Only a fluorocarbon peak at ∿291 eV appears in the C_{1s} ESCA spectra (Figure 3) when either higher temperatures or thinner coats are used in the coating process. On the other hand, with lower temperatures or thicker coats, a significant hydrocarbon peak appears at ∿284 eV, and the fluorocarbon peak is diminished. Also, the ESCA spectra of the latter samples showed an O_{1s} peak at ∿532 eV, which was absent in the former pair of samples. The hydrocarbon and oxygen peaks must derive from a residue of the surfactant used to stabilize the aqueous PTFE dispersion against settling. Unfortunately, a more detailed analysis of the structure of this hydrocarbon fraction is impossible because the sample is too small (a fraction of a surface layer <50Å thick) for any of the routine techniques.

Typical scanning electron micrographs representative of the pair of samples (higher temperature or thinner coats) that gave only a fluorocarbon peak in the C_{1s} ESCA spectra, are shown in Figure 4A (top). The surface appears as a uniform, dense packing of ridges about 0.1 by 0.5 microns in size. The unusually high molecular weight and crystallinity of poly-(tetrafluoroethylene)

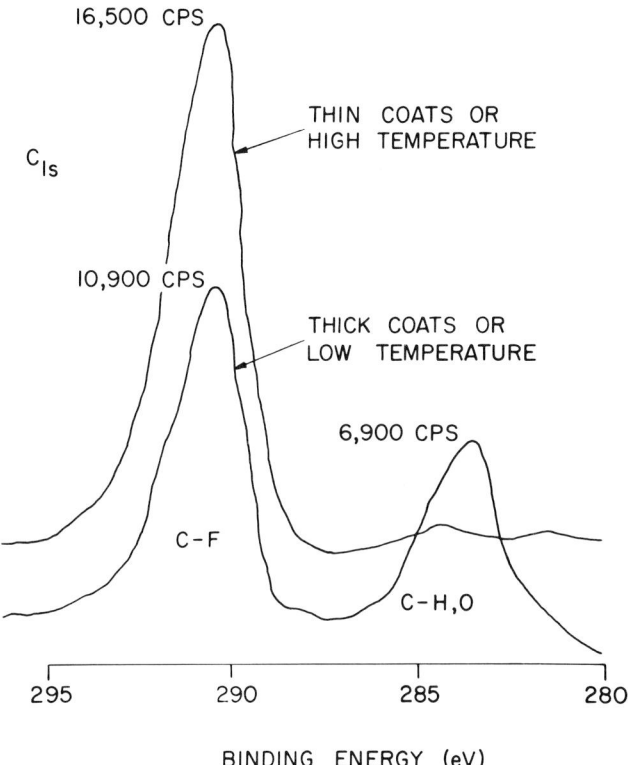

Fig. 3. Carbon 1s ESCA spectra from typical pairs of PTFE dispersion-coated fiber glass cloth samples. Different parameters employed during the coating process are indicated beside the spectra.

gives rise to a "stacked" lamellar morphology responsible for the observed surface structure. Transmission electron micrographs of fracture surface replicas of PTFE crystallized slowly from the melt[13] show a correspondence to our results.

The natural roughness of the PTFE surface lowers the receding contact angle so that the aqueous dispersion will cling as an unbroken film during its traverse through the drying oven. The residual surfactant (detected by ESCA as a hydrocarbon C_{1s} peak

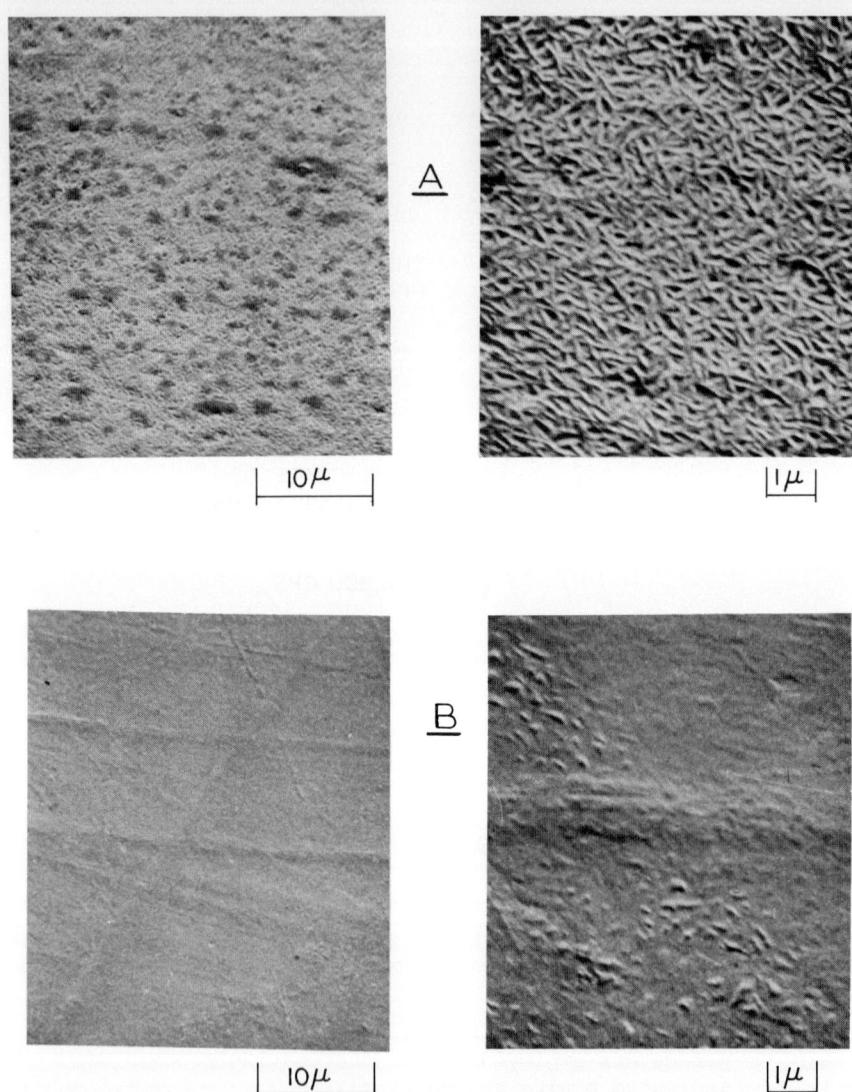

Fig. 4. SEM photomicrographs (2,500X and 10,000X) of the surface of PTFE film cast from aqueous dispersion. A. Ridge-like structure of homogeneous PTFE obtained when thinner coats of dispersion or higher oven temperatures were used during the coating procedure. B. Smooth surface of surfactant residue filling in between the PTFE ridges, representative of 30-50% of the samples for which thicker coats of dispersion or lower oven temperatures were used in the coating procedure.

on the thicker sample and the lower temperature sample) has the
effect of filling in the spaces between the PTFE ridges, as illustrated in Figure 4B. An estimated 30-50% of the surface of those
samples was covered with the relatively smooth areas shown, while
the remainder of the surfaces showed the same "unfilled" ridge
structure seen in Figure 4A. Note that the crude estimate by SEM
of hydrocarbon surface fraction is in agreement with the value of
39% calculated from ESCA peak intensities.

Fluoropolymer samples showing a hydrocarbon surface fraction
would be expected to be more wettable than the homogeneous fluoropolymer surface, and also to exhibit greater hysteresis, by analogy
with sodium-etched fluoropolymer films[1]. However, the contact-angle data in Table 1 show that it is the decrease in roughness
illustrated in the photomicrographs that is the governing factor.
Receding angles increase and advancing angles (and hysteresis)
decrease on the samples with the smooth hydrocarbon fraction.
These observations correspond to the theoretical predictions of
Johnson and Dettre[12] of the effect on contact angles of an energetically (chemically) homogeneous surface. Apparently, the physical
effect of the hydrocarbon fraction is the governing factor, dominating the tendency for its chemical nature to increase hysteresis.
The delicate balance of surface forces in this system may be unique:
a reduction of less than 5 dyne/cm in the surface tension of the
aqueous PTFE dispersion (by the addition of surfactant) facilitates
"re-wetting" by lowering the receding contact angle on the smoother
surface to zero. It should be emphasized that the advancing contact
angle is relatively unimportant in the coating process because the
film is submerged in the liquid, eliminating the requirement for
spontaneous spreading of the dispersion on the film. However, if
the equilibrium receding angle is much greater than zero. there
will be a tendency for the liquid film to rupture and form islands
or beads, i.e. "poor re-wetting" in the practical situation.

Skived Film and Annealing. The surface structures on PTFE
films prepared by skiving a rotating billet are much larger and
anisotropic compared with dispersion-cast films. When either type
of film is annealed, striking new surface features develop. A
comparison between dispersion-cast film before and after annealing
is made in the photomicrographs in Figure 5. As described earlier,
at higher magnification the as-cast film (Figure 5A, top) shows a
ridge-like structure in the 0.2 to 0.8μ range. At the level of
1-3μ, there appears to be a regular pattern of small, rounded
mounds and depressions. The dark striations or mars that appear
on this sample (A) at lower magnification are marks caused by
handling, illustrating how easily PTFE is deformed. After annealing, the surface structure undergoes a remarkable transformation.
The lower magnification view in Figure 5B shows an array of closely
packed bumps about 5μ in size. At higher magnification, the surface
appears to be an intertwined mass of very long fibers, 0.5 to 1.0μ

Table 1

Contact-Angle Hysteresis
on PTFE Dispersion-Coated Fiber Glass Cloth

Sample	Contact Angles (deg) at 24.5°C					
	Water			Hexadecane		
	Θ_a	Θ_r	$\Theta_a - \Theta_r$	Θ_a	Θ_r	$\Theta_a - \Theta_r$
Thin coats or High temperature	117	88	29	48	0	48
Thick coats or Low temperature	113	98	15	42	8	34

in diameter. At regular intervals, groups of fibers bend into loops projecting up from the surface creating "corrugated knobs" or what appear to be bumps at lower magnification. Apparently, the crystallizing forces operate during annealing in such a way as to organize groups of PTFE molecules into parallel alignment, while surface tension forces dictate a range of about 3,000 to 8,000 molecules to minimize the surface free energy of the system. Annealing appears to have increased the surface area considerably, indicating the magnitude of the crystallizing forces driving the system into a configuration that has excess surface energy.

As might be expected, the skiving process creates a rough, anisotropic PTFE surface with pronounced grooves and orientation in the skiving direction, seen clearly in the lower magnification photomicrograph, Figure 6A. On the right-hand side of the figure, the highly oriented nature of the surface is illustrated by the minute ($\sim 0.1\mu$) fibers drawn between two lumps of PTFE that appear to be polymer transferred by the skiving process.

When samples were prepared by annealing skived film, a looped, fiber-like topography appeared (Figure 6B), but the fibers are not quite as thick as seen previously in the dispersion-cast film (Figure 5B). The fibers make a sharper bend and form smaller, denser and more closely spaced knobs on the skived film. At lower magnification, the grosser effects of skiving are still apparent, and the knobs even seem to be aligned along the skiving direction, suggesting nucleation. This would be consistent with the greater number and smaller size of the surface features. At the level of $<20\mu$, the new surface structures completely dominate the effects of skiving.

Fluoropolymer Surface 323

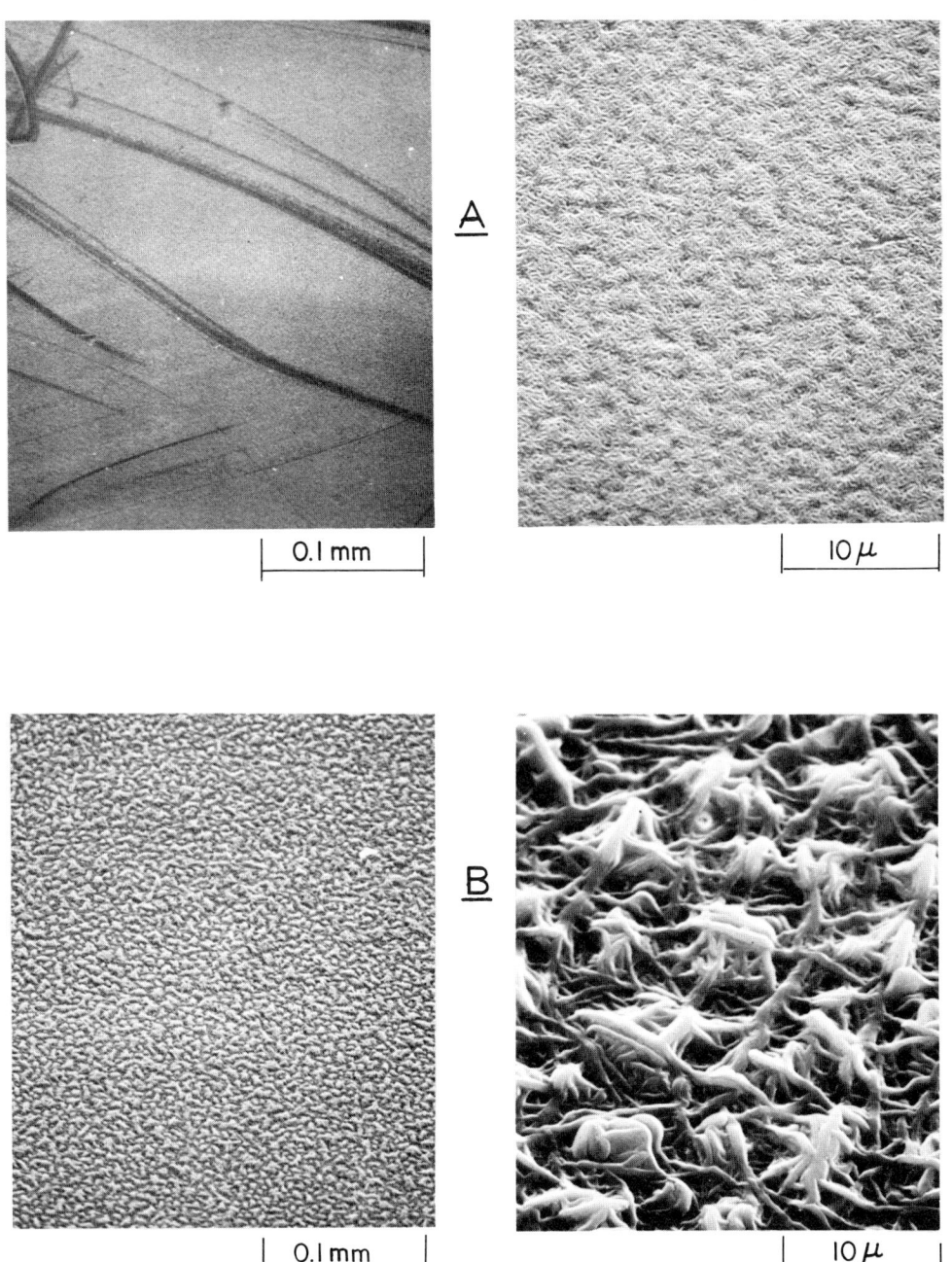

Fig. 5. SEM photomicrographs (300X and 3,000X) of the surface of
a homogeneous PTFE film cast from aqueous dispersion.
A. before, and B. after annealing.

This series of samples provides a unique set of homogeneous (i.e. only fluorocarbon and fluorine peaks in the ESCA spectra) PTFE surfaces of varying roughness. Water and hexadecane contact-angle hysteresis data were obtained on these samples by the goniometer method and are listed in Table 2. Note that these samples represent far more complicated forms of roughness than accounted for by theoretical models[12]. However, comparison of wettability before and after annealing reflects increased roughness: advancing contact angles and hysteresis increase markedly. As probed by water, some samples exhibited high receding contact angles indicative of a composite surface, where the liquid is no longer able to penetrate to the base of the rugosities -- a reasonable result in view of the fiber/knob topography of those samples. The other samples in this group showed the lowest receding angles with water, probably corresponding to the minimum in the theoretical curve of receding angle vs roughness, at the point where the composite surface first forms[5]. The dispersion-cast films after annealing showed an intermediate receding angle. A decrease in surface energy accompanies the introduction of perfluoromethyl side chains on the PTFE backbone, and should increase the advancing contact angle[15]. However, roughness effects predominate, so the smooth, extruded FEP film actually has the lowest advancing water contact angle and hysteresis.

Table 2

Contact-Angle Hysteresis on Dispersion-Coated
and Skived PTFE Films Before and After Annealing

Sample	Contact Angles (deg) at 24.5°C					
	Water			Hexadecane		
	Θ_a	Θ_r	$\Theta_a - \Theta_r$	Θ_a	Θ_r	$\Theta_a - \Theta_r$
Dispersion-coated	117	88	29	48	0	48
Dispersion-coated, after annealing	146	92	54	60	0	60
Skived	112	80	32	39	0	39
Skived, after annealing	148	76 & 120	72 & 28	60	0	60
Extruded FEP film	109	93	16	57	43	14

Fluoropolymer Surface 325

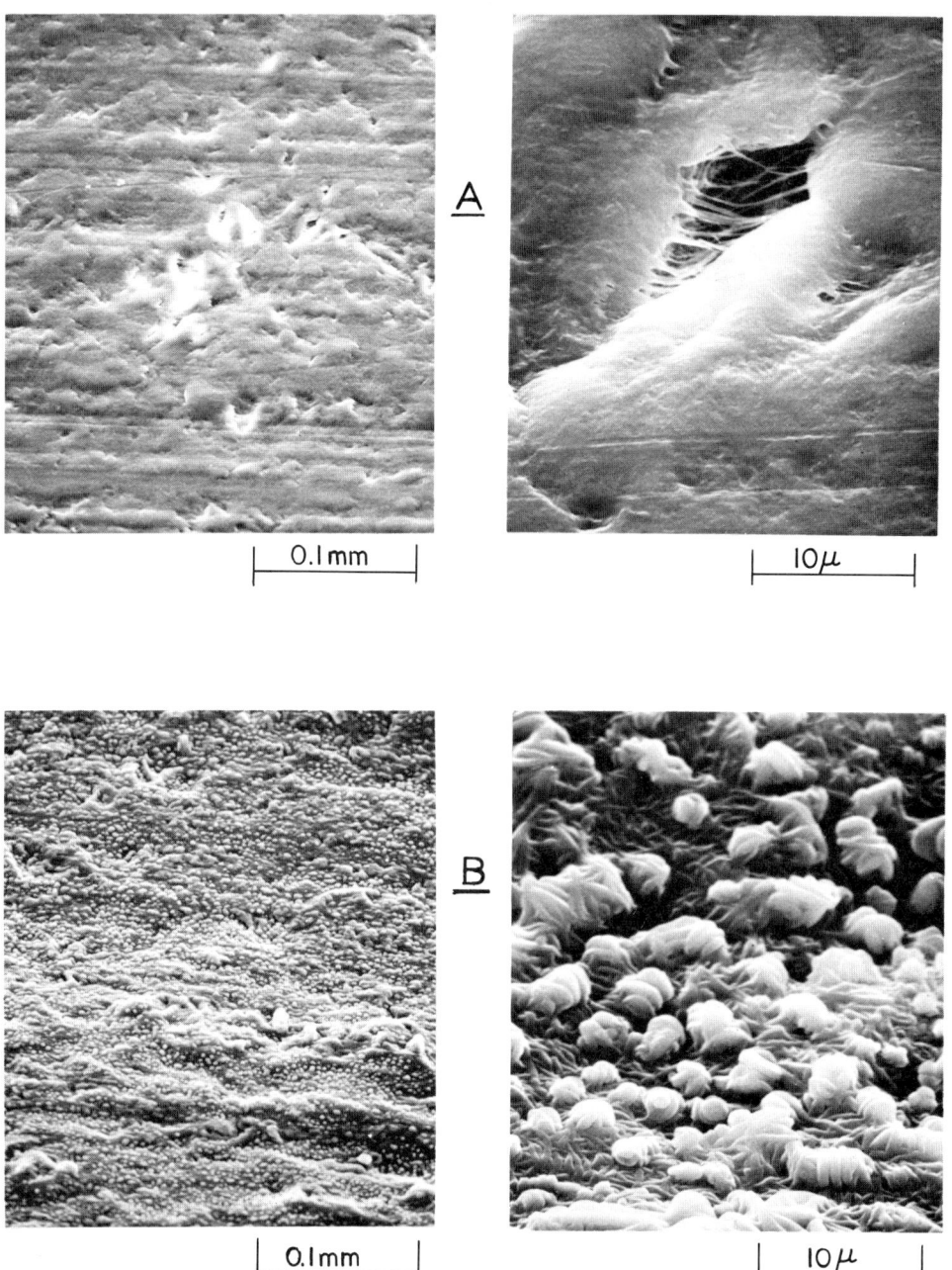

Fig. 6. SEM photomicrographs (300X and 3,000X) of a homogeneous PTFE film skived from a rotating billet. A. before, and B. after annealing.

Poly-(tetrafluoroethylene/hexafluoropropylene)

Our previous report[1] described the most common form (extruded film) of the perfluorinated copolymer "Teflon" FEP. Also this copolymer is blended into a spray enamel with thermosetting resins and solvents. After application to a substrate and baking, "that portion of film at the substrate interface is composed predominately of the auxillary material, while the other surface is either fused or particulate TFE/HFP copolymer[11]". This process is depicted schematically in Figure 7.

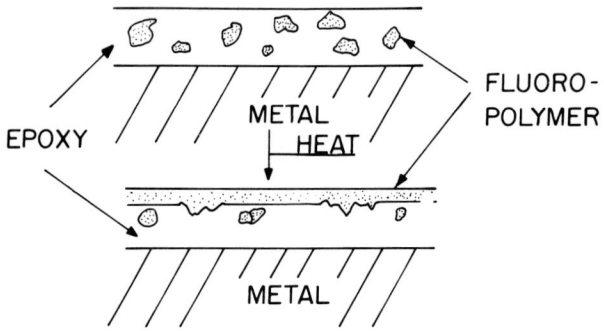

Fig. 7. Schematic diagram of the stratification thought to occur when a "Teflon-S' enamel is heat-cured.

We have obtained more detail on the cure mechanism in this system by ESCA and contact-angle analysis of two enamels using different types of FEP, each cured at three temperatures. To facilitate presentation of the ESCA data, peak intensity values were standardized with Wagner's sensitivity values[14] and listed in Table 3 as ratios to the fluorine concentration, a technique that effectively provides internal standardization, aiding semi-quantitative interpretation of the ratios.

The ratio of fluorocarbon to fluorine increases, suggesting a decreasing proportion of ($-CF_3$) side chains in the surface film as cure-temperature increases. The amine-cured epoxy ratios stay relatively constant at about C/O/N - 1.0/0.25/0.18.

The proportion of components in the surface region can be estimated from the fluorocarbon/hydrocarbon ratio calculated

Table 3

Variation in Atom Ratios
Calculated from ESCA Peak Heights
for "Teflon-S" Enamels Cured 15 min at Three Temperatures

Sample	ESCA Binding Energy (eV) and Atom Ratios					
	Cure Temp., °F	689eV F	293eV C-F	284eV C-H,O	532eV O	400eV N
"Teflon-S"						
Standard	350	1.0	0.34	0.54	0.14	0.10
Extracted		1.0	0.31	4.93	1.16	0.91
Standard	450	1.0	0.40	0.15	0.03	0.02
Extracted		1.0	0.39	0.40	0.09	0.11
Standard	500	1.0	0.43	0.07	0.01	0
Extracted		1.0	0.43	0.15	0.03	0.02
Extruded FEP Film	---	1.0	0.40	----	----	----

from ESCA peak intensities. For clarity, these results are plotted vs cure temperatures in the histogram in Figure 8. Two results are clear: the fluorocarbon surface fraction increases rapidly with cure temperature, and standard FEP contains more surface-film-forming fluorocarbons, especially apparent at lower temperatures.

Further detail on cure mechanism is provided by the trends in the peak widths-at-half-height, listed in Table 4. The fluorocarbon and fluorine peaks are significantly broader (0.4 to 1.1 eV) in the "Teflon-S" systems than in the FEP film, suggesting a wider distribution of fluorinated species in the former case. Peak widths are a maximum at the 450° bake, implying that the broadest distribution of fluorinated species populates the surface region under these conditions. At 500°F, the epoxy resin discolored and showed other signs of thermal-oxidative degradation or burning, and this probably explains the strongly broadening trend of the O_{1s} peak width.

The contact-angle hysteresis measurements (Table 5) allow further conclusions to be drawn about the surface processes during the cure of this blend. By comparison with the results on extruded FEP film, it appears that a complete fluorocarbon surface layer forms only with the standard "Teflon-S" composition after 450° and

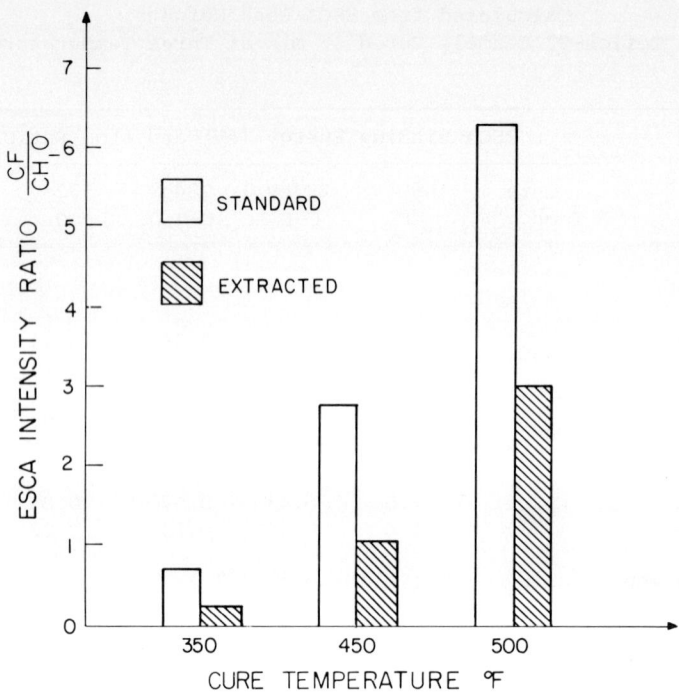

Fig. 8. Histogram illustrating the use of ESCA intensity ratios to follow the formation of fluorocarbon surface films from a blend of fluoropolymer and epoxy resins.

500°F. Although the ESCA data show reasonably strong fluorine and fluorocarbon signals from the "extracted" sample, the contact-angle data indicate that some patches of epoxy resin still occupy the surface even after the 500° cure. This series of experiments provides a good example of the danger of characterizing surfaces by the use of only advancing contact angles. Judging from those data alone would lead to the conclusion that most of the samples were essentially identical and composed entirely of perfluorinated material. Simply measuring the receding contact angles leads to the correct description.

The combined results indicate that this polymer blend acts like a chromatographic system during the cure: As the "Teflon" FEP component stratifies to the air interface, a separation occurs in the distribution of fluorinated species that comprise the

Table 4

ESCA Peak Width at Half-Height (eV)
for "Teflon-S" Enamels Cured 15 min at Three Temperatures

Sample	Cure Temp., °F	F	C-F	C-H,O	O	N
"Teflon-S"						
Standard	350	2.9	2.9	3.6	2.7	2.8
Extracted		2.6	---	3.0	2.5	2.6
Standard	450	3.3	3.2	3.9	3.5	3.2
Extracted		3.6	3.5	3.6	3.3	3.3
Standard	500	3.0	2.6	4.0	5.0	3.1
Extracted		3.0	2.5	3.3	4.8	3.2
Extruded FEP Film	—	2.2	2.2	---	---	---

Table 5

Contact-Angle Hysteresis
for "Teflon-S" Enamels Cured 15 min at Three Temperatures

Sample	Cure Temp, °F	Contact Angles (deg) at 24.5°C			
		Water		Hexadecane	
		Θ_a	Θ_r	Θ_a	Θ_r
"Teflon-S"					
Standard	350	117	68	54	0
Extracted		91	61	10	0
Standard	450	112	92	55	40
Extracted		107	78	48	25
Standard	500	109	93	55	40
Extracted		108	87	50	37
Extruded FEP Film	---	109	93	57	43

fluoropolymer, and the fluorocarbon molecules that are most mobile in the hydrocarbon matrix reach the surface first. Apparently, the ability to diffuse to the surface and form a film is related to low molecular weight and functionality, because the extracted fluoropolymer was much less effective, failing to form a complete surface layer even at temperatures that degrade the epoxy resin.

CONCLUSIONS

This paper presented the highlights of our characterization of the physics and chemistry of some common fluoropolymer film surfaces. A comprehensive view of the current state of the art is supplemented by our other reports[1,16]. Combinations of ESCA, SEM and contact-angle hysteresis data demonstrate exceptional power to elucidate subtle changes in the nature of minute surface layers in these systems. Significant changes (functions of time and temperature during preparation) were observed in the top few Angstrom units of the films. The importance of measuring the receding contact angle for surface characterization was clearly demonstrated; it was most sensitive to the presence of a high surface-energy fraction.

Residual surfactant remains on the surface of PTFE film cast from aqueous dispersion, unless the film is sintered at high temperatures or for long times. The surfactant residue has a leveling effect on surface topography, filling in between ridges that appear on the "uncontaminated" fluoropolymer surface. Annealing dispersion-cast or skived film produces a remarkable surface structure: an intertwined mass of very long, ∼0.5µ-wide fibers that form loops projecting up from the surface in a regular array of knobs. Comparison of contact-angle hysteresis on rough vs smooth fluoropolymer surfaces shows trends that agree qualitatively with the predictions of the Johnson-Dettre theory.

Stratification of fluorocarbons to the air surface occurs during cure of a blend of fluoropolymer and epoxy resins, and there were indications that low molecular weight, functional species are the primary film formers. Contact-angle hysteresis distinguishes between patchy and uniform fluorocarbon surface layers, even though the latter are less than ∼25Å thick (ESCA signal still visible from underlying elements).

ACKNOWLEDGMENTS

Experimental assistance of several co-workers at the DuPont Experimental Station, permission by the company to publish the data, and NASA support (Grant No. NSG-1124) during preparation of the manuscript are all gratefully acknowledged.

REFERENCES

1. D. W. Dwight and W. M. Riggs, J. Colloid and Interface Sci. $\underline{47}$(3), 650 (1974).
2. A. J. G. Allan and R. Roberts, J. Polym. Sci., $\underline{39}$, 1 (1959).
3. R. J. Good, J. A. Kvikstad and W. O. Bailey, J. Colloid and Interface Sci., $\underline{35}$(2), 314 (1971).
4. E. H. Cirlin and D. H. Kaelble, J. Polym. Sci., $\underline{11}$, 785 (1973).
5. R. E. Johnson, Jr. and R. H. Dettre, Advan. Chem. Series No. 43, 112 (1964), published by American Chemical Society.
6. D. T. Clark and D. Kilcast, Nature (Physical Science), $\underline{233}$, 77 (1971).
7. D. T. Clark, W. J. Feast, I. Ritchie, and W. K. R. Musgrave, J. Polymer Sci. (Chemistry), $\underline{12}$, 1049 (1974).
8. C. R. Ginnard and W. M. Riggs, Anal. Chem., $\underline{44}$, 1310 (1972).
9. M. M. Millard and A. E. Pavlath, Polymer Preprints, $\underline{16}$, 84 (1975).
10. D. F. O'Kane and D. W. Rice, Polymer Preprints, $\underline{16}$, 92 (1975).
11. J. C. Fang, (to DuPont) U.S. Patent 3,661,831 (May 9, 1972).
12. R. E. Johnson, Jr. and R. H. Dettre, in *Surface and Colloid Science*, Vol. 2, E. Matijevic, ed., Wiley-Interscience, New York, 1969, p. 121.
13. L. Melillo and B. Wunderlich, Kolloid-Z. u. Z. Polymere $\underline{250}$, 417 (1972).
14. C. D. Wagner, Anal. Chem. $\underline{44}$(6), 1050 (1972).
15. M. K. Bernett and W. A. Zisman, J. Phys. Chem. $\underline{64}$, 1292 (1960).
16. D. W. Dwight, "Surface Analysis and Adhesion in Fluoropolymers", Kendall Award Symposium, New York, April 1976, to be published in J. Colloid and Interface Sci.

Structural Characterization of Poly(N-vinylcarbazole)

C.H. Griffiths

Xerox Webster Research Center
Webster, New York 14580

The structure of poly(N-vinylcarbazole) has been studied in relatively monodisperse fractions covering the molecular weight (M_n) range of $3.7 \times 10^3 - 2.7 \times 10^6$. Folded chain crystalline lamellae were observed when the polymer was heated above Tg (227°). The crystallizability decreased with M_n, however, and was zero in fractions with a mean $M_n < 46,000$. The critical M_n for nucleation at 305°C appeared to be $\sim 10^5$. This behaviour follows that predicted by the nucleation theory of Hoffman and indicates that chain end free energy controls the stability of the chain folded nuclei. Diffraction analysis indicated that the lamellae had a paracrystalline structure with hexagonal symmetry about the chain axis and an interchain spacing of 12.00Å but some disorder down the chain axis. In the absence of sharp reflections due to chain axis periodicity this was calculated from the extrapolated crystalline density to be 2.16Å/monomer unit. This is consistent with an isotactic 3/1 screw axis and a basic trigonal structure with a = 12.00Å and c = 6.47Å.

INTRODUCTION

In recent years there has been considerable interest in the electronic properties of polymeric systems. Besides their established use as passive insulting materials a number of aromatic polymers have shown potential in active charge transport devices. Poly(N-vinylcarbazole) (PVK) has well established photoconductive properties which have led to commercial applications in photocopying devices[1]. A complete understanding of the electronic properties of this material depends on a knowledge of the interactions between the π electron systems of the large carbazole pendant groups. The intra and interchain π system interactions are in turn dependent on chain tacticity and conformation and on the degree of ordering of the chains. There has, therefore, been interest in the structure of PVK and in producing crystalline material in which the active groups are in well-defined positions in a three dimensional lattice.

$$-\left(CH-CH_2\right)_n-$$ Poly N-Vinylcarbazole

(with N-carbazole substituent on CH)

It has not been possible to unequivocally determine stereo-regularity in PVK using direct techniques such as NMR and IR spectroscopy. Crystallizability has, therefore, been used as a measure of this property. Solomon et al.[2] reported greater than 80% isotactic PVK using Ziegler-Natta type catalysts, however, Heller, Tieszen and Parkinson[3] were unable to obtain any evidence for crystallinity. Natta et al.[4] report synthesizing crystallizable PVK using cationic $(C_2H_5)_2AlCl$ type catalysts and Kimura and co-workers[5] obtained crystallizable PVK using radical, cationic, and Ziegler-Natta catalysts. The bulk of the evidence therefore indicates no significant influence of synthetic procedure on the crystallizability of PVK. It seems probable that the size of the carbazole group controls the geometry of the addition of monomer units to the growing chain. This leaves molecular weight to account for the differences in the observed properties of the material synthesized.

Molecular packing in crystalline PVK was first discussed by Kimura[5] who reported that drawn and thermally treated material exhibited a single crystalline diffraction peak at $2\theta = 8°15'$ or 10.7Å. This was interpreted at the $(10\bar{1}0)$ reflection of a pseudohexagonal lattice with rigid hexagonally packed chains. The chains were said to be composed of stereoblock arrangements of isotactic 3/1 and syndiotactic 2/1 helices. Crystal[6] derived a hexagonal structure similar to the isotactic chain with the 3/1 screw axis proposed by Kimura. The unit cell parameters, a = 12.30Å and c = 7.44Å, were calculated from electron diffraction data from solution grown crystals. The c axis dimension is much larger than that normally found in isotactic vinyl polymers (∿ 6.5Å) and the calculated density of 0.99g/cc is much lower than even the published value for amorphous PVK (1.2g/cc).[7] Such a structure would result in an 18° distortion in the tetrahedral C-C bond of the chain backbone.

In light of the contradictions concerning the properties of PVK and the unusual nature of the reported structure we have attempted to gain further insights into the molecular packing and the influence of molecular weight on this.

EXPERIMENTAL

Narrow molecular weight distribution samples of PVK were obtained over a molecular weight (M_n) range of 2.7×10^6 to 3.7×10^3 (chains of 14,000 to 19 monomer units) by fractional precipitation with methanol from benzene solution. The starting polymer in the high molecular weight range (M_n 2.7×10^6 to 0.5×10^6) was the commercially available BASF Luvican M170 PVK. The medium-range polymer (M_n 200×10^3 to 10×10^3) was synthesized from N-vinylcarbazole monomer with azobisisobutyronitrile (AIBN) as a free radical initiator. In the low molecular weight range (M_n from 10,000 to 3,000) the vinyl monomer was polymerized with t-butylperoxypivalate (TBPP) in n-butyraldehyde as described by Stolka.[8]

Prior to polymerization the N-vinylcarbazole monomer (BASF) was first refluxed in hexane solution over activated charcoal, filtered hot, then recrystallized from absolute methanol. N-vinylcarbazole was obtained as a pure white crystalline solid with a melting point of 64-66°C. The polymerizations were carried out under nitrogen at 45°C in the case of n-butyraldehyde solution (TBPP initiator) and under reflux in the case of the benzene solution (AIBN initiator). In both cases the polymer was first purified and then fractionated by precipitation with absolute methanol from a benzene solution. After multiple fractionations the final monodisperse fractions were dried at 40°C under vacuum for 12 hr.

The molecular weight and number average for each fraction were determined by gel permeation chromatography in tetrahydrofuran solvent using a Perkin Elmer 1210 liquid chromatograph. Glass-transition temperatures and overall thermal behavior were measured at 20°C/min. using a DuPont 900 thermal analyzer fitted with a differential scanning calorimeter (DSC) cell. The nuclear magnetic resonance spectra of each fraction was measured in deuterated chloroform solution using a JEOL 600 NMR spectrometer.

Films of PVK were prepared on evaporated carbon (200Å) on fused quartz, aluminum, Mylar and silver chloride substrates by coating a chloroform solution in a nitrogen atmosphere. Residual solvent was removed by heating at 100°C under vacuum for 6 hr. Samples of these films were then further heat-treated above and below the glass-transition temperature in a nitrogen atmosphere. The resultant structures were studied directly and as microtome sections and fracture replicas, using electron microscopy and diffraction. X-ray diffraction techniques including diffractometry and the Gandolfi[9] modification of the Debye-Scherrer method (to randomize preferred orientation) were also used to characterize and quantize crystallinity. The Gandolfi[9] camera was evacuated to eliminate air scattering and after development the exposed films were scanned with an optical densitometer to convert film density to x-ray intensities. Infrared spectra of films of each polymer fraction solution coated onto silver chloride were obtained with a Beckman IR-10 spectrophotometer. Densities of solvent cast films were measured in a 1 meter KBr/H_2O density gradient column with a range of 1.100 - 1.200 g/ml.

RESULTS

The PVK fractions isolated from the TBPP and AIBN syntheses and the two fractions from the high and low ends of the molecular weight range of the commercial PVK are shown in Table 1 along with their measured glass transition temperatures.

TABLE 1

PVK Fractions Isolated from Commercial Polymer and Free Radical Initiated Syntheses

M_n	M_w/M_n	T_g, °C	Density, g/cc
3,700 (TBPN)	1.30	152	-
8,900 (TBPN)	1.35	195	-
16,000 (AIBN)	1.39	215	-
26,000 (AIBN)	1.63	219	1.8153 ± 0.0005
46,000 (AIBN)	1.32	222	1.1850 ± 0.0002
83,000 (AIBN)	1.73	224	1.1851 ± 0.0002
157,000 (AIBN)	1.35	225	1.1848 ± 0.0002
500,000 (BASF)	1.20	225	1.1852 ± 0.0002
2.7×10^6 (BASF)	1.25	227	1.1850 ± 0.0002

The measured glass-transition temperature (Tg) shows a monotonic increase with increasing molecular weight until it reaches 227°C at $M_n = 2.7 \times 10^6$. This appears to be essentially the value at infinite molecular weight. The highest molecular weight fractions could be cycled through Tg with no change in the measured value (extrapolated onset) although there were small changes in the shape of the DSC trace. Below $M_n = 10,000$

cycling through Tg resulted in a reduction of the measured value. The addition of 5% vinylcarbazole monomer to the highest molecular weight fraction reduced Tg from 227°C to 170°C. Vinylcarbazole has been reported as the major thermal decomposition product of PVK up to 350°C.[10]

Above the glass transition the PVK thermograms were relatively featureless, regardless of molecular weight until at temperatures near 350°C there was a large exothermic excursion due to the rapid decomposition and volatilization of the decomposition products.

Attempts were made to detect differences in the structure of the PVK as a result of synthetic procedure and chain length. Neither the NMR spectra nor the IR absorption spectra, however, showed any appreciable change with molecular weight and were similar to the published data.[11,12] It was also not possible to detect specific chain end groupings or identify the polymer tacticity from this data.

PVK films with a thickness of 10-15 microns were prepared for X-ray and electron microscope examination and for density measurements. At the lower molecular weights film cracking and flaking tended to occur and, therefore, to increase substrate adhesion, all these films were prepared on roughened aluminum substrates.

The X-ray diffractometer scattering curves from the films as-prepared were found to be very similar over the full range of molecular weights. They consisted of the two major peaks at ca.8° and ca.20.5° shown in Figure 1. Previous work[5,6] indicated that the 8° peak was associated with the interchain distance and degree of chain parallelism. This data was, therefore, analyzed in terms of the exact position of the peak center at its half height (θ_1), the peak width at its half height ($\Delta\theta_1$) and the ratio of the heights of the 8° and 20° peaks.

The ratio of the heights of the 8° and 20° peaks was found to be rather inconsistent and also to depend on the adherence of the film to the substrate. PVK did not adhere to the Mylar substrates and the films were able to relax during drying. On roughened aluminum the films were very adherent and the tension produced by the contraction during solvent evaporation resulted in the alignment of polymer chains. The resultant orientation of the chains in the film plane was evidenced by a doubling in the 8°/20° peak height ratio.

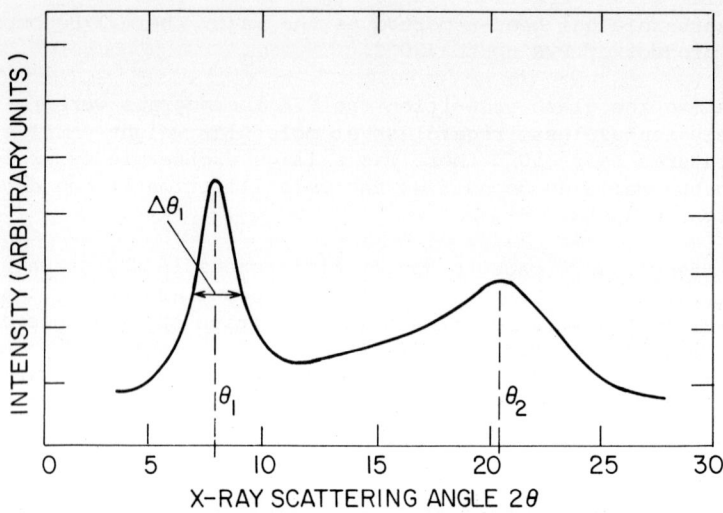

Fig. 1 X-ray scattering from amorphous PVK film.

The value of θ_1 varied from 7.5° at M_n = 3700 to 7.85 at M_n =2.7 x 10^6 and showed a general increase with increasing molecular weight. The value of $\Delta\theta_1$ varied from 3.3° to 2.9° generally decreasing with increasing M_n. In both cases, however, inconsistencies were present due apparently to anisotropic structure which interfered with the averaging of the scattered x-ray beam.

It is possible to relieve stress-induced alignment in polymers by heating above Tg, and heating the PVK films just above Tg of the high molecular weight polymer (at 240°C) did result in changes in x-ray scattering. The value of θ_1 after heating increased slightly at the higher molecular weights and more so at the lower molecular weights to bring all values into the 7.8° and 7.9° range. The value of $\Delta\theta_1$ decreased more at higher molecular weights to give a monotonically increasing $\Delta\theta_1$ from 2.10 to 3.30 with decreasing M_n. Prolonged heating of the highest molecular weight material at 240°C (24 hr.) also resulted in the appearance of a narrow crystalline X-ray scattering peak at 8.39°. This peak did not appear in the lower molecular weight fractions until higher temperatures and the amplitude varied systematically with molecular weight. At heating temperatures from 240-325°C the area under the 8.39° peak was greatest at M_n = 2.7 x 10^6 and decreased with molecular weight until below M_n = 46,000 the films did not crystallize at all. Along with the sharp peak at 8.39°

(10.39Å) there was increased resolution of a diffuse peak at 14° 2θ (6.3Å) but otherwise the scattering was basically unchanged.

The X-ray scattering from the amorphous PVK was not markedly different using the diffractometer or the Gandolfi techniques. In the heated materials giving the sharp 8.39° peak, however, the ratio of this peak intensity to the scattering at 21.5° was much greater in the diffractometer scan than in the Gandolfi pattern. This indicates that the plane giving rise to the 8.39° crystalline reflection was preferentially oriented in the plane of the film. The X-ray scattering curves are shown in Figure 2.

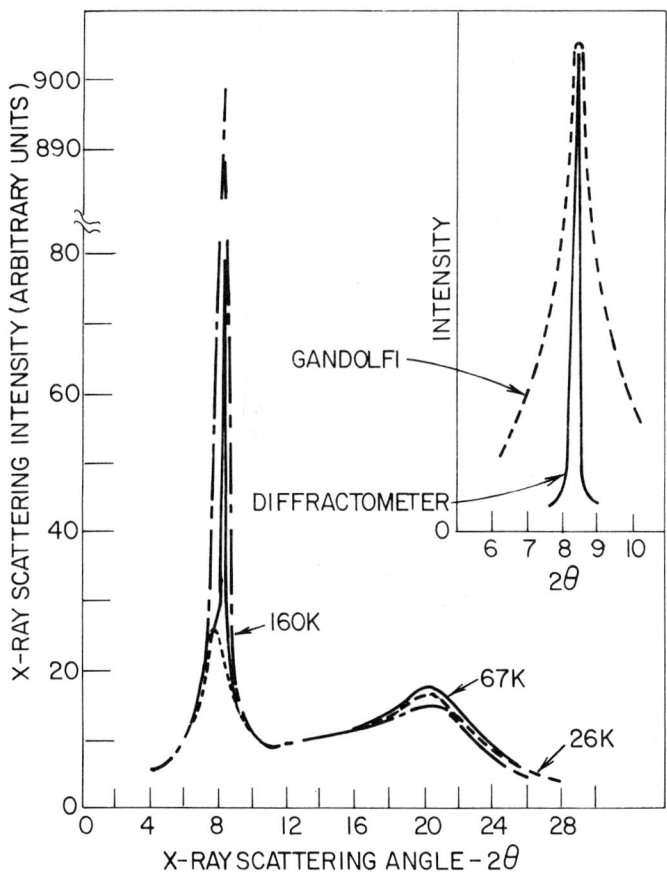

Fig. 2 X-ray scattering from crystallized PVK as a function of molecular weight and lamellae orientation (inset).

The 8.39° x-ray peak was found to be associated in the free standing films with the irregular stacked lamellar type of structure shown in Figure 3 ($M_n = 2.7 \times 10^6$ heated at 325°C for 1 hr).

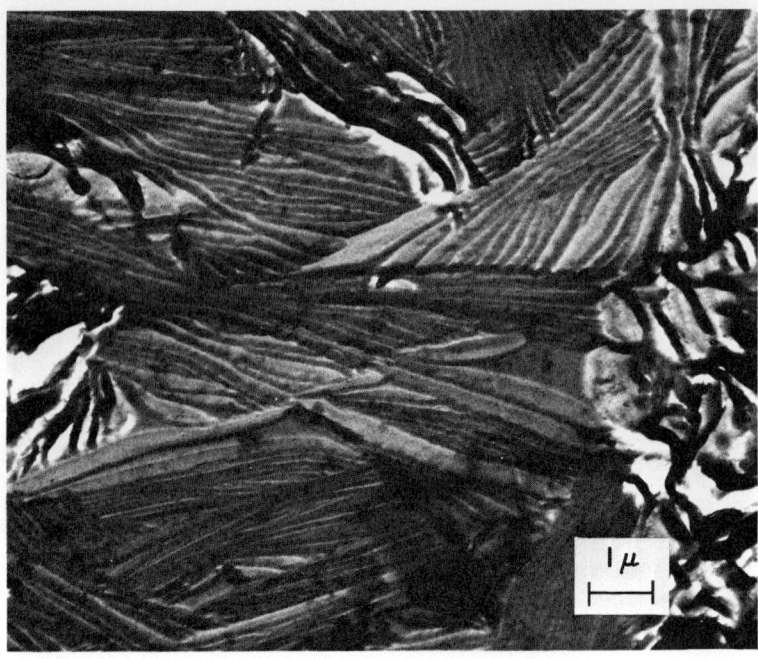

Fig. 3 Lamellar Structure of PVK heated at 325° for 1 hour.

True spherulitic morphology was not observed. Microtome sections taken normal and parallel to the plane of the high molecular weight films on aluminum, however, showed that nucleation had occurred almost entirely at the interface with the substrate and the free surface of the polymer. Further growth took place normal to these surfaces until the resulting lamellae met at the center of the film to give a very orderly structure of parallel, rectangular cross-section, stacked lamellae. The lamellar thickness varied from ca. 300Å after 1 hr. at 270°C to ca. 800Å at 330°C and the rectangular cross-section lamellar width was up to an order of magnitude larger. This type of structure produced at a temperature of 325°C is shown in Figure 4. The rectangular cross-section of the lamellae was clearly visible in microtome sections cut at 90° to that of Figure 4 an example of which is shown in Figure 5.

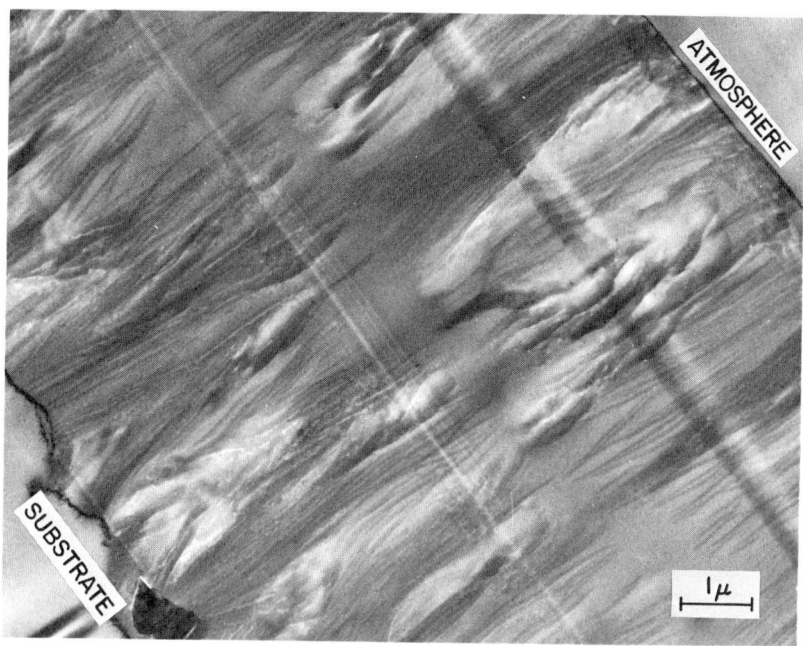

Fig. 4 Microtome section showing nucleation of paracrystalline PVK lamellae at the substrate and atmosphere interfaces of a 10 μ film ($M_n = 2.7 \times 10^6$).

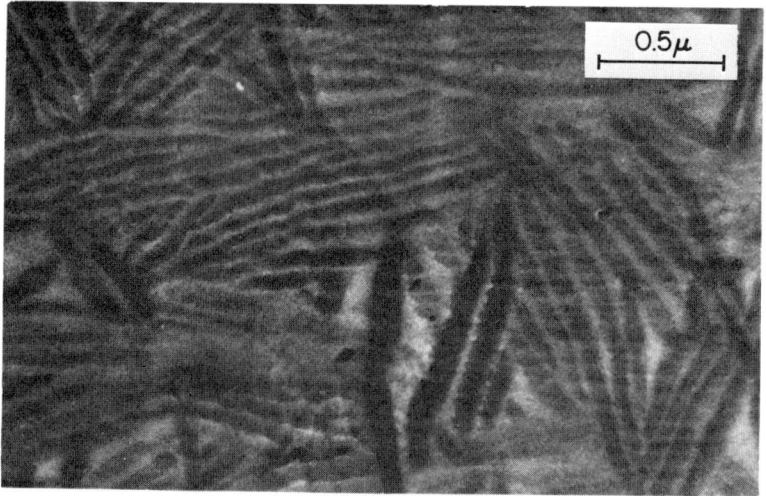

Fig. 5 Microtome section film shown in Figure 4 cut parallel to the substrate.

Note that the plane of the cleavage damage produced by the microtome knife in the lamellae is parallel to the small dimension of the cross-section rectangle.

With decreasing molecular weight the interfaces became noticeably less efficient nucleation sites. At M_n = 157,000 nucleation no longer occurred at the interface with the atmosphere and there was occasional branching of lamellae growing from the substrate interface. The secondary lamellae grew in a direction away from the normal to the interface and presented a barrier to primary lamellae growing close to the branching site. This resulted in areas of noncrystalline polymer being left between the barrier and the atmosphere interface. At M_n = 82,000 nucleation at the substrate was slightly less efficient and branching was much more prevalent. The secondary lamella growth direction was close to 90° to the primary lamella growth direction. This resulted in the termination of the growth of lamellae long before they reached the atmosphere interface and a structure containing a great deal of uncrystallized polymer. The trend of decreasing nucleation efficiency and increasing branching continued down through M_n = 67,000 resulting in more distinct and separated lamellae.

Figure 6 shows the lamellae in some detail. They have apparently a rectangular cross-section normal to the growth direction and this growth direction (long axis of the lamellae) is in, or at a small angle to, the plane of the microtome section. The major axis of the rectangular cross-section is at a variety of angles to the plane of the microtome section resulting in the differences in appearance of the various lamellae.

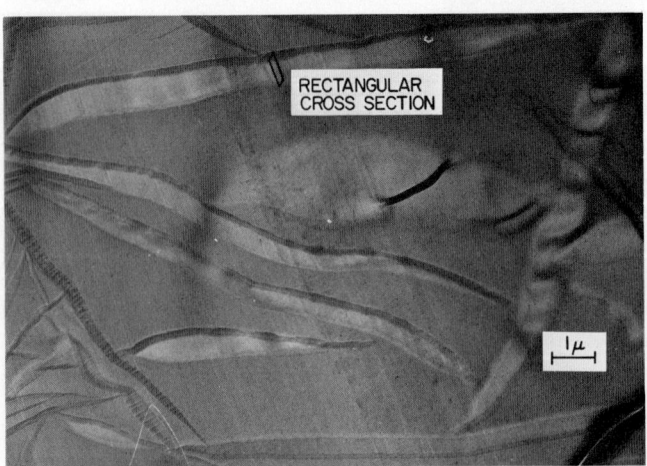

Fig. 6 Microtome section showing isolated PVK lamellae in an amorphous matrix (M_n = 67,000).

It should be emphasized that these are not complete lamellae but sections from lamellae and the lamellar dimensions are truncated where they intersect the top and bottom surfaces of the section. The dark strip running the length of one side of the lamellae images is apparently its area of intersection with the top surface of the microtome section and the adjacent lighter, wider parallel strip is the projection of the lamellar width onto the bottom surface of the section. In many lamellae the area of intersection with the bottom surface of the microtome section is also visible. As would be expected, this width is the same as that of the top surface intersection. The width of the lamellae images is, of course, a minimum when the minor axis of the rectangular cross section is in the plane of the microtome section.

Further reduction in chain length to M_n = 46,000 resulted in a dramatic decrease in nucleation efficiency of the aluminum substrate. Nucleation was very sparse indeed and occasional lamellae were observed to have apparently nucleated away from the surface. It is possible, however, that the link to the substrate interface was present, but was not in the plane of the microtome section. Below M_n = 46,000 no further nucleation of crystalline lamellae was observed and the PVK remained amorphous at all heating temperatures. Attempts to probe the structure of the individual lamellae discussed above by electron diffraction were unsuccessful due to the extreme deformation produced by the microtome blade. This structure was, therefore, studied in 700Å thick films of PVK on carbon. These films were heated at 325°C for 1 hr. and then floated off the quartz substrate and examined directly in the electron microscope.

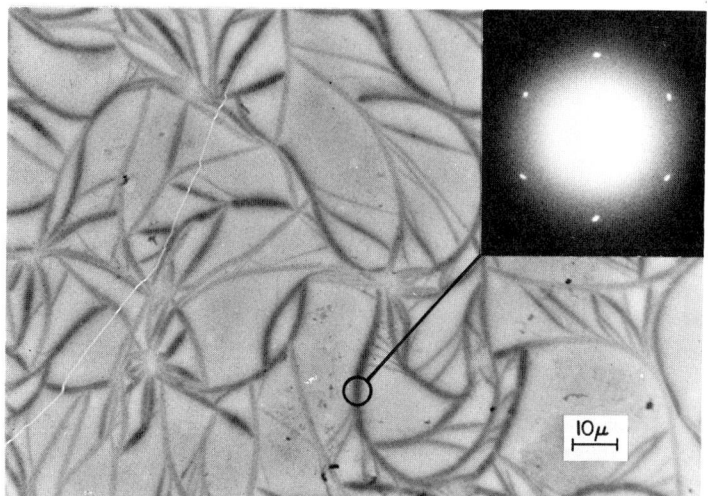

Fig. 7 Thin PVK film (700Å) heated at 325°C for 1 hr.

Crystal growth in the thin films on carbon produced a lamellar branching morphology with the long axis of the lamellae in the plane of the film (Fig. 7). Cracks propagating in the amorphous regions of the film changed direction abruptly at the lamellae and generally traversed the lamellae in what was apparently a cleavage plane at approximately 90° to the long axis. Transmission electron diffraction from these lamellae with the electron beam normal to the film plane gave a single sharp ring at a spacing of 10.4Å. This is similar to that observed in the x-ray diffraction data from thick films. Selected area diffraction from the marked single lamellae resolved this ring into the sharp single crystal spot pattern also shown in Figure 7. It was apparent that these lamellae were crystalline and that the observed diffraction spots were the same reflection shown by Kimura[5] to be due to the interchain spacing. It follows therefore, that the PVK chains were oriented normal to the plane of the carbon support film.

As the proportion of the lamellar phase increased, the intensity of the 8.39° peak also increased and the intensity of the ca. 8° diffuse peak, due to the interchain spacing in the amorphous phase, decreased. Using the method originated by Field,[13] the amorphous fraction was assumed to be proportional to the intensity of the amorphous halo and the crystalline fraction was determined by difference

$$X_c = 1 - X_a$$

The experimental Gandolfi scattering curves (preferred orientation eliminated) were corrected for incoherent scattering and normalized to constant scattering at the 21.5° peak not influenced by crystallinity. The crystalline and amorphous peaks at ca. 8° overlapped to the extent that the amorphous peak had to be resolved from the composite peak. The shape of the curve obtained from a sample heated for 1 hr. under nitrogen at 200°C (Tg = 227°C) was taken as the shape of the 100% amorphous peak. It was assumed that the 100% crystalline peak was Gaussian or at least symmetrical down to its half height. The amorphous component was then determined by subtracting off that weight of amorphous scattering required to leave a crystalline peak symmetrical at its half height. An example of such a composite peak and its separate components is shown in Figure 8. The maximum crystallinity found when the highest molecular weight PVK was heated at 325° for 24 hours, was ∿65%. Even after 1 hour at this temperature, however, the measured crystallinity was greater than 60%.

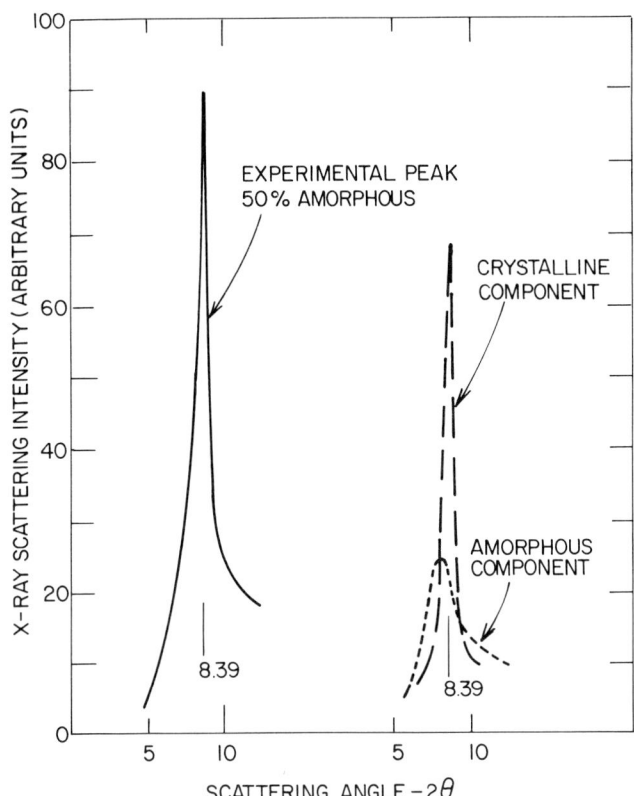

Fig. 8 X-ray scattering curve from PVK film resolved into crystalline and amorphous components.

It was found that the measured density of these crystalline films varied systematically with percentage crystallinity for films heated at 325° for different time periods. The measured densities are shown plotted against percentage crystallinity in Figure 9. This plot extrapolates to a value of 1.193 gm/cc at 100% crystallinity.

As stated previously the intensity of the 8.39° scattering peak was dependent on molecular weight for all heating temperatures in the 240-325°C range. The relative crystallinities resulting from heating for 1 hour at 305°C were calculated using a technique similar to that described above and the molecular weight dependence is shown in Figure 10.

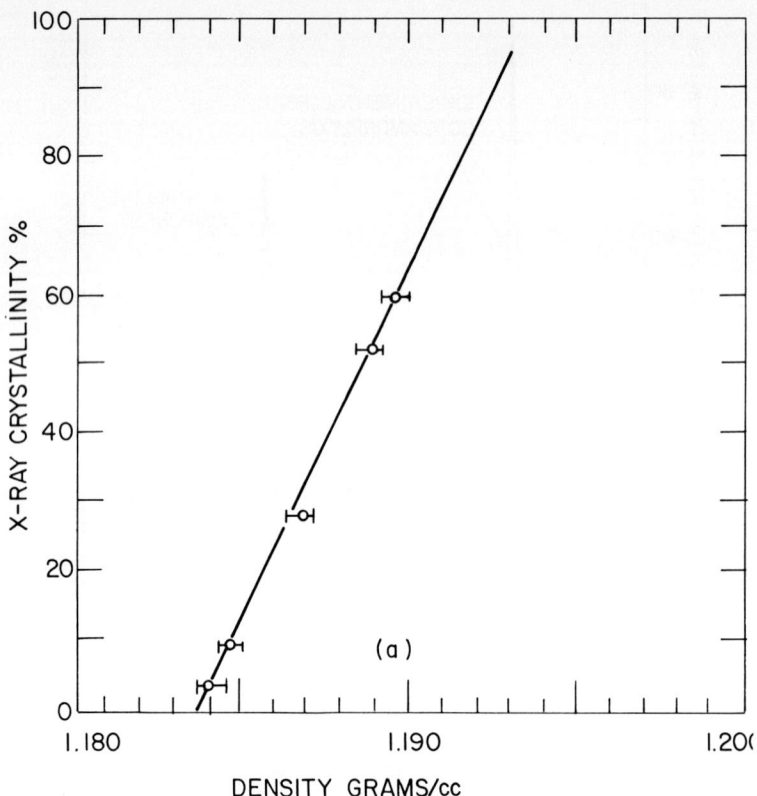

Fig. 9 Density of PVK as a function of X-ray crystallinity.

PVK was completely soluble in THF even after crystallization at 325°C and the molecular weight and dispersion of the M_n = 2.7 x 10^6 fraction were similar to the starting material. Tg was depressed slightly, due probably to the presence of a small amount of vinylcarbazole monomer decomposition product. A further sample heated at 350°C for 2 hr. was also completely soluble but M_n was reduced to 1.3 x 10^6 and M_w/M_n increased to 2.2.

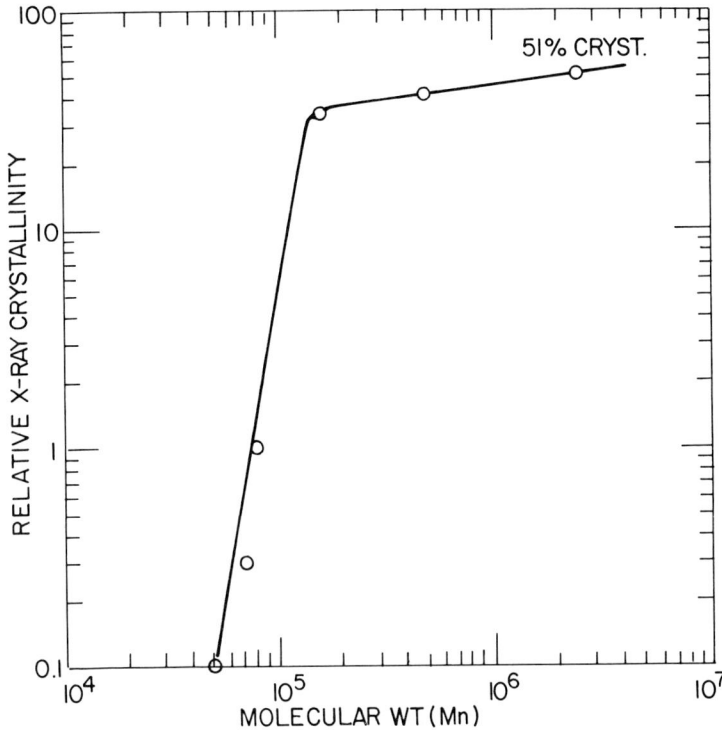

Fig. 10 Relative crystallinity of PVK films heated at 305°C for 1 hour from X-ray diffraction.

The direct relationship between chain length and crystallinty indicated that the concentration of chain termination groups and their perturbation of the structure might be important. In order to determine if the free volume associated with these groups was large, the density of the polymer was measured as a function of molecular weight. As shown in Table 1 the measured density stayed essentially constant at the high molecular weight value down to at least M_n = 26,000. The scatter in the data at 26,000 showed a significant increase over that at higher molecular weights due to the low physical strength and resultant cracking of the film. At molecular weights below 26,000 the material was so fragile and susceptible to cracking that no reproducible measurements could be made.

DISCUSSION

The physical properties other than Tg, of amorphous PVK measured in this study show very little dependence on either molecular weight or the synthetic procedure used in the polymerization. The synthetic procedure used in the commercial high molecular weight polymer is unknown; however, the medium and low molecular weight materials, where chain ends might be expected to be important were produced by two different synthetic routes. In the AIBN synthesis, the initiator is produced by the thermal decomposition

$$CH_3-\underset{\underset{CH_3}{|}}{\overset{\overset{CN}{|}}{C}}-N=N-\underset{\underset{CH_3}{|}}{\overset{\overset{CN}{|}}{C}}-CH_3 \rightarrow CH_3-\underset{\underset{CH_3}{|}}{\overset{\overset{CN}{|}}{C}}\cdot + N_2$$

and the possible termination groups are

$$-\underset{\underset{CH_3}{|}}{\overset{\overset{CN}{|}}{C}}-CH_3 \quad \text{and} \quad -CH_3$$

In the TBPP synthesis the thermal decomposition process produces two different radicals

$$CH_3-\underset{\underset{CH_3}{|}}{\overset{\overset{CH_3}{|}}{C}}-O-O-\overset{\overset{O}{\|}}{C}-\underset{\underset{H}{|}}{\overset{\overset{H}{|}}{C}}-\underset{\underset{CH_3}{|}}{\overset{\overset{CH_3}{|}}{C}}-CH_3 \rightarrow CH_3-\underset{\underset{CH_3}{|}}{\overset{\overset{CH_3}{|}}{C}}-O\cdot + \cdot O-\overset{\overset{O}{\|}}{C}-\underset{\underset{H}{|}}{\overset{\overset{H}{|}}{C}}-\underset{\underset{CH_3}{|}}{\overset{\overset{CH_3}{|}}{C}}-CH_3$$

and the possible chain termination groups are

$$-CH_3, \quad -\underset{\underset{H}{|}}{\overset{\overset{H}{|}}{C}}-\underset{\underset{CH_3}{|}}{\overset{\overset{CH_3}{|}}{C}}-CH_3 \quad \text{and} \quad -\overset{\overset{O}{\|}}{C}-\underset{\underset{H}{|}}{\overset{\overset{H}{|}}{C}}-\underset{\underset{CH_3}{|}}{\overset{\overset{CH_3}{|}}{C}}-CH_3$$

Neither the IR absorption nor the NMR spectra showed significant variation with molecular weight and, in particular, no absorption due to >C=O or -C≡N at 1720 and 2240 cm^{-1}, respectively, where PVK does not absorb. It is assumed, therefore, that the majority of chain ends are alkyl groups. While these groups are of small size compared to the carbazole group it is possible that the termination does produce a change in chain conformation close to the end group. The glass-transition temperature increased monotonically with increasing molecular weight and did not show discontinuities that might indicate differences in tacticity due to different synthetic procedure. In fact, the complete thermograms showed only small changes when the molecular weight was varied. It has been shown thet the glass transition in polystyrene, a polymer with similar chemical and physical structure to PVK only differs by 2° in the atactic and isotactic states.[14] In this case, therefore, Tg may not be a useful diagnostic tool for tacticity.

It has been suggested that the terminal groups of a polymer chain are associated with increased free volume and that an increase in the concentration of chain ends might be expected to reduce Tg. Assuming that all molecular weight fractions have equal free volume at their Tg, Fox and Flory[15,16] derived $T_g = T_g^\infty - K/M$, where T_g^∞ is the glass transition at infinite molecular weight, and K is a positive constant. A plot of $1/M_n$ against Tg should, therefore, give a straight line. This is shown to be valid for the PVK data in Figure 11. Conversely, it might also be expected that specific volume should increase with

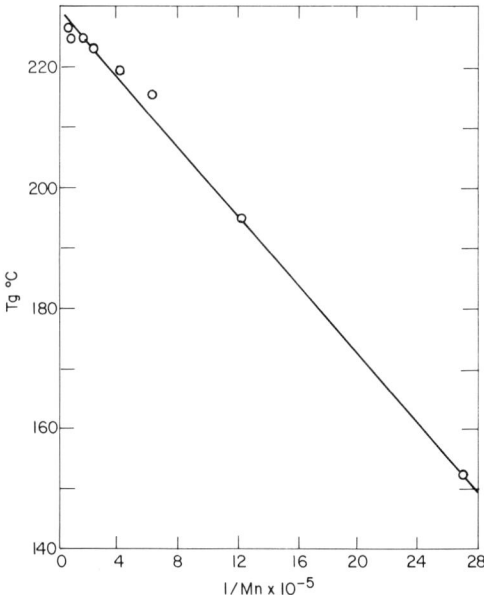

Fig. 11   Linear dependence of Tg on molecular weight in accordance with Fox-Flory model.

decreasing molecular weight and Tg. In the case of polystyrene an increase in specific volume of ca. 2% has been observed over the molecular weight range of 10^5 to 3×10^3.[15,16] No corresponding increase in specific volume (reciprocal density) was observed in PVK and, in fact, no change in density of even 0.1% was observed over a similar range of chain end concentrations. This seems to indicate that in PVK the increased free volume associated with the chain ends is small and that the chain-end free volume in itself should not be a large factor in determining crystallinity or ordering in the amorphous state. This is supported by the relatively small molecular weight dependence in the X-ray scattering data from the amorphous polymer outlined previously and shown in Figure 1. The position of the center of the ca. 8° peak, a measure of the average interchain spacing, changed only from 7.8 to 7.9° on going from a molecular weight of 3.7×10^3 to 2.7×10^6. The size of the shift relative to the width of the peak and the change in width with molecular weight makes any attempt to calculate interchain spacings rather suspect. Although some change has occurred in the interactions between chains its interpretation using this data is difficult.

The value of $\Delta\theta_1$, the half-width of the ca. 8° peak, changed much more dramatically with molecular weight. This value can be thought of as a measure of the dispersion of the interchain spacing or the size over which the average spacing is maintained. Scherrer[17] developed an equation relating diffraction peak broadening to crystallite size

$$L_{hk\ell} = \frac{K\lambda}{\beta_o \cos\theta} \tag{1}$$

where $L_{hk\ell}$ is the mean dimension of the crystal normal to the $hk\ell$ plane, β_o is the half-width of the diffraction peak ($\Delta\theta_1$) in radians, K is normally unity, and θ is the angle at the center of the peak.

While this formula is not strictly applicable to the present structure it does provide some useful insight into the meaning of the $\Delta\theta_1$ values. The formula gives a value of 50Å for $L_{hk\ell}$ at $M_n = 4,000$ and 81Å for $L_{hk\ell}$ at $M_n = 2.7 \times 10^6$. These distances are coherence lengths normal to the polymer chain axis and can be interpreted in terms of the presence of small volumes of pseudoparallel chains[18] with diameters of four interchain distances for $M_n = 4,000$ and seven interchain distances for $M_n = 2.7 \times 10^6$. Thus, possibly the mean interchain spacing but more particularly the distance over which this persists is found to be a function of molecular weight even though the measured densities indicate that the chain-end groupings do not contribute extra free volume to the chains.

The x-ray scattering patterns from the PVK heated above Tg to produce the lamellar morphology show the ca.8° peak is considerably narrower in half width. This suggests that in this material the chains are more strictly parallel and that this order persists over many interchain spacings. Selected area diffraction from single lamellae illustrated in Figure 7 showed that in fact, the 8.39° diffraction peak was resolvable into a hexagonal spot pattern indicating hexagonal or trigonal packing of the chains about the chain axis. It seems reasonable to interpret the 10.39Å reflection as the (10$\bar{1}$0) reflection of a pseudohexagonal lattice, that is, a lattice composed of PVK chains packed together in a hexagonal array with an interchain spacing, a, of 12.00Å. The individual lamellae in the thin films on carbon are therefore single crystals with the chain axis normal to the film plane which grow in the <210> direction. Since the film thickness is 700Å and the length of the PVK molecule calculated from M_n = 2.7 x 10^6 is approximately 2 microns, the PVK chain must fold at the top and bottom surfaces of the film. The x-ray diffraction data from the thicker films indicated that the planes giving rise to the 10.39Å reflection had a prefered orientation parallel to the film plane. That is the (10$\bar{1}$0) plane and the polymer chain axis were parallel to the film plane and the lamellar crystal growth direction was again <210>. It is apparent that these are the same folded chain crystals seen in the thin films. In Figure 5, the microtome section taken parallel to the film plane, the polymer chain axis must be parallel to microtome induced cleavage cracks which varied from ca.300Å after heating at 270°C to ca.800Å after heating at 330°C is, therefore, the lamellar fold period. This fold period did not vary with molecular weight.

The nucleation of folded-chain lamellae at aluminum and other high-energy interfaces and subsequent growth normal to these surfaces has previously been reported by Schonhorn[19] for other long-chain polymers. The driving force was said to be a minimizing of the interfacial energy by the orientation of the chain molecules parallel to the substrate.

Neither the electron nor x-ray diffraction data contained any sharp reflections indicative of the chain axis (c axis) period of the crystalline lamellae. It is possible to interpret the diffuse reflection at 14° 2θ (ca.6.3Å) in Figure 2 as the (0001) reflection of the crystal lattice. Such a spacing would be expected for an isotactic vinyl polymer chain with a 3/1 screw axis which had the trigonal symmetry suggested by the hexagonal packing. The (0001) reflection would not normally be an allowed reflection in an undistorted lattice but the lack of other than the {10$\bar{1}$0} reflection of the <001> zone and the diffuseness of the (0001) reflection indicate a distorted paracrystalline structure. The c axis spacing can actually be calculated independently from the area of the unit cell in the {0001} plane derived above and the volume of the unit cell calculated from the chemical formula and the crystalline density ρ_c (100% crystalline), where

$$\text{c axis spacing per monomer unit} = \frac{\text{wt. of monomer unit in grams}}{(a^2/\sqrt{3} \times 10^{24})\rho_c} \text{ Å}$$

Using the value of $\rho_c = 1.193$ g/cc derived in Figure 8 the above formula gives a value of 2.16Å per monomer unit or 6.47Å for an isotactic 3/1 helix. This value eliminates the possibility of a syndiotactic 2/1 chain which would require approximately 2.54Å per monomer unit and agrees fairly well with the 6.3Å diffuse peak in the crystalline x-ray scattering curve.

The measured change in density in going from the amorphous state (1.184 g/cc) to the paracrystalline state (1.193 g/cc) is unusually small at somewhat less than 1%. This would indicate a rather small structural change and is further evidence for an amorphous state which has considerable order. As a consequence of the small density change, the c axis dimension calculated from the crystalline density is insensitive to errors in the crystallinity measurements. The measured densities were reproducible to ± 0.0004 g/cc and it is to be expected that the most difficult measurement which enters the c axis calculation is in fact the crystallinity. Film recording and densitometer processing of the data can lead to distortion of the peaks and consequent errors. It is also possible that the shape of the X-ray scattering peak given by the noncrystallizable amorphous component is not identical to that given by the 100% amorphous sample which contains the crystallizable and noncrystallizable components. It should be pointed out therefore, that an error of 20% in the extrapolated 100% crystallinity is required to give an error of 0.01Å in the calculated c axis period. The c axis period in these folded-chain lamellae is certainly considerably shorter than that proposed by Crystal[6] and is in much better agreement with that of other isotactic vinyl polymers.

While the chain axis period calculated above is in the range expected for an isotactic vinyl polymer the only direct evidence for ordering along the chain axis is the diffuse X-ray peak at 6.3Å. This leads to the possibility of considerable disorder along this axis and indicates that the structure is most accurately described as paracrystalline. The idealized molecular packing for PVK as discussed above but without the disorder is shown in Figure 12. The reversal in direction of the noncentrosymmetric chain at the chain fold requires that the two segments of the chain have the bond attaching the pendant groups pointing up and down respectively with respect to a vertical chain axis. The exact position of reentry of polymer

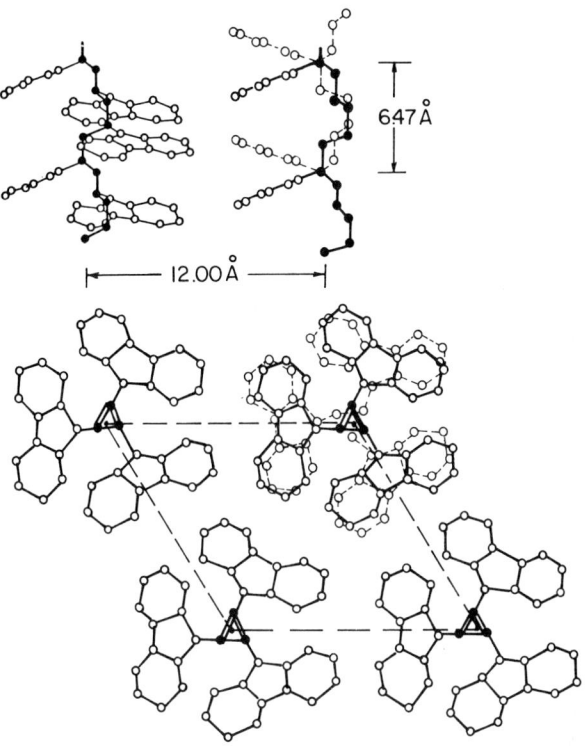

Fig. 12 Idealized structure model for crystalline PVK.

chains in melt-crystallized lamellae has received some attention and the bulk of the evidence suggests adjacent reentry as the dominant mechanism.[20] Assuming adjacent reentry in the case of PVK alternate chains will have a reversed sense of rotation. This chain conformation is shown drawn with dashed lines in Figure 12.

The proportion of the paracrystalline phase in the films heated to 305°C showed a definite dependence on M_n as illustrated in Figure 10. In this plot measured crystallinity decreases linearly with M_n down to M_n = 157,000 and then breaks sharply. The electron micrograph of the M_n = 67,000 (Figure 6) clearly confirms the lower crystallinity and below M_n = 46,000 no lamellar structure was observed. It is interesting to point out again that Heller et al.[3] were unable to obtain crystallizable PVK using Ziegler-Natta catalysts and could not confirm the work of Solomon et al.[2], Natta and coworkers[4] and Kimura et al.[5]

It seems reasonable to assume that, although Heller et al.[3] gave no molecular weight data, their polymer must have had a molecular weight of $M_n < 46,000$. There has been no evidence to suggest that any of the reported synthetic procedures produce differences in tacticity which would account for the lack of crystallinity. This decreasing crystallinity with decreasing molecular weight is at variance with the behavior usually found in polymers. It is normally expected that the smaller polymer chains in the lower molecular weight materials have greater mobility and find it easier to reorganize themselves into an ordered crystalline array.

The range of molecular weight fractions over which the crystallinity of PVK decreased sharply was 46,000 to 157,000. The fact that all the fractions contain a range of molecular weights suggests that the ability to crystallize might be lost at a particular molecular weight. Rapidly changing crystallinity would then reflect a changing proportion of polymer above and below the limit from one fraction to another.

The theory of the nucleation and growth of long-chain polymer crystals (where the nucleus is a single layer containing one or more folded chains similar to that shown in Figure 13) has been discussed by Hoffman.[21] In this treatment the free energy

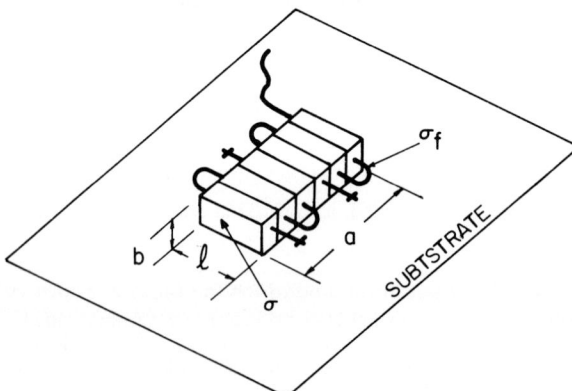

Fig. 13 Folded chain crystal nucleus (Hoffman[21]).

of formation of a single layer folded-chain nucleus $\Delta\phi$ is equal to the free energy of the created side fold surface and the created side linear-chain surface less the created crystalline volume multiplied by the energy of crystallization per unit volume: i.e.,

$$\Delta\phi = 2ab\sigma_f + 2\ell b\sigma - ab\ell\Delta Gv \qquad (2)$$

where ℓ is the fold period (determined by the crystallization temperature), a is the width of the molecule multiplied by the number of linear segments, b is the thickness of the monolayer, σ is the free energy of the side linear chain surface, $σ_f$ is the free energy of the side fold surface, and ΔGv is the bulk free energy difference between the supercooled liquid and the crystal. Because of the work required to fold the chains, $σ_f$ is expected to be considerably greater than σ. When the substrate and nucleus top surface free energies are not the same an additional term is involved. In the case of PVK on aluminum the initial monolayer nucleation is heterogeneous and crystal growth is determined by further nucleation of PVK on PVK so the eq. (2) is directly applicable.

The influence of molecular weight on nucleation and growth is through the changed population of chain ends. A chain end at or above the surface of a crystal has a lower free energy contribution than a chain end in the crystal interior but the free energy per unit surface area is still higher than that of a chain fold. The energy of a chain fold is spread over the surface area occupied by the cross section of two chains whereas that of the chain end is spread over one. The increase in the fold surface free energy term due to the chain ends may be written

$$\bar{σ}_f = σ_f \left[\frac{\emptyset + q_e/q_f}{\emptyset + 1} \right] \qquad (3)$$

where $\emptyset$ is the number of folds per chain and q_e/q_f is the ratio of free energies per unit area of a chain end versus a chain fold. There will be a lower limit of chain length for each crystallization temperature below which Δφ becomes positive due to the q_e contribution, and nucleation and growth will not occur.

In the case of PVK it seems likely that the critical value of M_n for folded chain nucleation at 305°C is in the region of 10^5 and the rapidly changing crystallinity below $M_n = 157,000$ does in fact reflect the rapidly decreasing proportion of the polymer which is above the critical value.

The length of the polymer chain at $M_n = 10^5$ is ca.1200Å and this critical value of M_n therefore corresponds to approximately three fold periods. This situation is illustrated in Figure 13 and assuming the chain ends are not incorporated into the body of the crystal, it can be seen that the fold surfaces are actually composed of equal number of chain ends and chain folds. No extended chain lamellae were observed but the likelihood of these nucleating from this and lower molecular weight fractions at this temperature is small, due to the increased free energy of the chain end versus the chain fold surface. No melting endotherm for the lamellar crystals was observed in the DSC scans so that the degree of supercooling (ΔT) represented by the maximum crystallization temperature of 330°C is not known. Using the generalism that Tg of polymer systems is expected to be ca. 2/3 of the melting temperature in degrees Kelvin it is possible to estimate the melting temperature at about 480°C. The maximum crystallization temperature limit set by decomposition of the polymer apparently represents the considerable supercooling of ca. 150°C. It is also interesting to reflect on the influence the large supercooling might have on the perfection of the folded chain crystals and to speculate that reducing Tm and Tg by suitable impurity addition might result in increased crystallographic perfection within the lamellae. Lowering the crystallization temperature might be expected to reduce the fold period and allow shorter chains to crystallize. Unfortunately this also results in an increase in the number of folds per unit volume of the nucleus and therefore, an increase in the $2ab\sigma_f$ parameter in equation (2). As would be expected from equation (2) the temperature below which crystal growth did not occur increased with molecular weight and at 280°C crystallization of the M_n = 46,000 fraction was no longer observed.

CONCLUSIONS

The physical properties of amorphous PVK, other than Tg, showed little dependence on molecular weight. Tg was linearly dependent on reciprocal molecular weight but density measurements indicated no measurable excess free volume associated with the chain ends as predicted by the Fox-Flory chain end free-volume model. There was no evidence for any influence of synthetic procedure on polymer tacticity and this is thought to be determined by the large size of the carbazole pendant group. Heating above Tg resulted in the formation of folded chain lamellar crystals whose fold period varied directly with heating temperature. These lamellae appear to have a basically trigonal structure with a = 12.00Å and c = 6.47Å. The c - axis period of 7.44Å measured by Crystal[6] was not observed and the measured densities of the thermally crystallized PVK would preclude such a spacing in these folded chain lamellae. The diffuseness of the diffraction peaks indicates that the lamellar crystals are badly disturbed and are, more precisely, paracrystals. The perfection of the crystals may be influenced by large degree of supercooling at which crystallization occurred but the lack of periodicity down the chain axis <001> may also be due to some lack of stereoregularity.

The intensity of the 8° diffraction peak from the amorphous phase and the small difference in density between this and the paracrystalline phase imply that there was some parallel ordering of the chains even in the amorphous state. This decreased with molecular weight. The nucleation efficiency and degree of crystallinity also decreased with molecular weight in the thermally crystallized polymer. The experimental observations can be explained in terms of the theory of nucleation and growth of polymer crystals outlined by Hoffman[21]. He suggests that chain ends reside at, and increase the free energy of, the fold surface. This renders the folded chain nucleus unstable when the chain end concentration reaches a critical value. At a crystallization temperature of 305°C this critical concentration was found to occur at a value of M_n approximating 10^5. The lack of crystallizability at low molecular weights probably explains the inability of Heller[3] to obtain crystalline PVK from the Ziegler-Natta synthesis.

ACKNOWLEDGEMENTS

The author would like to acknowledge the contributions of Ms. Anita VanLaeken and Mr. E.C. Williams and valuable discussions with W. Prest, T. Davidson and R. Penwell.

REFERENCES

1. R.M. Schaffert, IBM J. Res. Dev. 15, 75 (1971).
2. O.F. Solomon, M. Dimonie, K. Ambrozh, and M. Tomesku, J. Polym. Sci., 52, 205 (1961).
3. J. Heller, D.O. Tieszen and D.B. Parkinson, J. Polm. Sci., Pt. A. 1, 125 (1963).
4. G. Natta, G. Mazzanti, G. Dall'Asta and A. Casale, Belgian Patent No. 605,790 issued to Montecatini Societa Generale per L'Industria Mineraria e Chimica, August 31, 1961.
5. A. Kimura, A. Yoshimoto, Y. Akana, H. Hirata, S. Kusahoyashi, H. Mikawa, and N. Kasai, J. Polym. Sci., Pt A-2, 8, 643 (1970).
6. R.G. Crystal, Macromolecules, 4, 379 (1971).
7. E.H. Cornish, Plastics, 28, 61 (1963).
8. M. Stolka, to be published.
9. G. Gandolfi, Miner. Petrogr. Acta 13, 67 (1967).
10. J.M. Barrales-Rienda, J. Gonzales-Ramas, and M.I. Dabrio, Makromol. Chem., 43, 105 (1075).
11. D.J. Williams, Macromolecules 3, 602 (1970).
12. O.F. Solomon, N. Cobiana, V. Kucinschi, Die Makromol Chem. 89, 171 (1965).
13. J.E. Field, J. Appl. Phys., 12, 23 (1941).
14. F.E. Karasz, H.E. Bair, and J.M. O'Reilly, J. Phys. Chem., 69, 2657 (1965).
15. T.G. Fox and P.J. Flory, J. Appl. Phys., 21, 581 (1950).

16. T.G. Fox and P.J. Flory, J. Polym. Sci., 14, 315 (1954).
17. P. Scherrer, Göttinger Nachrichten, 2, 98 (1918).
18. G.S.Y. Yeh, J. Macromol. Sci., Phys., B6, 465 (1972).
19. H. Schonhorn, Macromolecules, 1, 145 (1968).
20. B. Wunderlich, *Macromolecular Physics*, Vol. 1, Academic Press, New York 1973.
21. J.D. Hoffman, SPE Trans., 4, 315 (1964).

Discussion

On the Paper by W.R. Goynes and J.H. Carra

Klara Kiss (*Stauffer Chemical Co.*): Why did you find it necessary to remove the embedding agent when it does not contain phosphorous or bromine which would interfere in X-ray images? Experimentally, how did you remove the embedding agent?

W.R. Goynes (*USDA Southern Regional Research Center*): It is often difficult to distinguish between fiber sections and embedding agent if the latter is not removed. It is especially difficult to see chemical deposits on the outer surface of the fiber section if the embedding agent is left intact. If necessary, the agent does not have to be removed for analysis, but interpretation of results is much more difficult in this case. Details of methods for removing embedments are pointed out in the paper.

R.M. Marshall (*Allied Chemical Fibers Division*): What carriers were used to carry the bromine compounds into the polyester? Were they applied from emulsions?

W.R. Goynes: A variety of solvents were used as carriers for the bromine containing compounds. These included xylene and perchlorethylene. Some finishes were emulsions, depending on the nature of the solvent.

C. Turnquist (*USCI (Division of C.R. Bard)*).

1. What accelerating voltage was used in your S.E.M. studies?

2. What observation angle?

W.R. Goynes:

1. 5 Kilovolts - A compromise voltage between resolution and charging.

2. 45°.

On the Paper by A.G. Causa

R.M. Marshall (*Allied Chemicals, Fibers Division*): What are the effects of 1. finish, 2. fabric construction, 3. tensilization conditions on cyclic fatigue results.

A.G. Causa (*Goodyear Co.*): All these three factors are well known to affect the lifetime of the composite under fatigue testing conditions. A discussion on this topic falls outside the scope of my presentation. Tensilization and heat treatment conditions insofar as they are able to alter the fiber microstructure may also alter the fracture topography developed upon rupture in a fatigue test.

Klara Kiss (*Stauffer Chemical Company*): Can you predict from the nature of fracture mechanism the tensile properties of cords?

A.G. Causa: Yes, if you have previously developed the background and expertise to do so. Of course, it is easier to generate the tensile data first, and then study the fractography and establish correlations.

Klara Kiss: What was the way of mounting the sample on the holder?

A.G. Causa: The broken cord tips were mounted using customary procedures. I recommend the reading of the book "The Use of the Scanning Electron Microscope" by J.W.S. Hearle, J.T. Sparrow and P.M. Cross, Pergamon Press, particularly Chapter 7.

On the Paper by D.W. Dwight

L.H. Lee (*Xerox Corp.*): Comments: This paper illustrates that ESCA can be used to study polymer surfaces in conjunction with other conventional techniques, e.g., contact angle measurements. Contact angle measurements, though convenient, frequently cannot provide fundamental answers to the actual condition of the surface. In the past, we observed that once a polymer was molded at high temperatures against a Teflon film, the critical surface tension of the polymer film decreased to the range of Teflon. At first, we thought that the former film was affected by the induced orientation due to Teflon film. Now, it appears possible to re-examine the phenomenon with ESCA to determine whether the contamination of the polymer film by Teflon had taken place.

On the Paper by C.H. Griffiths

L.H. Lee (*Xerox Corp.*): We would like to thank Dr. Griffiths for this post-conference contribution to this volume. Poly(N-vinylcarbazole) is an interesting organic photoconductor which has been used for electrophotography. Detailed results published in this paper should enable us to utilize this polymer properly by itself or in a polyblend.

The plenary lecture to this session by Dr. W.C. McCrone on "Surface Characterization by Light Microscopy" has been published elsewhere. Two other papers by Drs. Vanderhoff and Rowell are included in Part IV of this Volume.

PART IV

Surface-Chemical and Radiation Analyses

PART - IV

Surface-Tension and Radiation analysis

Characterization of Latexes by Ion Exchange and Conductometric Titration

J.W. Vanderhoff

Emulsion Polymers Institute
Lehigh University
Bethlehem, Pennsylvania 18015

Current emulsion polymerization technology uses a variety of comonomeric emulsifiers (e.g., acrylic acid or 2-sulfoethyl methacrylate) in the production of latexes. These conomomeric emulsifiers can polymerize in the aqueous phase, on the particle surface, or inside the particles; their mode of combination determines the performance of the latex in a given application. The loci of functional groups of these comonomeric emulsifiers can be determined by ion exchange and conductometric titration in combination with other analytical methods. The analytical procedure is reviewed and typical results on its practical application are presented.

INTRODUCTION

Monodisperse polystyrene latexes were first prepared at the Dow Chemical Company in 1947 and have since found a wide variety of research and diagnostic applications.[1] Despite the obvious utility of their very narrow particle-size distributions, until a few years ago these latexes had been used in only a few colloidal studies because little was known about the characteristics of their surfaces. This paper describes the characterization of these latexes as to the number and acid strength of their surface ionic groups and the development of an ion-exchange-and-conductometric-titration technique to give "clean" (i.e., emulsifier- and electrolyte-free) latexes that are useful, not only as model colloids, but also in other applications, e.g., as a substrate for adsorption studies of surfactants and proteins. Moreover, this technique is useful in preparing special (e.g., radioiodine-tagged) latexes and in "cleaning" other colloids.

OUTLINE OF PROCEDURE

The latexes were prepared either by conventional emulsion polymerization (i.e., by forming an emulsion of monomer in water with an anionic emulsifier and polymerizing with persulfate-ion free-radical initiator) or by "seeded" emulsion polymerization (i.e., by polymerizing additional monomer in a previously-prepared latex, to grow the particles to a larger size without initiating a new crop.) The details of the polymerization are described elsewhere.[2] Both types of polymerization give a colloidal dispersion of submicroscopic polymer spheres which bear a negative charge arising from the adsorbed emulsifier.

The classical representation of the latex is a system of hydrophobic polystyrene spheres, each stabilized by a double layer comprised of adsorbed emulsifier anions and their monovalent counterions. The negative charge of the particle is derived from the adsorbed emulsifier layer; if this emulsifier were removed, the particles would lose their charge and flocculate.

There is, however, another source of negative charge: the sulfate endgroups of the polymer molecules. These result from the initiation of polymerization by sulfate ion-radicals formed by the decomposition of persulfate ion in the aqueous phase.

$$S_2O_8^{2-} \rightarrow 2\ SO_4^-\cdot$$

Sulfate ion-radicals have no tendency to migrate to the negatively-charged particle-water interface, but instead undergo a reaction with the solute monomer to form an oligomeric radical.

$$SO_4^-\cdot + M \rightarrow SO_4^-M\cdot$$

$$SO_4^-M\cdot + M \rightarrow SO_4^-MM\cdot$$

$$SO_4^-MM\cdot + M \rightarrow SO_4^-MMM\cdot,\ \text{etc.}$$

The oligomeric radical soon grows to such a size that it becomes surface-active and adsorbs at the particle-water interface with the radical oriented toward the monomer-swollen polymer phase. The radical then grows into the particle by addition of monomer molecules, the supply of which is replenished by diffusion from the reservoir emulsion droplets or other monomer-swollen latex particles through the aqueous phase. The polymeric radical continues to grow until another oligomeric radical enters the particle and causes termination. Thus, this mechanism of initiation gives sulfate emulsifier groups chemically-bound to the particle surface.

Therefore, according to this mechanism, two sulfate endgroups should be found on the particle surface for each polystyrene molecule

formed assuming that:
1. the sulfate ion-radical does not undergo a transfer reaction to introduce a different endgroup into the polymer molecule.
2. none of the sulfate endgroups is buried inside the particle;
3. termination within the particle occurs by combination rather than by disproportionation;
4. transfer reactions with monomer, emulsifier, and other ingredients can be neglected.

Thus, the aqueous phase of the latex may contain emulsifier, residual initiator and its decomposition products, and electrolyte used as buffer. The latex particle surface may contain physically-absorbed emulsifier anions and chemically-bound anionic endgroups. The removal of the emulsifier and inorganic salts from the aqueous phase and particle surface would give a system of monodisperse polystyrene spheres stabilized only by chemically-bound strong-acid surface groups. If these chemically-bound groups were sufficient in number to stabilize the latex particles, and if this number were known, the latex would then be an ideal model colloid, i.e., comprised of monodisperse spheres with a constant and known surface charge density.

Two methods were tried for removing the emulsifier and electrolyte from the latexes: dialysis and ion exchange. With dialysis, the traditional method for purifying colloidal sols, the emulsifier was only partially removed, and the exchange of protons for the sodium or potassium counterions was incomplete. With ion exchange, the emulsifier and electrolyte were removed quantitatively, and the chemically-bound emulsifier groups were left in the H^+ form, so that their number could be determined by conductometric titration with base; however, this method presents the possibility of contaminating the colloidal sol with leached polyelectrolytes. This report describes the purification of the ion-exchange resin to avoid contamination by leached polyelectrolytes, the conductometric titration of the ion-exchanged latex to determine the surface charge density, the results obtained with typical monodisperse polystyrene latexes, the effect of the polymerization recipe on the bound surface groups, and the influence of the ion exchange on the stability and particle-size distribution of the latex.

CONDITIONING AND PURITY OF THE ION-EXCHANGE RESINS

Other workers have reported results that discouraged the use of ion-exchange resins in colloidal studies. Schenkel and Kitchener[3] found that deionized water contained a weakly-basic polyelectrolyte and warned against its use in experiments involving small surface areas, e.g., glass capillaries. Ottewill[4] found, in some cases, that the charge of colloidal sols was reversed upon treatment with ion-exchange resins. Therefore, in view of these experiences, it was of utmost importance to condition the ion-exchange resins so as to remove completely any soluble polyelectrolytes and to demonstrate that latex could be treated with these conditioned resins without significant alteration of its surface charge density.

The resins selected for this rigorous conditioning were 20-50 mesh Dowex 50W-X4(H^+ form)* and Dowex 1-X4(Cl^- form) resins. The Dowex 50W resin is the sulfonate salt of a 96:4 styrene-divinylbenzene copolymer with an ion-exchange capacity of 1.7 meq/ml wet resin; the Dowex 1 is the analogous trimethylammonium salt with a capacity of 1.33 meq/ml wet resin. The lower degree of crosslinking (4%) was used instead of the usual 8% to facilitate the pickup of soluble polyelectrolytes and emulsifier anions.

Possible sources of impurities in the resins are soluble low-molecular-weight sulfonated or aminated polymers, residual chemicals used in the preparation, additives such as screening agents used in sizing the beads, electrolytes used in the preconditioning, and very small, crushed, or fragmented beads.

After considerable work[5-8], the following procedure was developed for conditioning the resins. Each resin was washed with 85°C water until no more colored material was removed; it was then eluted consecutively with a threefold excess of 3M sodium hydroxide, hot water, methanol, cold water, a threefold excess of 3M hydrochloric acid, hot water, methanol, and cold water. This cycle was repeated four times. The last cycle was completed by the slow elution of the Dowex 50W resin with a five-fold excess of 3M hydrochloric acid to convert it to the H^+ form and likewise of the Dowex 1 with a five-fold excess of 3M sodium hydroxide to convert it to the OH^- form. The Dowex 1 was stored in the Cl^- form and converted to the OH^- form only shortly before use. The resins were rinsed copiously with doubly-distilled water and mixed in equivalent proportions under stirring. Before use they were washed under agitation and decanted or filtered..

The purity of the resins was controlled by observations of the color and odor, and by measurements of the conductance, surface tension, ultraviolet absorption (224 nm), and acid or base required for neutralization of the water in which the resin was agitated for 2 hrs. These results are summarized below, and typical values for the ultraviolet absorption and acid or base titration are given in Table 1. For comparison, results are also given for Bio-Rad AG 50W (H^+ form)** and Bio-Rad AG 1 (OH^- form) analytical-grade resins prepared by purification of the corresponding Dowex resins; these resins were used without further treatment.

---

*The Dow Chemical Company
**Bio-Rad Laboratories, Inc.

Table 1: Quality of Ion-Exchange Resins*

Resin	% Trans. (224 nm)	Acid or base, µeq/liter
Dowex 50W, as received	0.09	780
Dowex 50W, washed with water	30.8	—
Dowex 50W, purified	15.6	110
Dowex 50W, purified, washed with water	86.2	11
Bio-Rad AG 50W, as received	0.15	—
Bio-Rad AG 50W, washed with water	12.0	140
Dowex 1, as received	66.0	—
Dowex 1, washed with water	98.8	—
Dowex 1, washed with methanol	~100	<1
Bio-Rad AG 1, as received	37.5	120
Mixed Dowex resin, as received	~100	2.8
Mixed Dowex resin, washed with water	~100	<1
Mixed Bio-Rad AG resin, as received	~100	1.3
Mixed Bio-Rad AG resin, washed with water	~100	<1

*Ultraviolet absorption and titration values of deionized water treated with the indicated resin.

The color of the Dowex 50W resin was improved considerably by the conditioning; the red color which remained after washing with 85°C water and methanol was removed in the first two base-acid cycles (for comparison, the Bio-Rad AG 50W resin was light red as received, and some of the colored material dissolved in water upon agitation). The light-straw color of the Dowex 1 resin was improved only slightly by the conditioning. The purified Dowex 50W resin was odorless in both the H^+ and Na^+ forms, as was the Dowex 1 in the Cl^- form; however, the Dowex 1 in the OH^- form gave off an odor of amines upon standing, so this resin was stored in the Cl^- form.

The conductance of the wash water from the separate purified resins was 1-2 µmho, from the mixed resins 0.2-0.3 µmho. In our experience, values of 0.2-0.3 µmho are excellent, lower values being below the sensitivity of the conductivity bridge (however, Akelroyd and Kressman[9] measured values for pure water as low as 0.066 µmho, as compared with the theoretical value of 0.055 µmho). No polyelectrolytes were detected by other methods in samples with conductances in the 0.2-0.3 µmho range. The surface tension (DuNouy ring method) of the wash water from the separate purified resins was 70-72 dyn/cm, from the mixed resins 71-73 dyn/cm. The polyelectrolytes leached from the resins do not lower the surface tension appreciably, and, therefore, this method does not give a sensitive measure of their presence.

Boyd and Bunzl[10] found that the polyelectrolytes leached from Dowex ion-exchange resins displayed a strong ultraviolet absorption at 224nm. Table 1 gives the values of the percent transmission (relative to doubly-distilled water) obtained by shaking 8-15 g resin in 150 ml water for 2 hrs, then filtering through a sintered-glass Büchner funnel. An appreciable absorption was observed in the wash water from the separate Dowex resins before conditioning and the separate Bio-Rad AG resins. Repeated washing completely removed the leachable polyelectrolyte from the Dowex 1 resin. It was much more difficult to remove from the Dowex 50W resin, however, and a small amount was even found in the wash water from the purified resin. The mixed resins, however, showed only a negligible absorption, even without conditioning, as did the mixed Bio-Rad AG resins. Apparently, the polyelectrolyte released by one resin is picked up readily by the other. For comparison, the laboratory deionized water supply* gave values in the range 90-100% transmission, indicating that occasionally it contained contaminants.

The wash water from the resins was also titrated conductometrically with 0.01 M sodium hydroxide or hydrochloric acid to determine the acid or base content. Blank titrations were carried out on distilled water or distilled water boiled out and titrated under nitrogen. By this method, acid or base can be detected in concentrations of less than 10^{-6}M. Table 1 shows that the results of this analysis confirm those of the ultraviolet absorption. Repeated washing of the Dowex 1 resin reduced the base content of the wash water to a very low value (much less than that of the Bio-Rad AG 1 resin); the slope of the conductometric titration curve showed that the material leached was a weak base. The Dowex 50W resin was much more difficult to purify--a small concentration of a strong acid was leached even from the purified resin. The wash water from the mixed resins, however, contained no strong acid within the limits of detection; also, no base was detected (Schenkel and Kitchener[3] suggested that their deionized water contained a weakly-basic polyelectrolyte leached from the unstable anion-exchange resin; apparently, this polyelectrolyte was too large to be picked up by the 8X-cation exchange resin used by these authors, because in this work no base was detected using the 4X-resin). Thus, both the conductometric titration and the ultraviolet absorption at 224nm are sensitive measures of the purity of the conditioned resins.

The purity which can be achieved routinely using the foregoing conditioning technique is shown by a typical analysis of the wash water in which the mixed resin was agitated for 2 hrs: 72 dyn/cm

---
*1702 Bldg., The Dow Chemical Co., Midland, Michigan.

surface tension; 0.2 μmho specific conductance; 100% transmission (within the limits of error) at 224nm; <0.1 μeq/100 ml acid (limit of detection). Thus, in a 2.5%-solids sample of latex A-2 which by titration showed a surface charge of 5×10^{-5} eq/100 ml, the contamination from the mixed resin would amount at most to about 10^{-7} eq of acid, or about 0.2% of the total charge. This demonstrates that the ion-exchange resins can be purified to such an extent that they can at least be used in experiments of large surface area without introducing an appreciable amount of contamination.

This conditioning procedure is generally satisfactory for reducing the concentration of leachable polyelectrolyte to a level that does not interfere with the characterization of the latex particle surfaces; however, ion exchange resins show batch-to-batch variations, and some batches cannot be purified using this procedure--polyelectrolytes can still be leached from these resins even after the conditioning procedure has been repeated many times.[8] This conditioning failure is attributed to the ion exchange resin itself--incomplete crosslinking of the original styrene-divinylbenzene copolymer beads, side reactions in the sulfonation and other chemical reactions used to prepare the resins, and breaking of chemical bonds by swelling or de-swelling--rather than to the conditioning procedure. The remedy is to discard the batch that cannot be conditioned and seek another more satisfactory batch. Preliminary experiments with a small amount of resin can be used to determine whether it can be conditioned satisfactorily.

ION-EXCHANGE TECHNIQUE

The ion-exchange technique finally developed consisted of mixing 300-500 ml latex with a fivefold excess (estimated from the emulsifier and electrolyte content) of mixed resin under agitation for 2 hrs, filtering through a sintered-glass Büchner funnel to remove the resin beads, and titrating. This process is repeated until a constant value is obtained for the surface charge. Earlier batch experiments used only a twofold excess of resin and a 3-hr contact time; this proved less efficient than the foregoing procedure. The first ion exchanges were carried out in small columns, either a single column of mixed resin or separate columns of Dowex 50W and Dowex 1 resins, rather than in batch; both variations proved less efficient than the batch technique.

The latexes were ion-exchanged at low concentrations, usually 5% or less, depending upon the particle diameter. After 15-30 min contact with the ion-exchange resin, the viscosity of the latexes increased twofold or more, the conductance greatly decreased, and, in some samples, interference colors became apparent. These changes are indications of the reduction of the electrolyte concentration to a low level and the resulting expansion of the double layer.

DETERMINATION OF SURFACE CHARGE DENSITY

After ion exchange, the latex particles are stabilized only by the sulfate endgroups of the polymer molecules, all of which are in the H^+ form. Thus, their number can be determined by titration with base. Generally, the titrations were followed conductometrically. In the first experiments, small increments of 0.1 N sodium hydroxide were added with a microsyringe to the latex in a dip-type cell under nitrogen, and the resistance was measured (1000 Hz) after each addition by balancing the conductivity bridge. Later, the conductometric titrations were carried out in dip-type cells, adding 0.01 N sodium hydroxide continuously with a constant-rate burette. In some titrations, the resistance was measured (1000 Hz; General Radio model 1650A conductivity bridge) at various intervals while the sodium hydroxide was being added continuously. In others, the voltage drop across a 10-Ω resistor in series with the cell and in parallel with a 6-V transformer was recorded continuously. All titrations were carried out at 25.0 ± 0.1°C.

Figures 1 and 2 show conductometric titration curves for ion-exchanged latexes using the measured-at-intervals and continuous-recorder-trace methods, respectively, as the variation of specific conductance with added sodium hydroxide. Both the descending and ascending legs are linear, and their extrapolation to intersection gives the equivalence point. The rounded minimum is due principally to the slow attainment of equilibrium near the equivalence point. This shape of curve is typical of that found in the titration of strong-acid surface groups with a strong base. The descending leg is linear because the strong-acid surface groups have an equal preference for Na^+ and H^+ counterions; however, because of the limited mobility of these counterions in the double layer, its slope is much smaller than for a strong acid in solution. For example, the ratios of the slopes of the descending and ascending legs are 0.216 and 0.23 for latex A-2 (Figs. 1 and 2, respectively) and 0.21 for latex D-4 (Fig. 2), as compared with the theoretical value of 1.213 for the titration of a strong acid with a strong base at 25° and the experimental value of 1.215 for the titration of hydrochloric acid with sodium hydroxide. These smaller values are of the same order as those found earlier for acidified silver iodide sols, 0.179 by de Bruyn and Overbeek[11] and 0.24 by van Os[12]. Beyond the equivalence point, the specific conductance increases linearly with excess sodium hydroxide. An equivalence point of 3.38 ml 0.01 N sodium hydroxide for 1.458 g latex polymer (Fig. 1) corresponds to a surface charge of 3.60 µC/cm^2 (note that "surface charge" or "surface charge density" is defined here as the total charge measured by titration; this is distinguished from the "effective charge" or diffuse charge, which takes into account any undissociated negative groups and any counter-ions residing in the Stern layer; thus the effective charge is the product of the surface charge and the "apparent degree of dissociation" α;

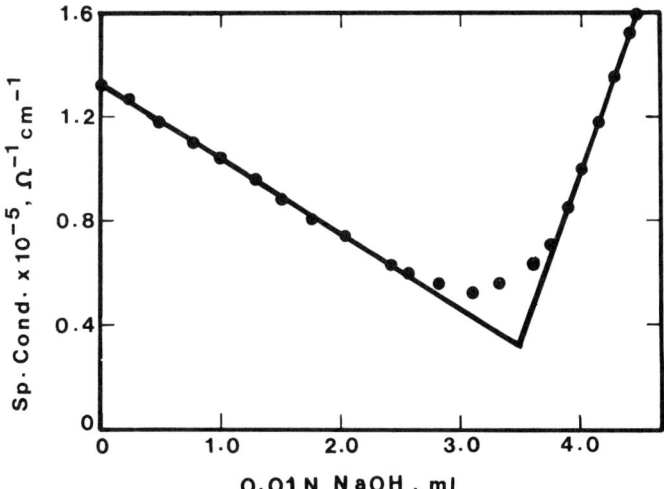

Fig. 1. Conductometric titration of ion-exchanged latex A-2 with sodium hydroxide.[6]

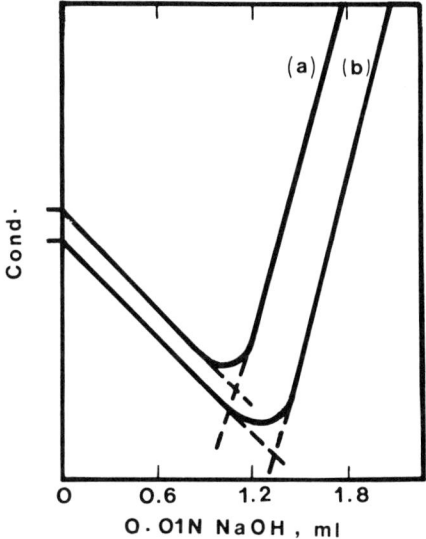

Fig. 2. Conductometric titration of ion-exchanged latexes with sodium hydroxide: (a) Latex A-2, (b) Latex D-4.[2]

as will be seen later, this effective charge is much smaller than the surface charge).

Figure 3 shows the results of conductometric titration with barium hydroxide of the same sample of latex A-2 shown in Fig. 1. In this case, only the ascending leg is linear; the descending leg is curved initially, but eventually becomes linear, so that extrapolation to intersection at 1.84 ml gives the equivalence point. The descending leg is curved because the innermost H^+ ions in the double layer, i.e., those that contribute least to the conductivity of the latex, are replaced first by the Ba^{2+} ions, and only later are the outermost faster-moving H^+ ions exchanged. Although the mobilities of the Ba^{2+} and H^+ ions are markedly different, the initial addition of Ba^{2+} ions gives only a slight decrease in the conductivity, but further additions give an increasing decrement in the conductivity until the descending leg becomes linear. An equivalence point of 1.84 ml 0.009 N barium hydroxide for 0.735 g latex polymer corresponds to a surface charge of 3.47 $\mu C/cm^2$, which is in good agreement with the value obtained with sodium hydroxide.

The reproducibility of this analysis is shown by the values of 3.60 and 3.47 $\mu C/cm^2$ obtained for the same sample of latex A-2 titrated with sodium and barium hydroxides, respectively. This is confirmed by the values of 3.58 and 3.61 $\mu C/cm^2$ obtained in duplicate titrations with sodium hydroxide carried out on the same day and the values of 3.60 and 3.50 $\mu C/cm^2$ from duplicate titrations carried out 14 days apart. Although the close agreement of these values demonstrates the reproducibility of the titrations, the average value of 3.56 $\mu C/cm^2$ does not accurately represent the surface charge of latex A-2. This sample was ion-exchanged at an early stage of the work, before the conditioning technique was fully developed. Later experiments with latex A-2 using mixed resins rigorously-purified according to the best technique gave a slightly smaller value for the surface charge; the average of twenty determinations was 3.3 ± 0.1 $\mu C/cm^2$.

The reproducibility of the ion-exchange-and-conductometric-titration technique is also shown by the results for latex A-3. Early experiments using the column technique of ion exchange gave a value of 1.97 $\mu C/cm$ for the surface charge. Later experiments with another sample of the same latex using the batch technique of ion exchange and the rigorously-purified resins gave a value of 2.02 $\mu C/cm^2$. Although the final values are in good agreement, it took several passes through the ion-exchange columns to achieve the same surface charge obtained after one batch treatment. Thus these results demonstrate the reproducibility, not only of the conductometric titration, but also of the ion-exchange technique.

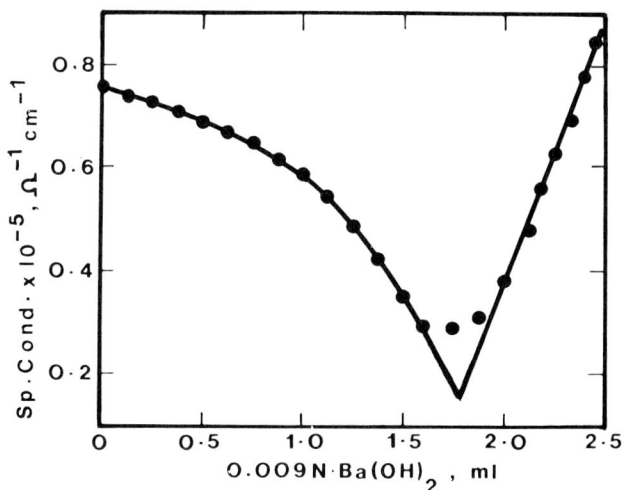

Fig. 3. Conductometric titration of ion-exchanged latex A-2 with barium hydroxide.[6]

This technique can also be used to determine weak-acid surface groups. Figure 4 shows the conductometric titration curve (continuous recorder trace) of a polystyrene latex prepared in the presence of acrylic acid using persulfate-ion initiator. The two distinct regions of this curve are attributed to the sulfate and carboxyl groups, respectively. Thus, it is possible to determine the number of both groups in the same titration, so that the absence of bound carboxyl groups (as was found for polystyrene latexes by Ottewill and Shaw[13], and Shaw and Marshall[14]) and residual carboxylate emulsifier (where used) can be demonstrated.

The surface charge can also be determined by potentiometric titration. Figure 5 shows typical curves for latexes A-2 and D-4, which also indicate that only strong-acid groups are present.[2] These potentiometric titration curves, however, are more difficult to analyze than the corresponding conductometric titration curves and, therefore, this method was not used extensively.

EFFICACY OF THE ION-EXCHANGE TECHNIQUE

Table 2 shows the variation of the surface charge of latex A-2 with the number of ion-exchange cycles. This series of experiments used the earlier batch ion-exchange procedure, i.e., a twofold excess of resin and a 3-hr contact time (thus necessitating more ion-exchange cycles to achieve a constant surface charge than if

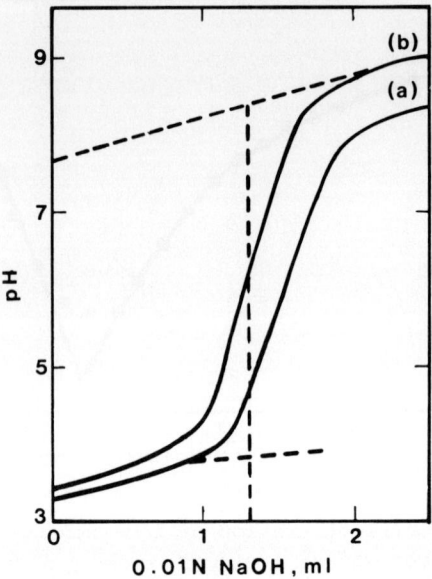

Fig. 4. Potentiometric titration of ion-exchanged latexes with sodium hydroxide: (a) latex A-2, (b) latex D-4.[2]

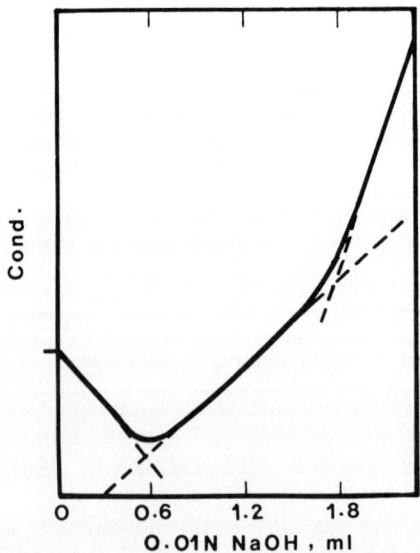

Fig. 5. Conductometric titration of ion-exchanged sytrene-acrylic acid copolymer latex with sodium hydroxide.[2]

the newer procedure were used). The surface charge reached s constant value of 3.3 µC/cm^2 after three ion-exchange cycles; however, it reached a minimum value of 2.86 µC/cm^2 after the first cycle and then increased to the constant value. It is tempting to take this minimum value as the actual surface charge of latex A-2 and attribute the increase in charge upon further ion exchange to contamination by polyelectrolytes leached from the resins. In fact, however, this minimum results from the incomplete exchange of Na$^+$ and K$^+$ ions for H$^+$ ions. This is shown by the constancy of the surface charge at a higher value of 3.3 µC/cm^2, whereas additional ion-exchange cycles should add to the contamination and thus gradually increase the surface charge.

This incompleteness of the ion exchange is also confirmed by experiments with Aerosal MA* emulsifier in which the same phenomenon was observed in the absence of latex particle surfaces. In these experiments, the specific conductance of a given volume of emulsifier solution in contact with 15 g mixed resin (0.125 meq) was measured as a function of time. The amount of emulsifier was varied from 0.125 to 10^{-4} meq, to give equivalent ratios of ion-exchange capacity/emulsifier ranging from 1 to 900. For ratios in the range 10-900, the specific conductance decreased sharply to a constant value equivalent to that of pure water within about one hour. For the ratio 3.2, however, the specific conductance increased twofold within about 15 min, then decreased to a constant value after 8 hr; this increase was attributed to the incomplete exchange of Na$^+$ ions for H$^+$ ions, since titration of the 24-hr sample with 0.01 N sodium hydroxide showed that 44% of the initial 3.7 x 10^{-2} meq emulsifier had been removed and that 84% of that remaining was in the Na$^+$ form. Thus, the completeness of the ion exchange is limited by the exchange of H$^+$ ions for Na$^+$ ions, as well as that of OH$^-$ ions for emulsifier anions, independent of whether the emulsifier is in solution or adsorbed on the latex particle surface.

Table 2 shows that the surface charge of latex A-2 reached a constant value of 3.3 µC/cm^2 after three ion-exchange cycles. That this value represents the number of chemically-bound negative surface groups remaining after complete desorption of the emulsifier was shown by several additional experiments. A sample of this rigorously ion-exchanged latex dialyzed in a cellophane bag for 5 days at 45°C with at least twenty changes of water showed upon titration a surface charge of 3.4 µC/cm^2, indicating that dialysis after ion exchange did not remove any additional surface groups. Another sample of this rigorously ion-exchanged latex was subjected to four more ion-exchange cycles for a total contact time of 40 hr; the final surface charge of this sample was also 3.4 µC/cm^2, indi-

---

*American Cyanamid Company

Table 2: Variation of Surface Charge of Latex A-2 with Successive Ion Exchange Cycles

Wt. of latex, g	Wt. of wet resin[a], g	Contact time, hr	Surface charge, $\mu C/cm^2$	Surface area per negative charge, Å^2
original	——	——	10.7	150-160
180	100	16	2.86	563
500	50	3	3.11	517
500	50	3	3.28	490
500	50	4.5	3.30	487
500	50	16	3.23	498

[a] about 50% water.

cating that additional ion-exchange cycles neither increased the surface charge by contamination nor decreased it by removing more surface groups. In other experiments, five different emulsifiers (Sipon WD,[*] Aerosal MA, Aerosol OT[**], potassium oleate, Dresinate 214[***]) were added to samples of the rigorously ion-exchanged latex in such concentrations as to exceed the critical micelle concentration in the aqueous phase; these samples were then ion-exchanged to remove this added emulsifier and titrated; in all five cases, the added emulsifier was removed completely after two ion-exchange cycles, and the average of the final surface charges was 3.4 $\mu C/cm^2$. Thus, the emulsifier is desorbed completely by the ion exchange without introducing enough ionic contaminants to significantly affect the particle charge.

The surface charge of the latex is also influenced by the purity of the resin used for the ion exchange. Table 3 gives the values for latex A-2 after one ion-exchange cycle (fivefold excess; 2-hr contact time) for various mixed resins. Different batches of the rigorously-purified Dowex mixed resin gave the same value for the surface charge, 3.3 $\mu C/cm^2$, but the same resin without purification gave a somewhat higher value. The Bio-Rad AG mixed resin gave a higher value without purification, and a lower value after washing, than the rigorously-purified Dowex resin. Thus, the use of unpurified resins gives a higher value for the surface charge, although these values are only 8% higher than those obtained using rigorously-purified resins.

---

[*] Alcolac, Inc.
[**] American Cyanamid Company
[***] Hercules, Inc.

Table 3: Effect of Resin Purity on Surface Charge of Latex A-2 After One Ion-Exchange Cycle[a]

Mixed resin	Surface charge, $\mu C/cm^2$	Surface area per negative charge, $Å^2$
Dowex, as received	3.53	454
Dowex, purified	3.34	480
Dowex, purified and washed	3.34	480
Dowex, purified and washed (second batch)	3.29	487
Bio-Rad AG, as received	3.58	448
Bio-Rad AG, washed	3.18	504

[a]Fivefold excess of resin; 2-hr contact time.

COMPARISON WITH DIALYSIS

Dialysis is the traditional method for purifying colloidal sols; however, in our experience, this method does not give complete removal of emulsifier or complete exchange of Na^+ and K^+ ions for H^+ ions. Table 4 gives a comparison of ion exchange with dialysis at 25° and 45°. The values of the surface charge were measured by conductometric titration and the residual concentration of Na^+ and K^+ ions by atomic absorption. The surface charge measured by conductometric titration was 3.4 $\mu C/cm^2$ for the ion-exchanged sample, which is in good agreement with other determinations of this latex. The values for both dialyzed samples, however, were considerably smaller. This might have been construed as evidence for the more complete removal of emulsifier anions by dialysis were it not for the fact that these dialyzed samples also contained a much higher concentration of Na^+ and K^+ ions than the ion-exchanged sample. The conductometric titration measures only the number of H^+ ions; any anion which has a Na^+ or K^+ counterion will go undetected by this technique. Therefore, the total concentration of anions (i.e., the chemically-bound negative groups plus the residual emulsifier and inorganic anions) is the sum of the concentrations of the H^+, Na^+, and K^+ ions; this is shown in the last column of Table 4 as the total surface charge. These values are significantly greater for the dialyzed samples than for the ion-exchanged sample. Moreover, the total concentration of Na^+ and K^+ ions in the ion-exchanged sample is negligible compared to the concentration of H^+ ions, but in the dialyzed samples it is so much greater that, if neglected, it can lead to a serious error in the determination of the surface charge. Thus, dialysis is less effective than ion exchange, both in removing emulsifier and inorganic anions and in exchanging H^+ ions for Na^+ and K^+ ions.

Table 4: Comparison of Dialysis with Ion Exchange For Latex A-2

Treatment	Surface charge of polymer by conductometric titration		Total concn. of Na^+ and K^+ ions in polymer, µeq/g	Total surface charge, µC/cm^2
	µC/cm^2	µeq/g		
Ion exchanged	3.4	21	<1	3.4
Dialyzed at 25°	2.1	13	∿20	∿5.3
Dialyzed at 45°	1.7	10	∿30	∿5.5

Other workers have also found that dialysis does not completely remove emulsifier from latexes. Brodnyan and Kelley[15] found that C^{14}-tagged sodium lauryl sulfate solutions in dialysis bags equilibrated with the surrounding medium in less than 24 hr, but that only 9.5 and 22% of the same emulsifier was removed from two polystyrene latexes under the same conditions. Edelhauser[16] found that the rate of dialysis of sodium lauryl sulfate against water was initially high, but soon dropped off to a virtual standstill without reaching an equilibrium (several earlier examples of this type are also cited in ref. 16); at this point, if the solution was transferred to a fresh dialysis bag, the rate was unaffected, but if the emulsifier concentration inside the bag was increased by further addition, the rate increased markedly; the driving force for dialysis is apparently the concentration gradient across the membrane; when this gradient is too small, the dialysis comes to a standstill. Force et al.[17] found with a butadiene-styrene copolymer latex stabilized with Dresinate 214 that only about 50% of the emulsifier was removed by dialysis for 160 days. On the other hand, Ottewill and Shaw[13] found from measurements of the electrophoretic mobility and from dialysis experiments using C^{14}-tagged emulsifier that the emulsifier was completely removed from polystyrene latexes by dialysis, or at least that a constant surface charge was attained.

SURFACE CHARGE OF MONODISPERSE POLYSTYRENE LATEXES

Table 5 summarizes the results for the monodisperse polystyrene latexes ion-exchanged to constant surface charge. In several cases, the surface charge was constant after only one ion-exchange cycle; however, latex A-1, which had the smallest particle diameter and was prepared with the highest concentration of emulsifier, required at least three cycles. The latexes prepared with potassium

Table 5: Monodisperse Polystyrene Latexes: Surface Charge and Number of Sulfate Endgroups/Molecule.

Latex	Particle diameter, Å	M_n,x10^{-4}	Surface charge, $\mu C/cm^2$	Surface area per negative charge, Å^2	Sulfate endgroups / molecule added	surface	total
A-1	250	——	0.5	3100	(406)	(12.4)	(13.0)
A-2	880	4.3	3.3	500	3.3	0.98	1.57
A-3	2340	17.3	2.0	800	5.2	0.87	1.59
B-1	1580	8.1	4.2	390	2.8	1.26	1.21
B-2	2480	7.5	5.7	280	2.6	1.02	1.23
C-1	2540	7.6	5.4	300	2.6	0.95	0.94
D-1	1090	10.6	1.6	1000	3.8	0.95	——
D-2	1870	10.6	1.8	910	5.5	0.90	——
D-3	2850	10.1	4.9	330	3.5	1.03	1.63
D-4	4470	8.5	8.1	200	3.1	0.93	2.07

A = Aerosol MA (bis-1,3-dimethylbutyl sodium sulfosuccinate);
B = Sipon WD (sodium lauryl sulfate); C = Aerosol OT (di-2-ethylhexyl sodium sulfosuccinate); D = potassium oleate.

oleate were first ion-exchanged with the mixed resin in the Na$^+$/OH$^-$ form to keep the emulsifier completely dissociated; the Na$^+$ ions were then exchanged for H$^+$ ion in the subsequent cycles. These ten latexes, prepared with four different emulsifiers, represent a particle-diameter range of 250-4470Å. The particle diameters are the number-average diameters determined by electron microscopy.[18] The values of the number-average molecular weight $\bar{M}_n$ were determined by osmometry of tetrahydrofuran solutions of polymer (recovered by coagulating the ion-exchanged latex by freezing, washing with methanol, dissolving in benzene or tetrahydrofuran, reprecipitating with methanol, and drying under vacuum). The values for the surface charge were determined by conductometric titration. The area per charge was calculated using the values of the average particle volume and surface area determined by electron microscopy. The number of surface charges per molecule was calculated using the values of the surface charge, average particle volume, and $\bar{M}_n$. To determine the total number of charges per molecule, a portion of the polymer recovered for osmometry was dissolved in dioxane-water mixture, ion-exchanged, and titrated again to give the total amount of sulfate (in some cases, these values were corroborated by X-ray fluorescence). For comparison, the total amount of sulfate ion added to the polymerization in the form of persulfate ion is also given. The value of M_n was not determined for latex A-1; in this case, the values given in parentheses are in μmol/g polymer.

The values for the surface charge varied from 0.5 to 8.1 µC/cm^2 and generally increased with increasing particle diameter. This increase is to be expected if all polymer endgroups are on the surface of the particles, because the values of $\bar{M}_n$ lie within a relatively narrow range while the surface/volume ratio decreases with increasing particle diameter, e.g., in the D-series, in which the values of $\bar{M}_n$ are in the range $8.5\text{-}10.6 \times 10^4$, the surface charge increased about fivefold and the total surface area decreased about fourfold over the particle-size range 1090-4470Å.

The values for the number of surface charges per molecule were all close to one (average 0.99) instead of the expected value of two. This discrepancy might result if some sulfate endgroups were buried within the particle. The values for the total number of charges per molecule show that this is the case for some latexes, but not for others, e.g., latex D-4 contains about 0.9 surface charges and about 1.1 buried charges per molecule, but latexes B-1 and C-1 contain no buried sulfate endgroups. The total number of charges per molecule were corroborated in some cases by sulfur analysis by X-ray fluorescence: latex A-2 contained 0.11 ± 0.01% sulfur by X-ray fluorescence and 0.117% by titration; latex D-3 0.052 ± 0.004% by X-ray fluorescence and 0.048% by titration. The values for the total number of charges per molecule were much smaller than the values corresponding to the total amount of persulfate used in the polymerization, indicating that some persulfate is lost through side reactions or remains undecomposed after the polymerization is completed.

PRESENCE OF HYDROXYL GROUPS

Except for the latex of largest particle size (latex D-4), the total number of sulfate endgroups per molecule was less than two. Other investigators have found similar values, e.g., about 1.2 endgroups per molecule for the Mutual recipe at 80°C using radioactive persulfate[19] and about 0.8 for 0.21% potassium persulfate and 0.5% sodium laurate at 35° and 60° using the dye-interaction and dye-partition tests.[20] One possible explanation for this discrepancy is that termination occurs by disproportionation (one sulfate endgroup per molecule) rather than by combination (two endgroups per molecule); however, there is strong evidence[21-25] that, for styrene, termination occurs principally by combination. Another possibility is the incorporation of hydroxyl endgroups according to the following reaction suggested by Kolthoff and Miller.[26]

$$SO_4^- \cdot + H_2O \rightarrow HO\cdot + HSO_4^-$$

The hydroxyl radicals thus formed would presumably initiate polymerization in a manner similar to that of the sulfate ion-radicals (except that the negatively-charged surface of the latex particles would not repel the uncharged hydroxyl radical as it might the sulfate ion-radical) and the nonionic hydroxyl endgroups thus incorporated into the polymer molecules would not be detected by the foregoing conductometric-titration analysis. If the hydroxyl radical-initiated polymerization occurs, the number of sulfate endgroups per molecule would be smaller than the expected value of two, and the discrepancy would be greater at low pH because the generation of hydroxyl radicals increases with decreasing pH.[26]

One possible method of detecting hydroxyl endgroups is to oxidize them to the carboxyl form, which can be titrated. The persulfate ion used for polymerization initiation is also a powerful oxidizing agent with a redox potential of -2V. The oxidation is usually slow, but can be catalyzed by traces of such metal ions as Ag^+.[27]

To test this possibility, ion-exchanged latex A-2 was heated for 6-hr at 90°C with various amounts of potassium persulfate and 10^{-5}M silver nitrate as catalyst; the samples were then cooled, ion-exchanged again, and titrated.[2] The titration curves showed a decrease in the number of sulfate endgroups and the appearance of weak-acid endgroups (presumably carboxyls). The results are given in Table 6. The values in the last column were calculated from the number of sulfate endgroups that were converted to carboxyls. The number of hydroxyl endgroups thus determined added to the number of sulfate endgroups gives a total number of per molecule of 2.06. Similar experiments with latex A-3 gave a value of 2.28. Moreover, the infrared spectra of latex A-2 polymer after oxidation showed the appearance of carbonyl attributable to carboxyl groups; however, the polystyrene (or styrene) is oxidized only to benzaldehyde groups.

Also, a series of polymerizations (Aerosol MA emulsifier) were carried out in which the emulsion was buffered at various pH values.[2] Table 7 shows that the number of sulfate end groups per molecule was much smaller at low pH than at high pH. As the production of hydroxyl radicals increases with decreasing pH, the decrease in the number of sulfate endgroups can be explained by the increasing proportion of hydroxyl radical-initiated polymerization.

COMPARISON WITH OTHER WORK

The total number of sulfate endgroups was also determined by the dye-partition method as described by Palit et al.[20,28], using benzene or chloroform as the polymer solvent and methylene blue

Table 6: Oxidation of Sulfate and Hydroxyl Endgroups in Latex A-2

$K_2S_2O_8$, % based on polymer	Number of endgroups, μmol/g polymer		
	$-SO_4^-$	-COOH	-OH
None (original)	22.5	0	0
10	11.7	22.0	11.2
15	8.0	24.7	10.2
30	6.4	26.3	10.2

Table 7: Monodisperse Polystyrene Latexes: Effect of pH of Polymerization Medium

pH	Particle diameter, Å	$\bar{M}_n$, X10^{-4}	Endgroups per molecule		
			Surface sulfate	Total sulfate	Hydroxyl plus sulfate
1.9	1430	11.6	0.19	0.20	1.84
3.3	1320	9.1	0.55	1.00	1.75
4.4	1330	7.9	0.62	1.05	2.00
5.0	1280	8.9	0.71	1.10	2.00
6.4	1300	9.2	0.46	0.55	2.01
7.8	1290	11.6	1.40	2.07	2.07
11.6	1620	7.9	0.62	1.00	1.42

in 10^{-2}N hydrochloric acid as the dye. The number of sulfate endgroups was determined by two methods: 1. The optical density of the solvent phase after shaking with the dye was compared with a calibration curve obtained with sodium lauryl sulfate instead of polymer[29]; 2. the optical density of the aqueous phase was determined, and the amount of dye that left the water phase to combine with the polymer endgroups was calculated. With both methods, the results were dependent upon the time of agitation, the solvent, and the polymer concentration. Figure 6 shows the effect of polymer concentration. Table 8 shows that the number of sulfate endgroups, even when extrapolated to zero concentration, was always much lower than the number determined by titration or X-ray fluorescence.

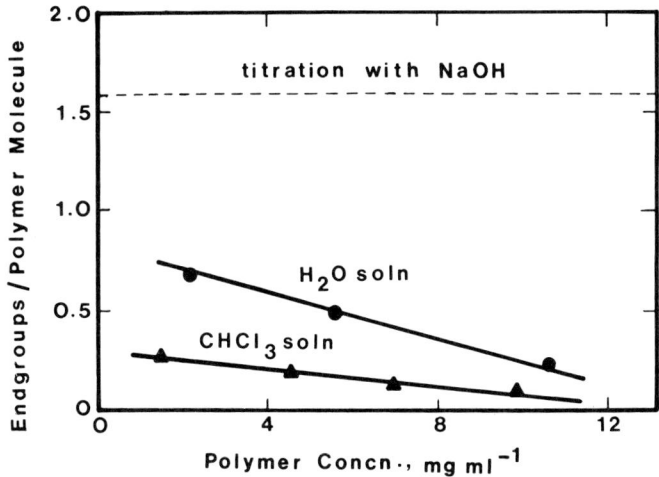

Fig. 6. Comparison of different methods for determination of sulfate endgroups of polystyrene from ion-exchanged latex A-2: (▲) dye partition, measured in chloroform solution; (●) dye partition, measured in aqueous phase, polymer in benzene; (---) conductometric titration.

Table 8: Comparison of Endgroup Determinations

Method	Number of sulfate endgroups per polymer molecule	
	Latex A-2	Latex D-4
Latex surface titration	0.98	0.93
Polymer solution titration	1.57	2.07
X-ray fluorescence	1.5	2.0
Dye partition	0.08-0.30	0.06-0.12

Thus, the total number of sulfate endgroups determined by conductometric titration in dioxane-water mixtures is in good agreement with that determined by X-ray fluorescence; however, the dye method gave values which were consistently lower and dependent upon the experimental conditions. Similar results were reported by Bitsch[30], who found that the number of endgroups determined for polystyrene by the dye-partition method was always much lower than that determined by other methods; however, for the more polar polymethyl methacrylate, dye partition gave the expected value of two endgroups per polymer molecule.

Apparently, because of the necessarily-poor solubility of the dye in the organic solvent, the dye-endgroup ion-pair must be formed in the water-solvent interface. Because of the apolar nature of polystyrene, few endgroups actually come to this interface and, therefore, the ion-pair formation is not quantitative. As the polymer-polymer interaction increases at higher concentrations, the extent of ion-pair formation decreases. This explanation is confirmed by Huber and Thies[31], who studied the adsorption of toluene-soluble polymers at the toluene-water interface; they concluded that polystyrene has little affinity for adsorption at this interface, but that polymethyl methacrylate absorbs significantly at concentrations as low as 10^{-4} g/100 ml; that the sulfate (and other polar groups) are not sufficient to bring the polymer to the interface is shown by the fact that an 89:11 ethylene-vinyl acetate copolymer adsorbs to only a slightly greater extent.

Another point is the presence of carboxyl groups. Ottewill and Shaw[13], and Shaw and Marshall[14], found a large proportion of these groups in polystyrene latexes prepared with hydrogen peroxide initiator and sodium laurate emulsifier, and cleaned by dialysis. Later, Ottewill et al.[32] found carboxyl groups in polystyrene latexes prepared without emulsifier using persulfate-ion initiator, and others[33] found carboxyl groups even with emulsifier.

In comparison, our experiments showed no significant concentrations of carboxyl groups in the rigorously ion-exchanged latexes, whether prepared with sulfonate, sulfate, or carboxylate emulsifier, as evidenced by both conductometric or potentiometric titrations, infrared spectra of the polymer, and electrophoretic mobility of the particles (however, the infrared spectra gave evidence for hydroxyl groups). To reconcile the discrepancy, polymerization experiments were carried out to determine the effect of inorganic buffering electrolytes on the sulface endgroups[34] because the emulsifier-free system[35] uses persulfate ion alone without buffering electrolyte so that the pH decreases form 7-8 to low values during the polymerization. Table 9[34] confirms, not only that the emulsifier-free system with persulfate ion alone gives both sulfate and carboxyl endgroups, but also the potassium persulfate-sodium bicarbonate systems (which was used to prepare most of the latexes described in this paper) gives only sulfate groups; similarly, the ammonium persulfate-ammonium hydroxide system gives only sulfate groups. The pH variation during the polymerization is not the only factor in determining whether carboxyl groups are formed, however, because potassium dihydrogenphosphate or sodium tetraborate buffers give both sulfate and weak-acid groups, even in the presence of sodium bicarbonate; in these systems, the

Table 9: Monodisperse Polystyrene Latexes: Effect of Polymerization Recipe on Surface Groups.[34]

Latex	Emulsifier	Initiator	Buffer	Endgroup Concentration µeq/g polymer sulfate	carboxyl	pH
41	Aerosol MA	$K_2S_2O_8$	$NaHCO_3$	6.99	None	7.0
W23	Aerosol MA	$K_2S_2O_8$	$NaHCO_3$	7.71	None	8.3
42	Aerosol MA	$K_2S_2O_8$	$NaHCO_3$, KH_2PO_4	6.95	1.41	7.0
43	Aerosol MA	$K_2S_2O_8$	$NaCHO_3$, $Na_2B_4O_7$	6.80	4.60*	8.7
44	Aerosol MA	$K_2S_2O_8$	$NaHCO_3$, $AgNO_3$	5.63	3.15	8.5
W1	Aerosol MA	$K_2S_2O_8$	NaOH, KH_2PO_4	4.01	9.70	7.0
W2	Aerosol MA	$K_2S_2O_8$	$NaHCO_3$, NaOH, KH_2PO_4	3.12	3.57	8.7
520	None	$K_2S_2O_8$	None	9.34	1.45	2.0
W21	Aerosol MA	$(NH_4)_2S_2O_8$	NH_4OH	7.28	None	2.7
W22	Aerosol MA	$(NH_4)_2S_2O_8$	NH_4OH	7.89	None	2.3
716D	Aerosol MA	$(NH_4)_2S_2O_8$	None	5.01	2.36	2.2

*Two different weak-acid endgroups, 2.51 and 2.09 µeq/g polymer.

weak-acid group may be introduced by chain transfer. Oxidation may play a part in the formation of carboxyl groups because the addition of a small amount of silver nitrate (a catalyst for the oxidation of hydroxyl groups to carboxyl groups) to the sodium bicarbonate system gives some carboxyl groups. The sodium hydroxide-potassium dihydrogenphosphate buffer system also gives carboxyl groups, even when added to the sodium bicarbonate system. Thus, these results show that even minor components of the polymerization recipe have a significant effect on the surface endgroups produced, but there are two systems ---- potassium persulfate-sodium bicarbonate and ammonium persulfate-ammonium hydroxide ---- which gave only sulfate groups.

LATEX STABILITY

Most of the monodisperse polystyrene latexes described in Table 5 have a sufficient number of surface sulfate groups to give at least adequate colloidal stability after ion exchange. For example, ion-exchanged latex A-2 (880Å diameter, 3.3 $\mu C/cm^2$ surface charge) displayed critical coagulation concentrations for sodium chloride, calcium chloride, aluminum chloride (pH3), and aluminum chloride (pH7) of 180, 18.5, 0.37, and 0.15 mM, respectively.[6] Such latexes are ideal model colloids, i.e., they comprise uniform-size spheres stabilized with a known number of chemically-bound strong-acid surface groups, and they have been used for this purpose in investigations of stability, viscosity, adsorption, conductance, interference colors, ultracentrifugation[5-7], and electrophoresis.[36]

Some of the foregoing latexes, however, were too unstable to be used for this purpose, e.g., latex A-1 (250Å; 0.5 $\mu C/cm^2$) flocculated at least partially during ion exchange and completely upon titration with sodium hydroxide[6], presumably because of its combination of small particle diameter and low surface charge. Also, latexes D-1 (1090Å; 1.6 $\mu C/cm^2$) and D-2 (1870Å; 1.8 $\mu C/cm^2$) flocculated slowly upon standing after titration in an excess of 10^{-4}N sodium hydroxide, presumably because of their low surface charge.

There are several possible mechanisms for the flocculation of latex particles upon ion exchange: (1) the anionic latex particles deposit on the surface of the cationic resin beads; (2) the emulsifier is removed from the surface of particles which have too-few residual surface groups to give stability; (3) the particles are flocculated by polyelectrolytes leached from the ion-exchange resin.

The deposition of latex particles on the resin beads should occur for both stable and unstable latexes; however, the amount of polymer lost in this way is likely to be small, and the stability of the particles remaining in dispersion is unaffected. The flocculation of latex particles by leached polyelectrolytes is minimized by using rigorously-purified ion-exchange resins. Therefore, the flocculation observed in the foregoing cases is most likely due to small surface charge.

The partial flocculation of a monodisperse latex upon ion exchange, of course, precludes its use as a model colloid; however, it does not alter the particle size distribution, nor does it affect the determination of the surface charge, assuming that all particles bear the same charge. For latexes with broader particle-size distributions, however, the preferential flocculation of either the smaller or larger sizes could alter the shape of the particle-size distribution curve and lead to appreciable errors in the determination of the surface charge. Therefore, the effect of ion exchange on latex stability was investigated,[37] particularly with reference to particle size and surface charge.

In addition to purified and unpurified Dowex resins, two other mixed resins were used: Amberlite Monobed MB-1* and Amberlite Monobed MB-3. The orange-brown Monobed MB-1 resin is a mixture of the sulfonate (H^+ form) and trimethylammonium (OH^- form) salts of styrene-divinylbenzene copolymer beads, with a capacity of 7.06 meq. $CaCO_3$/ml wet resin. The grayish-black Monobed MB-3 resin is a mixture of the sulfonate (H^+ form) and dimethylhydroxyethylammonium (OH^- form) salts, with a capacity of 8.48 meq. $CaCO_3$/ml wet resin. The most significant difference between the two Monobed resins is that the Monobed MB-3 contains an indicator dye that shows when the resin is exhausted. Both resins were used as received.

This work used three monodisperse polystyrene latexes with the following values for the particle diameter and surface charge: 1760Å (1.7 $\mu C/cm^2$); 330Å (1.1 $\mu C/cm^2$); 640Å (not determined, but probably in the same range as the other two latexes). These latexes were combined to form two mixtures of bimodal particle-size distribution: mixture I with 13:1 number ratio of 330Å/1760Å particles; mixture II with 9.8:1 number ratio of 640Å/1760Å particles. The particle number ratios were calculated from the weights of polymer and the number-average particle volumes determined by electron microscopy.

These mixtures were shaken in a separatory funnel with Monobed MB-1 resin, sampled at various times by filtration through a sintered-glass disk, and examined by electron microscopy. Table 10 gives the particle number ratios for different contact times as determined by electron-microscopic particle counts. In both cases, the number ratio of small/large particles decreased with increasing contact time, about sixtyfold for mixture I and about fivefold for mixture II after 48 hr. Thus, upon ion exchange with Monobed MB-1 resin, the smaller particles of both mixtures flocculated preferentially, the 330Å-diameter particles to a greater extent than the 640Å-diameter particles.

These three latexes were also ion-exchanged individually for 16 hr, either in separatory funnels or 8-oz** screw-cap bottles.

---
*Rohm & Haas Company
**1 oz = 28.4 cm^3.

Table 10: Effect of Ion Exchange with Amberlite Monobed MB-1
Resin of the Particle-Size Distribution of Latex
Mixtures

Mixture	Contact time, hr.	Number ratio small/large particles	
		1st Series	2nd Series
I	0	4.8/1	———
I	9	———	5.8/1
I	16	2.8/1	2.6/1
I	24	0.20/1	1.2/1
I	32	———	0.11/1
I	40	0.23/1	———
I	48	———	0.08/1
II	0	10/1	———
II	9	———	5.4/1
II	16	5.7/1	4.6/1
II	24	3.3/1	2.3/1
II	32	———	1.9/1
II	40	5.5/1	2.0/1
II	48	———	2.2/1

Table 11 shows the effect of ion exchange with various resins on the latex solids content and average particle diameter as measured by dissymmetry of light scattering[38]; this latter technique does not give an accurate value for the average particle diameter, but merely indicates the presence or absence of submicroscopic aggregates of latex particles. Ion exchange in separatory funnels produced visible flocs of polymer in all three latexes; however, the degree of visible flocculation of the 1760Å-size was significantly less for the purified Dowex resin than for the unpurified Dowex or Monobed MB-1 resins. In all cases, however, the dissymmetry measurements showed the presence of submicroscopic aggregates. From these results, the extent of flocculation was greatest for the 640Å-size, less for the 330Å-size, and least for the 1760Å-size.

Different results were obtained in the 8-oz bottles. None of the samples displayed flocs of polymer, and all samples with the exception of the 1760Å-size ion-exchanged with the purified Dowex resin or used as a "control" (i.e., shaken in the bottle without resin) showed the presence of submicroscopic aggregates by dissymmetry. Moreover, in several cases, the solid content increased upon ion exchange, the increase being greatest for the 1760Å-size ion-exchanged with the purified Dowex resin; this sam-

Table 11: Effect of Ion Exchange on Latex Solids Content and Particle Size

Latex Particle diameter, Å	Resin	After ion exchange	
		% solids	Average particle diameter,* Å
Separatory funnel			
330	Monobed MB-1	0.033	1300
640	Monobed MB-1	0.016	1320
1760	Monobed MB-1	0.069	∿ 2500
1760	Dowex, unpurified	0.066	1800-2200
1760	Dowex, purified	0.090	1700-2000
8-oz bottles			
330	Monobed MB-1	0.109	2600
330	Monobed MB-3	0.116	∿ 2600
330	Dowex, purified	0.122	1900-2500
640	Monobed MB-1	0.035	2600
640	Monobed MB-3	0.064	1900-2500
640	Dowex, purified	0.117	1950
1760	Monobed MB-1	0.028	1900-2400
1760	Monobed MB-3	0.054	1800-2400
1760	Dowex, purified	0.135	1500
1760	control	0.096	1500
Original			
330	———	0.100	315
640	———	0.100	690
1760	———	0.100	1500

*dissymmetry of light scattering.

ple showed no evidence of flocculation. This increase in solid content is attributed to the absorption of water by the ion-exchange resins, which were not equilibrated with water before use; if the resins absorb water from the aqueous phase of the latex, the latex particle concentration increases; whether the solid content increases or decreases upon ion exchange then depends upon the relative values of the solid-decrease due to flocculation and the solid-increase due to water absorption by the resin. Using the results for the 1760Å-purified Dowex resin combination as a reference point, the degree of flocculation was greatest with Monobed MB-1, less with the Monobed MB-3, and least with the purified Dowex resins. With the purified Dowex resin, the most stable latex was the 1760Å-size, followed by the 330Å- and 640Å-sizes in that order. With the unpurified Monobed MB-1 and MB-3 resins, however, the most stable latex was the 330Å-size, followed by the 640Å- and 1760Å-sizes in that order.

Thus, these results show that the flocculation of a latex upon ion exchange depends, not only upon the relationship between particle size and surface charge, but also upon the conditions of the ion exchange: size and shape of the vessel; the intensity of agitation; purity of the resin. The ion exchange of the latex mixtures with Monobed MB-1 resin, which showed preferential flocculation of the smaller sizes, is consistent with both mechanisms of flocculation: destabilization because of emulsifier removal and flocculation by leached polyelectrolytes. The ion exchange of the individual latexes with various resins, however, distinguishes between the two mechanisms and shows that, in certain cases, both are operative. The degree of flocculation in all cases was minimized using the rigorously-purified Dowex resin.

The ion exchange of latexes to remove adsorbed emulsifier and solute electrolyte is a useful technique for characterizing the particle surfaces and preparing model colloids. However, caution must be exercised in ion-exchanging the latex, particularly those comprised of broad or bimodal particle-size distributions. Flocculation upon ion exchange has more serious consequences for a polydisperse latex than for a monodisperse latex. The partial flocculation of a monodisperse latex, of course, precludes it use as a model colloid, but does not affect the characterization of the surface charge or the determination of the average particle size and particle-size distribution by electron microscopy. The partial flocculation of a latex comprised of a broad or bimodal particle-size distribtution, however, can lead to serious errors in the average particle size, particle-size distribution, and surface charge; also, such flocculation may be quite difficult to detect.

The use of rigorously-purified resins is also recommended for avoiding errors in the determination of the surface charge. The fact that very low values are obtained for the conductance of deionized water from unpurified resins does not necessarily mean that a latex ion-exchanged with this resin will not be contaminated by leached polyelectrolytes. For example, the wash water from unpurified Dowex mixed resin showed no evidence of leached polyelectrolytes, while that from the individual resins showed ample evidence of their presence. This suggests that in the mixed resin the leached polyelectrolytes are picked up by the resin of opposite charge. If this is the case, however, each latex particle must be considered as a tiny ion-exchanger which competes with the resin for the adsorbable polyelectrolytes according to its surface area and surface characteristics. In a typical ion exchange, the total surface area of the latex particles is substantial relative to that of the resin, and their surface characteristics are similar to those of the Dowex 50W resin. The adsorption of cationic polyelectrolytes would destabilize the latex because of charge neutralization (unless, of course, the polyelectrolyte concentration is great enough to give charge reversal, in which case the latex would be stable and cationic[39]). The adsorption of anionic polyelectrolytes would either stabilize or destabilize the latex according to the composition (charge/mass ratio), molecular weight, and concentration of the polyelectrolyte (the influence of these parameters on the adsorption and its subsequent effect on the stability is not yet well understood). In either case, however, the ion exchange would give a very low value for the conductance of the latex, but the surface charge would be altered, perhaps irreversibly, by the adsorbed polyelectrolytes, thus obviating any efforts to correlate the colloidal properties of the latex with theory. This is borne out by the significant difference in surface charge obtained with the purified and unpurified resins (Table 3).

FUTURE WORK

The general thrust of this work comprises three parts:

1. the preparation of monodisperse latexes;
2. the characterization of their surfaces;
3. the use of these well-characterized monodisperse latexes as model colloids.

The first part was accomplished many years ago in the Dow Chemical Company and, more recently, in several other laboratories. The second part is accomplished in this work. For the third part, some progress has been made---preliminary experiments using the well-characterized monodisperse latexes as model colloids give interesting and consistent results---but for definitive work, the model colloid must contain only one type of surface group. With a few exceptions, the latexes described here contain more than one type of surface group, e.g., sulfates and hydroxyls. The techniques of latex preparation must now be combined with the techniques of latex characterization to produce model colloids of different particle size, with variable numbers of one type of surface group. It is to this problem that we must now turn our attention.

SUMMARY

Monodisperse polystyrene latexes prepared with persulfate-ion initiator can be ion-exchanged to remove the adsorbed emulsifier and solute electrolyte. Rigorous purification of the ion-exchange resin is necessary to avoid contamination by leached polyelectrolytes. These ion-exchanged latexes are stabilized with the residual sulfate endgroups of the polymer molecules, the number of which can be determined by conductometric titration. The result is a dispersion of monodisperse spheres with a constant and known surface charge due to chemically-bound strong-acid surface groups. These latexes are ideal models for colloidal studies and preliminary experiments of their use in stability, adsorption, viscosity, sedimentation, interference colors, conductance, and electrophoresis studies (reported elsewhere[5-7,36]) give consistent results. Various sources of error in the ion-exchange and surface characterization are discussed.

REFERENCES

1. J.W. Vanderhoff, Preprints, Org. Coating and Plastics Chem. Div., A.C.S. 24(2), (1964) 223.
2. H.J. van den Hul and J.W. Vanderhoff, Br. Polym. J., 2, (1970) 121.
3. J.H. Schenkel and J.A. Kitchener, Nature, 182, (1958) 131.
4. R.H. Ottewill, private communication, 1966.
5. H.J. van den Hul and J.W. Vanderhoff, J. Colloid Interface Sci., 28, (1968) 336.
6. J.W. Vanderhoff, H.J. van den Hul, R.J.M. Tausk and J. Th. G. Overbeek in G. Goldfinger (Ed.), *Clean Surfaces: Their Preparation and Characterization for Interfacial Studies*, Marcel Dekker, New York, 1970, p. 15.
7. H.J. van den Hul and J.W. Vanderhoff in R.M. Fitch (Ed.), *Polymer Colloids*, Plenum Press, New York, 1971, p. 1.
8. H.J. van den Hul and J.W. Vanderhoff, J. Electroanal. Chem., 37 (1972), 161.
9. E.I. Akelroyd and R. Kressman, Chem. Ind. (Lond.), (1950) 189.
10. G.E. Boyd and K. Bunzl, J. Amer. Chem. Soc., 89 (1967) 1776.
11. H. de Bruyn and J. Th. G. Overbeek, Kolloid-Z., 84 (1938) 186.
12. G.A. van Os, Thesis, University of Utrecht, 1943.
13. J.N. Shaw, Thesis, University of Cambridge, 1965; R.H. Ottewill and J.N. Shaw, Kolloid-Z.Z. Polym., 215 (1967) 161.
14. J.N. Shaw and M.C. Marshall, J. Polym. Sci., A1 6 (1968) 449.
15. J.G. Brodnyan and E.L. Kelley, J. Colloid Sci., 20 (1965) 7.
16. H.A. Edelhauser, J. Polym. Sci., C27 (1969) 291.
17. C.G. Force, E. Matijevic and J.P. Kratohvil, Kolloid-Z.Z. Polym., 223 (1968) 31.
18. E.B. Bradford and J.W. Vanderhoff, J. Appl. Phys., 26 (1955) 864.
19. I.M. Kolthoff, P.R. O'Connor and J.L. Hansen, J. Polym. Sci., 15 (1955) 459.
20. P. Ghosh, S.C. Chadha, A.R. Mukherjee and S.R. Palit, J. Polym. Sci., A2 (1964) 4433.

21. F.R. Mayo, R.A. Gregg and M.S. Matheson, J. Amer. Chem. Soc., 73 (1951) 1691.
22. L.M. Arnett and J.H. Peterson, J. Amer. Chem. Soc., 74 (1952) 2031.
23. J.C. Bevington, H.W. Melville and R.P. Taylor, J. Polym. Sci., 12 (1954) 449; 14 (1954) 463.
24. C.H. Bamford and A.D. Jenkins, Nature, 176 (1955) 78.
25. C.G. Overberger and A.B. Finestone, J. Amer. Chem. Soc., 78 (1956) 1638.
26. I.M. Kolthoff and I.K. Miller, J. Amer. Chem. Soc., 73 (1951) 3055.
27. D. A. House, Chem. Rev., 62 (1962) 185.
28. S.R. Palit, Pure Appl. Chem., 4 (1962) 451; S.R. Palit and B.N. Mandal, J. Macromol. Sci.-Rev. Macromol. Chem.,C2 (1968) 225.
29. P. Mukherjee, Anal. Chem., 28 (1956) 870.
30. B. Bitsch, Thesis, University of Strasbourg, 1968; B. Bitsch, G. Parmeland, G. Riess and A. Banderet, Prepr. IUPAC Int. Symp. Macromol. Chem., Budapest, 1969, Vol. II, p. 49.
31. H.F. Huber and C. Thies, J. Polym. Sci., A2 8 (1970) 71.
32. J.W. Goodwin, J. Hearn, C.C. Ho and R.H. Ottewill, Br. Polym. J., 5 (1973) 347; J. Hearn, R.H. Ottewill and J.N. Shaw, Br. Polym. J., 2 (1970) 116.
33. R.H. Ottewill et al., I.M. Krieger et al., R.M. Fitch et al., A. Homola et al., M.E. Gultepe et al., Discussions at the NATO/ASI Meeting "Polymer Colloids", Trondheim, Norway, June 30 - July 11, 1975.
34. W.C. Wu, M.S. El-Aasser and J.W. Vanderhoff, unpublished research results, Lehigh University, 1976.
35. A. Kotera, K. Furusawa, and Y. Takeda, Kolloid-Z. Z. Polym., 240 (1970) 667; A. Kotera, etc., see manuscript.
36. G.D. McCann, J.W. Vanderhoff, A. Strickler and T.I. Sachs, Separation Purification Methods,2 (1), (1973) 153.
37. G.D. McCann, E.B. Bradford, H.J. van den Hul and J.W. Vanderhoff in R.M. Fitch (Ed.), *Polymer Colloids*, Plenum Press, New York, 1971, p. 29.
38. M.E. Elder, Thesis, Case Institute of Technology, 1951.
39. J.W. Vanderhoff and E.F. Gurnee, Tappi,39 (1956) 71.

Surface Area of Polymer Latexes by Angular Light Scattering

Robert L. Rowell and Raymond S. Farinato

Department of Chemistry
University of Massachusetts
Amherst, Massachusetts 01002

Angular light scattering may be used to determine particle size distribution, refractive index and number concentration of a polymer latex. An overview is given of the theory of angular light scattering including the relations for determining particle size distribution, number concentration and surface area. The dependence of angular light scattering patterns on size parameters is shown in terms of computer-drawn intensity surfaces. Methods of determining surface area are discussed and a new multi-dimensional fit is proposed. Results in terms of particle size distribution, number concentration and surface area of several latexes are discussed. It is shown that angular light scattering is a sensitive measure of the specific surface area.

INTRODUCTION

The examination of a polymer latex using electromagnetic radiation may take on several forms such as microscopy, angular light scattering, turbidity spectra or Rayleigh linewidth spectroscopy. We report some recent progress in the application of angular light scattering to the study of polymer latex suspensions. Our attention was drawn to this method because of two general advantages: 1) an *in situ* technique, and 2) a non-destructive technique.

Angular light scattering has a long history but has been reviewed recently in Kerker's treatise on light scattering[1] and is a subject of continuing interest in many laboratories[2-18].

Fundamentally, there are four pieces of information obtainable from angular light scattering: 1) average or modal size, 2) breadth or spread of particle size, 3) refractive index of the polymer latex, and 4) number concentration of particles. Clearly, if the particle size distribution and number concentration are determined, the surface area of the polymer latex is readily calculated.

We should like to point out that there are several important limitations of angular light scattering as a general method, but fortunately these do not limit a wide application. The principal limitation is a restriction to systems of low polydispersity. Light scattering theory has been longest known and most widely applied to spheres although recently significant progress has been made for ellipsoids[1,14-16,19]. Particle size distributions susceptible to analysis should not be too polydisperse, and the particles should be of homogeneous composition, i.e., chemical composition should be represented by a single refractive index. However, it should be pointed out that considerable progress has been made in the study of coated spheres and other non-spherical shapes[1].

In this paper, we should like to present an overview of light scattering calculations, suggest some new directions, and report some preliminary results. This overview will be carried out by representation of light scattering calculations in terms of computer-drawn intensity functions which will serve to illustrate the problem of determining particle size distribution and number concentration from angular light scattering.

THEORY

The theory describing the scattering of electromagnetic radiation by a dielectric sphere has been widely attributed to Mie[20], although it has recently been shown that the roots extend to the work of Lorenz[1]. The Lorenz-Mie theory is covered, in detail, in the context of a general consideration of the scattering of light in Kerker's treatise[1]. In general, we may write that the light scattered by an assembly of spheres is a function of five variables

$$I = I(\alpha, \sigma, m, \theta, N) \tag{1}$$

The key variables are particle size parameter $\alpha = 2\pi r/\lambda$ which is the ratio of the circumference of the particle to the wavelength of light in the medium; σ the breadth parameter of the particle size distribution; m the refractive index of the particles, and θ the scattering angle between the direction of observation and the direction of propagation of the illuminating light beam. If absolute intensity measurements are made, then the scattering is proportional to the total number of particles present so that the number concentration N may be determined. This assumes that measurements are carried out on a dilute latex so that mutual interference and multiple scattering are absent and that the total scattering is additive.

One can use polarized or unpolarized light in a variety of ways as noted below. As an example, we consider the case of vertically polarized incident light with observation of the vertically polarized scattered light, V_v scattering, to illustrate the relationship between light scattering and surface area[21].

$$V_v = N(\lambda/2\pi)^2 \int_0^\infty G(\alpha)(i_1)_\theta d\alpha \qquad (2)$$

Here, V_v is the measured Rayleigh ratio for vertically polarized light; λ is the wavelength in the medium; $G(\alpha)$ is the particle size distribution function; $(i_1)_\theta$ is the theoretical intensity coefficient calculated from the Lorentz-Mie theory, and α is the size parameter given above. In practice, the integral is replaced by a sum to sufficiently approximate the particle size distribution. The form of the distribution is unknown so that the usual practice is to assume a two-parameter distribution[6] and accept the best-fit results as a close approximation to the true distribution. This error is small for nearly monodisperse systems where the distribution is neither very broad nor very assymetric but becomes more of a limitation for broader distributions. With the size distribution known, the surface area is calculated from

$$S = 4\pi \Sigma N_i r_i^2 \qquad (3)$$

The total surface area is an extensive property but very often an intensive property such as the specific surface is desired. The specific surface S_w (usually in m^2/g) is obtained with a knowledge of the density and the total volume V given by

$$V = (4/3)\pi \Sigma N_i r_i^3 \qquad (4)$$

The specific surface becomes

$$S_w = S/V\rho = (3/\rho)(\Sigma N_i r_i^2)/(\Sigma N_i r_i^3) \qquad (5)$$

which is essentially the ratio of the second moment to the third moment of the size distribution. It is instructive to consider the important special case of a monodisperse system which results in a simple expression for the specific surface.

$$S = 3/\rho r = 6\pi n/\rho \lambda_o \alpha \qquad (6)$$

Here, we have also expressed the surface in terms of the dimensionless light scattering size parameter α, the vacuum wavelength λ_o of the incident light and the refractive index n of the medium. It is important to note that the specific surface of a monodisperse colloid becomes independent of the number concentration and refractive index of the spherical particles.

ANGULAR LIGHT SCATTERING PATTERNS

The usual unit of information available in angular light scattering is a single scan of intensity versus scattering angle which may exhibit one or more maxima and minima. The number of angular location of maxima and minima vary with particle size distribution. To illustrate this, we have prepared computer-drawn graphs of the vertically polarized scattering for a fixed range of angles from 45° to 115° and for a fixed range of size parameter varying from 2.0 to 6.9.

Figure 1 shows the V_V intensity surface for a refractive index of 1.200 characteristic of many polymer latex systems. The regularity of the "wave-like" structure is clearly shown as a function of scattering angle facing left in the figure and as a function of size parameter facing right in the figure. The intensity scale is logarithmic and varies over four orders of magnitude so that the oscillations shown in the figure are of large magnitude. The figure clearly distinguishes regions of pronounced structure from regions that tend to be featureless presenting a concise overview of the domains most likely to reveal size distribution parameters.

Figure 2 shows the V_V intensity surface for the same angle-size domain but for a refractive incex of 1.486. One can imagine that with higher refractive index the simple wavelike structure of the 1.200 refractive index surface is replaced by an interference pattern of two "waves" giving rise to a more complicated variation in the pattern of maxima and minima. This surface helps us to see at a glance how an angular scattering pattern showing a single maximum at small particle size can split to yield two maxima at a larger size.

Figure 3 shows the V_V intensity surface for a refractive index of 1.510 which appears to be identical with the surface for a refractive index of 1.486 shown in Figure 2.

The differences between Figures 2 and 3 are shown in Figure 4 where the computer has plotted the matrix obtained by element-by-element subtraction of the V_V intensity data used to generate Figures 2 and 3. Figure 4 is a difference surface for the intensities of refractive index 1.486 minus the intensities of refractive index 1.510. The figure shows that even though the refractive index difference is very small there are some regions for which the light scattering differences are very large. Note that the intensity scale in the figure is logarithmic.

Surface Area of Polymer Latexes 401

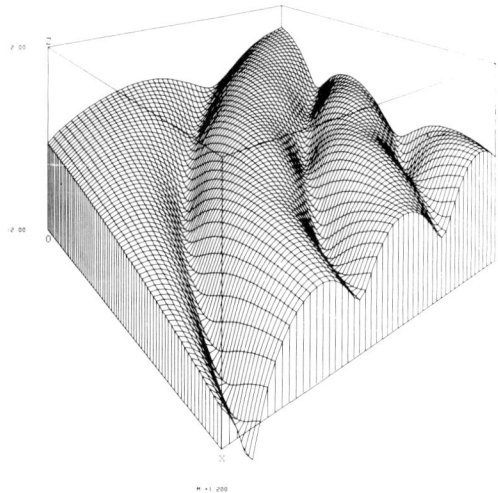

Fig. 1. V_v Surface, m = 1.200

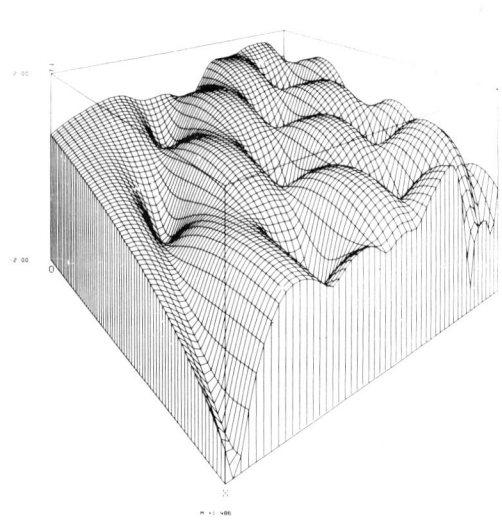

Fig. 2. V_v Surface, m = 1.486

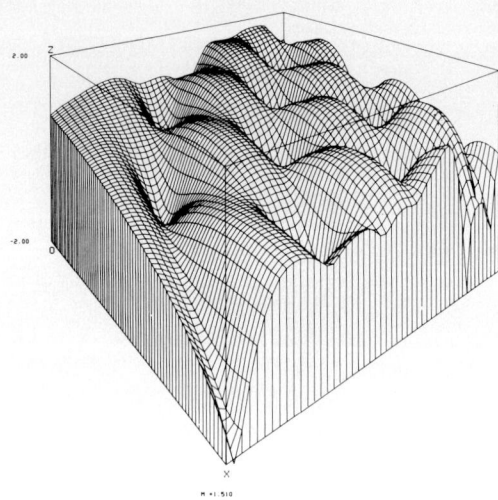

Fig. 3. V_v Surface, m = 1.510

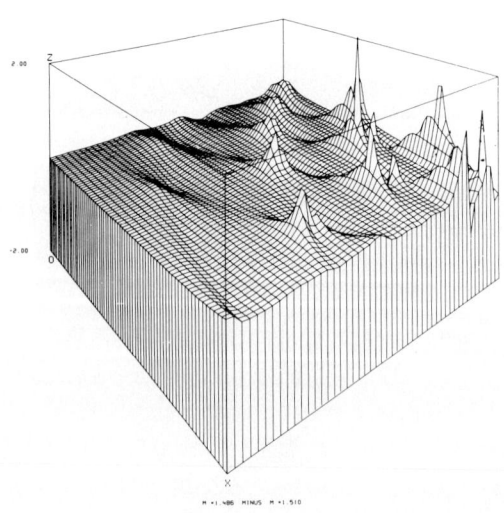

Fig. 4. V_v Difference Surface, m = 1.486 - 1.510

A difference surface with a great deal of structure is shown in Figure 5 which gives the difference for the V_V intensity surface of refractive index 1.200 minus the intensity surface of refractive index 1.510. The large variations in structure over nearly all the angle-size domain demonstrate the sensitivity of the angular light scattering to refractive index.

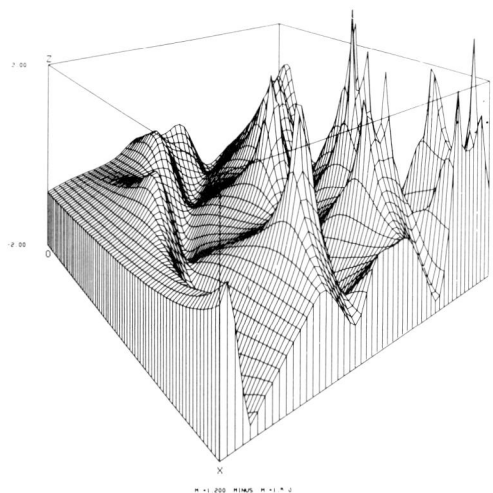

Fig. 5. V_V Difference Surface, m = 1.200 - 1.510

With this overview of the angular light scattering information in mind, we can re-examine the series of intensity surfaces discussed above for the case of a polydisperse system. The computed surfaces shown above are the patterns generated by a single particle of varying size. For N identical particles the scale of the surface is shifted but the pattern is the same since a logarithmic scale was used. However, for a distribution of particle sizes about some modal size, the features would be less pronounced. The effect of polydispersity on the patterns shown in Figures 1-5 is to smooth out the structure to a degree dependent on the extent of polydispersity. It is therefore easy to see that some domains of parameters are more useful than others in the determination of particle size distribution. In the following paragraphs, we discuss these problems in more detail.

DETERMINATION OF SURFACE AREA

The problem of the determination of the surface area of a

polymer latex by angular light scattering is twofold: 1) the determination of the characteristic size distribution parameters: modal size, distribution width, and refractive index; and 2) the determination of the scale parameter or number concentration. The latter problem requires either an absolute intensity measurement or a measurement that has been calibrated against some standard.

Let us consider the problem of the determination of the characteristic size distribution parameters. This is also known as the inversion problem since it is in essence a problem of relating an observed angular scattering pattern to a set of size distribution parameters. The problem has been worked on extensively by several laboratories as indicated above. Various characteristics of the scattering surface have been used in the inversion such as angular location of the minima, relation of minima to maxima, etc., or simply a least-squares type of best-fit procedure using one of several matching parameters such as the intensities V_v, H_h or $V_v + H_h$, the depolarization ratio H_h/V_v or the polarization $P = (H_h - V_v)/(H_h + V_v)$. In all of these approaches, however, one has in essence attempted to find a one-dimensional fit, i.e., to find that one dimensional curve on the scattering surface that gives the best fit to a one-dimensional experimental template, the angular scattering pattern. Such an experimental template is shown for example on a linear scale in Figure 6.

We should like to present an extension of the one-dimensional inversion procedure. Instead of using a single template to find a best-fit between experiment and theory, we suggest that two or more dimensions be simultaneously used in comparing experiment and theory. It is obvious from an examination of the sample V_v intensity surfaces we have presented here that a two-dimensional fit involving matching a surface to a surface will have a better chance at a unique solution than a one-dimensional fit wherein a one-dimensional experimental template is matched to a library of machine computations.

We are not limited to a two-dimensional fit between experiment and theory since by varying simultaneously the experimental wavelength and polarization, i.e., multichannel collection of experimental data, it is easy to generate a multi-dimensional comparison matrix.

RESULTS AND DISCUSSION

We have been working on formulating, programming and testing the multi-dimensional approach to the inversion of angular light scattering. In the following paragraphs, we report on some early test results using a simplified one-dimensional fit and calibration procedures already available in the literature.

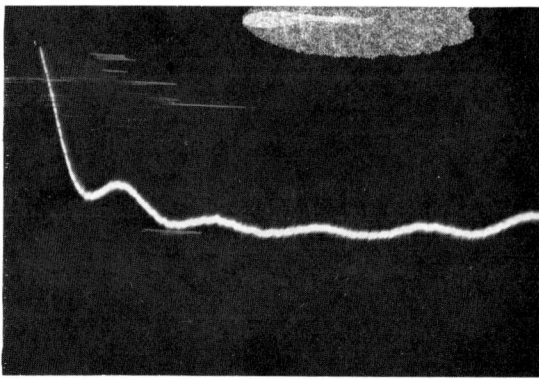

Fig. 6. Oscilloscope trace of angular light scattering from minicomputer memory.

Our experimental data is rapidly collected using a minicomputer interfaced to a standard Brice-Phoenix light scattering photometer that has been automated to improve the speed of data collection[17]. Figure 6 shows a photograph of the oscilloscope display of the digitized angular scattering pattern. The minicomputer, a Nicolet 1072 has a 12-bit memory word and a 4000 word memory so that high resolution of the experimental data is possible.

Several polystyrene latex samples of low polydispersity have been measured by angular light scattering using a log-normal distribution function and finding the best fit following the procedure of Rowell and Levit[17]. The results are summarized in Table 1 which gives the log-normal size distribution parameters from light scattering, α_M the modal size and σ_o the distribution width. Since absolute intensities were measured, it was also possible to obtain the total number concentration N at which the light scattering measurements were made according to Eq. (2). From these data and the dilution factor, we calculated the light scattering (LS) percent solids in the parent sample and compare it with a direct gravimetric determination (G) in Table 1. Finally, in Table 1, we show the specific surface as obtained from Eq. (5) and compare it with the contribution to the specific surface S_M arising only from particles of the modal size as computed from Eq. (6).

In all cases, the light scattering gave higher values than the gravimetric percent solids. We are continuing the investigation of the discrepancy.

A discrepancy in the number concentration obtained from light scattering was also noted by Nicolaon et al. in a study of the Brownian coagulation of submicron aerosols[9]. In their work, it was possible to compare the particle concentration obtained from light scattering with the particle concentration calculated from the known light scattering size and the mass of aerosol collected in a thermal precipitator. They obtained an aerosol count of 1.20×10^7/cc from thermal precipitation compared to 3.3×10^6 by light scattering.

It is clear that the light scattering method on an absolute base is subject to considerable uncertainty. This may be contrasted with earlier work in which <u>relative changes</u> in the surface area of a growing polydisperse sol could be observed with much greater precision[17].

Table 1

Size Parameters and Specific Surface*
of Polystyrene Latexes

Sample	α_M	σ_o	N (10^6/ml)	LS (% Solids)	G (% Solids)	S_W (m^2/g)	S_M (m^2/g)
w-235	2.20	0.08	62.2	9.4	8.8	24.5	24.9
w-424	4.10	0.05	12.6	11.6	8.6	13.1	13.4
w-540	5.00	0.08	9.3	16.6	9.8	10.8	11.0
w-670	6.60	0.06	1.96	7.9	3.6	8.2	8.3

*The specific surfaces have been calculated from Eqns. (5) and (6) with $n = 1.34$, $\lambda_o = 0.4358 \times 10^{-4}$ cm, and $\rho = 1.057$[22,23].

As we have shown above, the specific surface of a colloid of low polydispersity is expected to be less sensitive to the absolute number concentration of particles. This is demonstrated in the specific surface S_W of the latexes given in Table 1 which in every case is practically identical with the modal specific surface S_M. The low polydispersity of the latexes is also shown in the parameter σ_o, the zeroth order log-normal standard deviation. For narrow distributions, σ_o is related to the Gaussian standard deviation σ by $\sigma \approx \alpha_M \sigma_o$ so that σ_o is a measure of the fractional spread in particle size.

We should like to emphasize that in systems of low polydispersity the specific surface is essentially determined by the particle size as shown by Eq. (6). The reliability of our measurements of specific surface is governed mainly by the reliability of the determination of modal size. The samples analyzed in this work were prepared in the laboratory of Professor Thomas P. Wallace of the Department of Chemistry of Rochester Institute of Technology.

The sample code-labels in Table 1 are 100 times the modal size parameter obtained by Professor Wallace in his independent evaluation of particle size by light scattering. The agreement in the determination of modal size ranges from 1.5 to 8% and may be used as a measure of the uncertainty in the specific surface as determined by light scattering.

ACKNOWLEDGEMENT

The calculations were carried out in the University of Massachusetts Computing Center. The preparation of the computer-drawn figures was done by Dr. Melvin Prueitt using his program PICTURE at the Los Alamos Scientific Laboratory of the University of California. The general investigation was supported in part by a grant from the National Science Foundation. We should also like to acknowledge the assistance of Mr. Stephen Vasconcellos in the final preparation of the manuscript.

REFERENCES

1. M. Kerker, *The Scattering of Light*, Academic Press, New York, 1969.
2. L. J. Stryker and E. Matijevic, J. Colloid Interface Sci. 31, 39 (1969).
3. T. P. Wallace and J. P. Kratohvil, J. Polymer Sci. A-2, 8, 1425 (1970).
4. J. P. Kratohvil and T. P. Wallace, J. Phys. D Appl. Phys. 3, 221 (1970).
5. W. Heller and J. Witeczek, J. Phys. Chem. 74, 4241 (1970).
6. R. L. Rowell and A. B. Levit, J. Colloid Interface Sci. 34, 585 (1970).
7. T. P. Wallace, J. Polymer Sci. A-2, 9, 595 (1971).
8. T. P. Wallace and W. B. Scott, J. Polymer Sci. A-2, 10, 527 (1972).
9. G. Nicolaon, M. Kerker, D. D. Cooke and E. Matijevic, J. Colloid Interface Sci. 38, 460 (1972).
10. M. Kerker, J. Colloid Interface Sci. 39, 2 (1972).
11. F. T. Gucker, J. Tuma, H.-M. Lin, C.-M. Huang, S. C. Ems and T. R. Marshall, J. Aerosol Sci. 4, 389 (1973).
12. P. McFaden and E. Matijevic, J. Colloid Interface Sci. 44, 95 (1973).
13. D. D. Cooke and M. Kerker, J. Colloid Interface Sci. 42, 150 (1973).
14. W. Heller and M. Nakagaki, J. Chem. Phys. 60, 3889 (1974).
15. M. Nakagaki and W. Heller, J. Chem. Phys. 61, 3297 (1974).
16. W. Heller and M. Nakagaki, J. Chem. Phys. 61, 3619 (1974).
17. A. B. Levit and R. L. Rowell, J. Colloid Interface Sci. 50, 162 (1975).

18. T. P. Wallace, R. J. Cembrola, A. J. Migliore and D. E. DuCann, J. Colloid Interface Sci. 51, 283 (1975).
19. S. Asano and G. Yamamato, Applied Optics 14, 29 (1975).
20. G. Mie, Ann. Physik 25, 377 (1908).
21. R. L. Rowell, T. P. Wallace and J. P. Kratohvil, J. Colloid Interface Sci. 26, 494 (1968).
22. T. L. Pugh and W. Heller, J. Colloid Sci. 12, 173 (1957).
23. W. Heller and T. L. Pugh, J. Colloid Sci. 12, 294 (1957).

Evaporative Rate Analysis: Its First Decade

John Lynde Anderson

ERA Systems, Inc.
4048 Brookfield Circle
Chattanooga, Tennessee 37412

This is a review paper summarizing all known technical and professional publications in the field of Evaporative Rate Analysis. Even though the published literature extends back to 1963, the first commercial sales took place in late 1965. A comprehensive annotated bibliography is appended which includes, but is not restricted to, all papers read during the three Symposia on ERA sponsored by the ACS Division of Organic Coatings and Plastics Chemistry. These Symposia are identified as ERA I (NYC, Sept. 1969), ERA II (Chicago, Sept. 1970), and ERA III (Dallas, April 1973). The bibliography includes two recent ASTM methods and all U.S. patents in the field (foreign equivalents in Canada, England, West Germany and Japan are not listed). The articles are listed chronologically within general categories to facilitate use: General and Descriptive (GD), Residues and Lubricants (RL) which also includes cleaning and cleanliness phenomena, Cure and Polymer Characterization (CP), Adsorption/Desorption Phenomena (A/D), Miscellaneous (M), and Patents (P).

INTRODUCTION

ERA is an analytical technique used in a number of industrial and governmental facilities (within the United States and abroad) both for research and development purposes as well as for quality and production control. The technique differs from most other methods of surface analysis in that the characterization of the surfaces and surface related phenomena is carried out by the addition of at least one chemical detector to the test surface followed by observing the rate at which this detector disappears from the surface.

UNDERLYING PRINCIPLE

The underlying principle on which ERA is based is that a monolayer equivalent of a high boiling but volatile compound tends to be retained by the surface on which it is deposited as a function of a number of different surface properties. Thus a particular compound may essentially react with a surface (the process of chemisorption) and be retained for very long periods; it may rapidly evaporate within just a few seconds as, for example, from a clean non-retentive surface; it may diffuse into and out of a surface (such as the action of a solvent on a polymer surface); or it may be retained by other molecular forces for greater or lesser periods of time. The balance of these tendencies determines just how long the added volatile compound will remain on the surface under standardized evaporative conditions. The extent of the interactions or retentive forces may be used as a means of evaluating or characterizing the pre-existing surface properties merely by sensing the rate of evaporation of the added high boiling, but volatile compound. The ERA technique, as presently practiced, usually takes only one to three minutes per test surface and is carried out at ambient temperature and pressure in the laboratory atmosphere (but with metered dry, gaseous nitrogen flowing over the deposited compound on the test surface). A large number of high boiling, but volatile compounds may be used in these determinations of surface properties, but as of this writing, only seven compounds have been studied and used extensively as are discussed in detail below.

Each particular high boiling, but volatile compound in ERA use has been made highly radioactive by earlier synthesis using pure $BaC^{14}O_3$ so that each molecule has at least one carbon-14 nucleus. Thus, by placing a thin end-window Geiger Mueller detector adjacent to the test surface, sufficient beta emissions may be detected to provide accurate data representing the rate at which the added compound evaporates or otherwise disappears from "view". The aforementioned dry gaseous nitrogen sweeps the already evaporated radioactive molecules out from under the detector window and the rate of evaporation is determined by counting the number of emissions from the retained radiochemical on a time-phased basis.

CHEMICAL DETECTOR

As mentioned above, one chemical detector is normally employed in the ERA method, being the already described volatile radiochemical. In addition, a solvent or solvent mixture of approximately 50°C boiling point is used in ca. 100,000:1 solvent:radiochemical ratio. The solvent has two primary purposes, namely: To permit accurate and reproducible deposition of the approximately 5×10^{14} molecules of radiochemical used in each test and to enhance the particular surface properties being evaluated. An inert solvent is used for adsorptive/desorptive measurements; a participating solvent is selected to temporarily increase the fractional free volume of polymer surfaces thus permitting increased diffusion of the radiochemical molecules or to solvate residues, lubricants or contaminants in those measurements. The solvent also tends to reduce temporarily the number of detected carbon-14 emissions since the solvent absorbs a portion of the relatively weak betas before they are detected by the G.M. tube. Although the solvent disappearance normally takes place within twenty or thirty seconds from a non-participating metal surface, it may be retained for considerably longer periods where diffusion related phenomena are involved.

The solutions of volatile radiochemical in low boiling solvent are known as Test Solutions and are distributed exclusively by ERA Systems, Inc. as preformulated, standardized compositions in sealed-in-glass ampules to preserve their integrity and titre. The Test Solution formulations so far used have been based on tetrabromoethane-C14 (A), tridecane-C14 (B), diethyl succinate-C14 (C), N,N-dimethyl-n-decylamine-C14 (D), 2-ethylbutyric acid-C14 (E), beta-cyclohexylethanol-C14 (F), and alpha,omega-dibromononane-C14 (G) all of which (except G) exhibit rather remarkably similar rates of evaporation from non-reactive, non-retentive surfaces. For solvents, trifluorotrichloroethane (J) and cyclopentane (K) have been used as primary solvents with chloroform (L), methanol (M), methyl acetate (N), and tetrahydrofuran (P) being used as secondary solvents. The letter designations are those currently used for coding the various Test Solution formulations (the second solvent, if present, being preceded by the percentage of that solvent). Thus TSAJ is the Test Solution consisting of tetrabromoethane-C14 in trifluorotrichloroethane while TSBK20P is the Test Solution consisting of tridecane-C14 in cyclopentane:tetrahydrofuran in an 80:20 ratio.

The precision dispensers used in ERA technology (described in more detail below in the discussion of the MESERAN Surface Analyzers) permit the accurate and highly reproducible deposition of a nominal 20 microliters of Test Solution per test. Each deposition consists of approximately 0.05 microcuries of carbon-14. There are 2.50 ± 0.05 microcuries per milliliter of Test Solution. Thus since each Test Solution ampule contains a nominal 1.6 ml of solution, there are 4.0 microcuries per sealed-in-glass ampule. (Note: These figures apply strictly to C, D, E, F, and G radiochemicals whereas B is formulated at 2.70 ± 0.05 microcuries per ml, while A is always formualted at 3.25 ± 0.05 microcuries per ml.)

Ths uses of the MESERAN Test Solutions supplied by ERA Systems, Inc. are exempt from licensing requirements under present U.S. Nuclear Regulatory Commission (and so-called "Agreement States") requirements, due to the extremely small amounts of radioactive material involved.

From a standpoint of safety in the use of the radiochemicals, the rules of the U.S. Nuclear Regulatory Commission provide that up to 10^{-7} microcuries of bound carbon-14 may be released to unrestricted areas per milliliter of air (the word "bound" refers to carbon-14 incorporated in chemicals as compared to non-bound carbon in carbon dioxide where the limit is 10^{-6}). Thus in a "typical" laboratory of 3 x 4 x 5 meters with 10 air changes per hour, there are 600 cubic meters of air per hour. Since each ERA test releases only 0.05 microcuries, it is obvious that even with as many as 20 tests per hour, the release rate will never exceed 1/60 of the limits to unrestricted areas. The recommended use of hoods or other supplemental venting to the outside atmosphere greatly reduces even this minimal exposure.

Insofar as the radiological aspects of ERA technology is concerned, the amount of radioactivity per test (or in the aggregate) is without demonstrable hazard.

Among the several Test Solution components, several are listed as toxic or hazardous in the list of toxic materials promulgated by the U.S. Government. Thus tetrabromoethane (listed as acetylene-tetrabromide) is considered to be carcinogenic and recent studies show that chloroform also has a degree of carcinogeneity. Methyl acetate is toxic to a lesser degree and certain other components are also shown to be "hazardous" under high exposure conditions.

The listed maximal permissible limit for tetrabromoethane is 1 part per million of air. There is never any possibility that that limit can ever even be approached in ERA technology since each deposition consists of only 3×10^{14} molecules (specific activity of the carbon-14 is 114 mCi/mMole) which is equal to 0.00001 ml of gas. Even if an entire ampule is broken and lost to the atmosphere by evaporation into, say, 1 cubic meter of air the concentration of tetrabromoethane will not exceed 1 part per billion. Again, the recommended use of hooding or other exhausting significantly lowers even that extremely low exposure.

Chloroform used in some Test Solution formulations is, of course, present in significantly higher concentrations than is the radioactive tetrabromoethane. However, even in the case of TSAJ20L (tetrabromoethane in trifluorotrichloroethane:chloroform of 80:20 volume ratio) each 20 lambda (microliter) deposition corresponds to only ca. 1 ml of chloroform gas which is 1 ppm in a cubic meter of air, again almost completely insignificant with respect to the volume of air in the "typical" laboratory, particularly when hoods or other special exhausting apparatus are employed.

MESERAN SURFACE ANALYZERS

As far as is known all published work in the ERA field have involved the MESERAN Surface Analyzers and MESERAN Test Solutions, marketed exclusively by ERA Systems, Inc. (Chattanooga, Tennessee).

The term MESERAN is a registered U. S. Trade Mark and is an acronym for Measurement and Evaluation of Surfaces by Evaporative Rate Analysis; it is also registered in a number of countries foreign to the United States.

Figure 1 is an illustration of the MESERAN Model 720 system. It shows the electronic control and analysis unit which also provides control of the nitrogen flow, the precision dispenser and the mechanical apparatus which permits reproducible positioning of the detector with respect to each test specimen. The various test specimens may be positioned and held firmly in a number of ways including a vacuum stage for holding the specimen from below (a device which has proved useful for holding coatings panels, steel aluminum and other metal sheets and foils, plastic films, paper and the like) in many cases providing a slight concavity in the test specimen surface; also available is a magnetic stage which is convenient for steel sheets and foil.

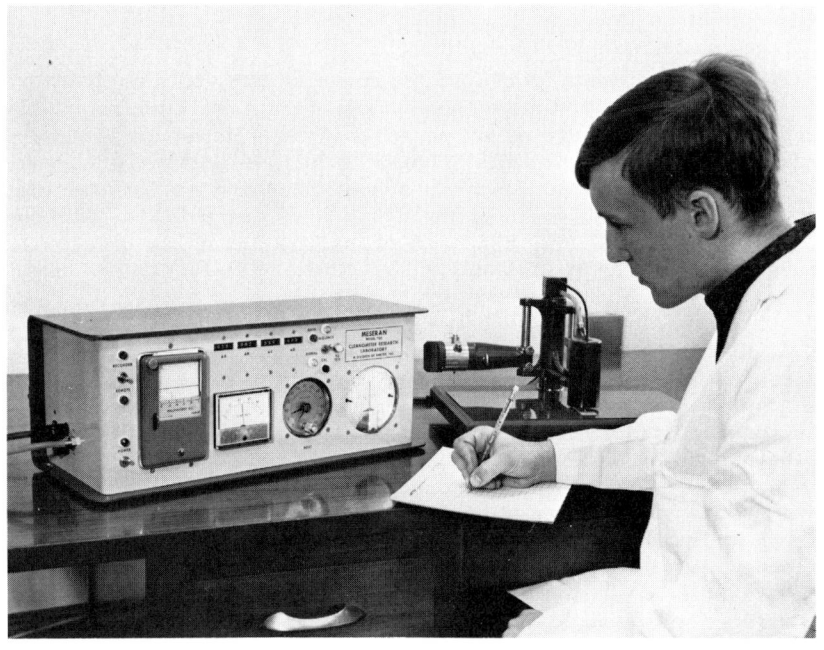

Fig. 1. The MESERAN ® Model 720 System.

Since a liquid droplet is deposited in the ERA technique, it is helpful (to insure constant geometry from test to test of the test surface/detector relationship) that the surface be capable of being positioned with a slight concavity such as occurs when a coatings panel (or foil or film) is held on the vacuum stage. Such a concavity discourages the droplet from meandering about the surface during evaporation of the solvent and thereby altering the relationship of droplet to Geiger Mueller detector tube. When it is impossible to attain a concavity, flat surfaces may be examined if care is taken to insure that the surfaces are level during testing.

The precision dispenser is used for reproducible deposition of the nominal 20 microliter amount of Test Solution used in each test. It is normally qualified to have a $\pm$ two sigma error (mechanical) of less than 3%. The dispenser is designed to minimize any possible change in titre of the Test Solution composition during use.

Not shown in Figure 1 but recommended for use in ERA testing is an exhaust and isolation box which prevents extraneous air currents from disturbing the evaporating Test Solution and also provides means for exhausting the vapor phase radioactive material and other Test Solution components.

In the equipment, rates of evaporation are expressed numerically as a function of radioactive counts per unit time over sequential counting periods initiated by positioning the detector assembly over the test specimen. The MESERAN Model 720 Surface Analyzer, for instance, uses four fourteen second counting periods during each minute of counting known as delta A to delta D and expresses one eighth of the actual detected emissions per delta area. These values are known as M (for MESERAN) counts or values since they differ from the actual counts by the 8 factor. (Note: The Model 720 system employs a 60 Hz cam timer and the additional 4 seconds per minute (60 - 4 x 14) are used for various switching functions.) For expressing M-values over longer periods of time, a reset button is either manually or automatically depressed. Following Test Solution deposition and detector positioning, an adjustable but deliberate delay time of several seconds is normally used prior to the onset of counting thereby discarding that portion of the evaporative process associated with the evaporation of the low boiling solvent (normally of less value than the following evaporation of radiochemical).

Oftentimes, a distinct series of delta M-values will be accumulated (such as sigma (delta A - delta D) or sigma (delta C_1 - delta D_2) to give a single numerical value as an expression of the observed evaporative rate. These values are known as sigma M-values or counts. Such single values do not differentiate well among various <u>shapes</u> of evaporative curves but such differentiation is oftentimes not of significant value in routine or repetitive measurements.

The MESERAN Model 820 system, not illustrated, is designed for quality and production control applications. In this system all counts are accumulated for a prespecified period of time, usually factory preset, for a predetermined end-use application. In general, the actual conditions will have been examined and determined earlier using the more versatile Model 720 system.

For repeat accuracy, a number of variables are routinely controlled in the MESERAN systems. These include but are not limited to: (1) Reproducible and standardized control of the test surface/detector geometry, (2) Holding the temperature constant or at least known and calibrated (since evaporative phenomena are a function of the temperature), (3) Controlling the rate and quality of the gaseous nitrogen which is permitted to flow between the detector and the test surface, and (4) Insuring that the added radioactive Test Solution is substantially identical both as to composition and as to deposited quantity from test to test.

MECHANISMS THROUGH WHICH ERA OPERATES

In more detail the mechanisms through which the ERA method permits characterization of surfaces are:

(1) When little or no participation of solvent or radiochemical occurs (i.e., when the surface system does not serve to retain the radiochemical and the solvent is inert), Figure 2 is illustrative of the evaporative progress as shown by the detector placed just above a test surface of aluminum foil using TSAJ (tetrabromoethane-C14 in trifluorotrichloroethane) and on which aluminum foil a monolayer of adsorbed triglyceride had previously been adsorbed. In this case, A-B represents the evaporation of the bulk droplet with relatively little effect from the surface while B represents the point at which substantially all of the low boiling solvent has visually disappeared from the surface and the maximal amount of the residual emitted radiation reaches the detector. B-C is the rate at which the residual radiochemical disappears from the surface under the conditions of the test. C represents a value of approximately 10^{12} molecules of radioactive material (distributed over the approximately 0.7 cm^2 area covered by the original Test Solution droplet, this point represents that amount of residual radioactivity when the thin-end window detector can no longer adequately differentiate the residual radiation from background. The illustrated evaporative process normally takes place with an overall elapsed time of approximately 38 seconds at 23°C. In general, the smoother the surface the slower the observed evaporation rate; rougher surfaces with greater surface areas provide more rapid heat

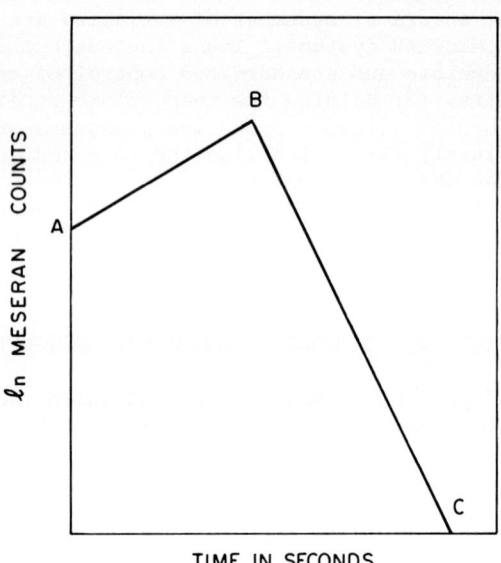

Fig. 2. Evaporation from a non-retentive surface.

transfer and therefore faster evaporation. Surface complexities such as pores or crevices with undercuts tend to force retention for longer periods. From this discussion, it is apparent that surface roughness does affect observed evaporation rates to a degree but, in general, the effect is relatively smaller than is observed in mechanisms (2) to (4).

(2) When the solvent and the radiochemical (in this case TSAJ05N--tetrabromoethane-C14 in trifluorotrichloroethane with 5% of methyl acetate) interact primarily thru solution forces with soluble residues on the surface such as occurs when dioctyl sebacate (DOS) has been applied to steel or aluminum can stock (at a nominal level of approximately 10 to 20 molecular layers --0.6 to 1.2 micrograms per square centimeter), the rate at which the solvent evaporates is slowed somewhat and the rate

at which the radiochemical evaporates is slowed considerably, with the observed rates being functions of the amount of DOS on the surface. Figure 3 illustrates these processes in which ABC is repeated from Figure 2, A'B'C' is the rate due to a lower level of DOS and A"B"C" is the rate due to a higher level. The rationale of using a preset dead time (t_0 to t_1) or non-counting time is apparent from inspection of Figure 3 since increased differentiation results from comparing the total counts between t_1 and t_2. This same reasoning applies for all such determinations of soluble residues whether the residues are deliberate (normally lubricants or waxes) or non-deliberate (contaminants). In general, the higher the molecular weight of the residue, the less the differentiation of the amount of residue, other factors being equal.

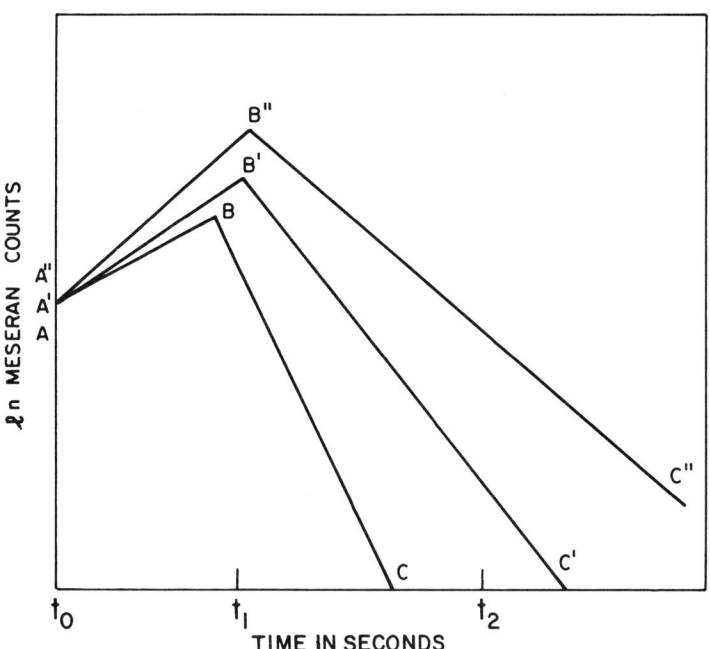

Fig. 3. Increasing Amounts of Residue ABC → A'B'C → A"B"C".

(3) When the solvent is essentially inactive or non-reactive but when the radiochemical is actively retained (as for example when the radiochemical contains an organic functional group), the extent of chemisorption of the functional group on the available surface sites can readily be determined. For this type of surface characterization, the TSAK, TSBK, TSCK, TSDK, TSEK, and TSFK Test Solutions have proved valuable since the evaporative rates from nonretentive surfaces are practically identical and the radiochemical varies from a halide, to a hydrocarbon, to a diester, to an amine, to an acid, and to an alcohol. Adsorption has been observed to occur primarily at the retreating interface of the evaporating droplet (not at the bulk droplet/surface interface); the adsorbed molecules appear to lie relatively flat on the surface (of aluminum at least) rather than in the standup configuration of Langmuir-Blodgett monolayers.

Since the number of radiochemical molecules deposited is known to approximately ± 3% accuracy and since the measurement equipment and geometrical considerations during counting are rigidly standardized, the number of adsorbed molecules can almost literally be counted. From this information, the number and availability and types of surface adsorptive sites can be rather unequivocally derived. Further, energy considerations of site/adsorbate interactions can also be derived by determining rates of desorption. The effects of changes in the environment on surfaces and on the typical adsorbed monolayers can also readily be evaluated. Figure 4 illustrates typical chemisorptive retention (AB'C') compared to non-retentive evaporation (ABC) (from Figure 2).

(4) When the solvent and the radiochemical both tend to diffuse into and out of a polymer surface, a considerable degree of polymer characterization may be achieved. The solvent effect in this case serves primarily to increase the effective fractional free volume of the polymer surface layers so that increased amounts of the radiochemical can diffuse into and out of the polymer surface. In this way, even crosslinked polymers may be characterized with a good deal of sensitivity as is shown in Figure 5 which indicates the effect of non-crosslinked conditions (A), a partially crosslinked condition (B) and a complete chemical crosslinked condition (C). By increasing the solvent strength for a given coating, for example, increased differentiation in evaporative responses may be obtained for increased levels of crosslinked density. Thus while TSBK10P often serves to differentiate the under-cure levels of a particular epoxy coating, increasing to TSBK20P permits markedly improved differentiation of the cured levels. When a solvent/radiochemical composition which does not invade the polymeric substrate is used, it is possible to measure levels of residues on the surface of the substrate

such as occur in migration of plasticizers or lubricants. Further in noncrosslinking coatings, the loss of traces of solvents may be followed since the presence of the solvents also tends to increase the fractional free volume and therefore the invasion of the radioactive test chemical.

APPLICATIONS

As noted earlier the bibliography is divided chronologically into the major areas of ERA applications and is annotated. There appears to be little to be gained by repeating those notes and comments in this section. The reader is referred to the several references. Obviously in many of the listed references, more than one area of application is mentioned and particularly in those with the General and Descriptive classification.

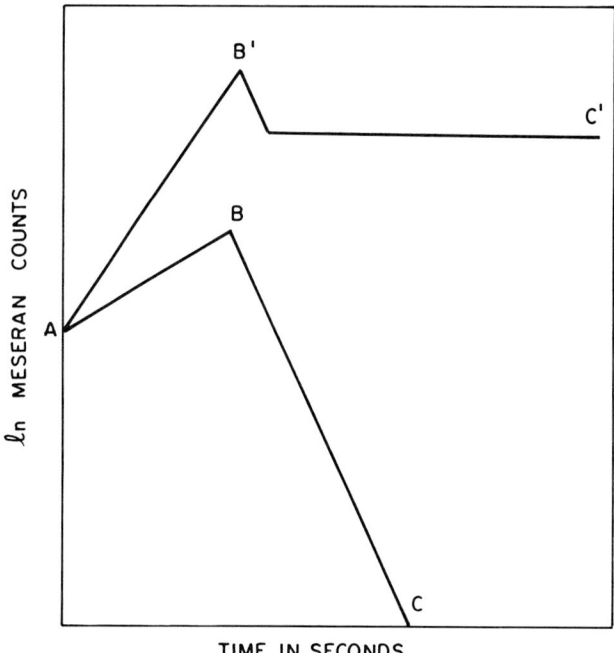

Fig. 4. Adsorptive Retention (AB'C')

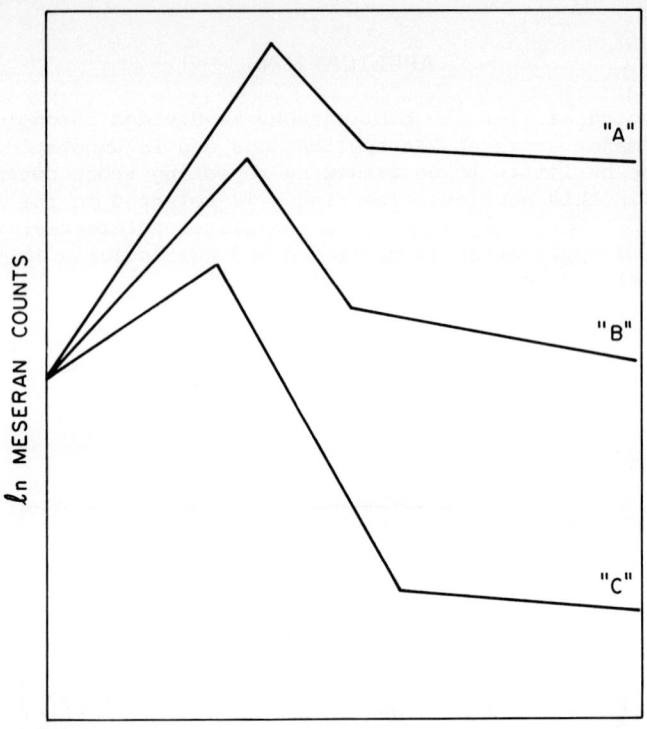

Fig. 5. Degree of Crosslinking
   A - Undercured
   B - Partially Cured
   C - Cured

SUMMARY

Evaporative Rate Analysis is a method for the chemical characterization of surfaces involving the use of an added radioactive labeled chemical followed by observing the rate at which the added material evaporates or otherwise disappears from the surface.

Depending on the surface being tested and on the particular Test Solution employed, the ERA method may productively be used to determine (primarily by comparing the results with other independent calibrating methods and/or by establishing suitable calibration standards and/or through use of the basic knowledge of what is being measured):

(1) The amount of lubricant or solvent miscible organic residue on a surface.
(2) Certain surface topographical characteristics such as porosity or surface complexity.
(3) The degree of cure of thermo- or other setting (crosslinking) materials.
(4) The identity of lot to lot and batch to batch materials.
(5) Drying properties of coatings.
(6) Migration of lubricants and/or plasticizers.
(7) Surface changes caused by aging or other factors.
(8) Degree of crystallization of polymeric films.
(9) Certain of the properties that relate to adhesion.
(10) Levels of wax or other additives.
(11) The ability of surfaces to adsorb and to desorb compounds possessing functional groups and from the rates of desorption to derive valuable information with respect to such sorptive processes.

Figure 6 illustrates the wide range of sensitivity obtainable using the ERA technology in measuring adsorbed monolayers to residues and other soluble materials on surfaces. Six to seven orders of magnitude of such materials may be determined to approximately 3 per cent error for a single determination.

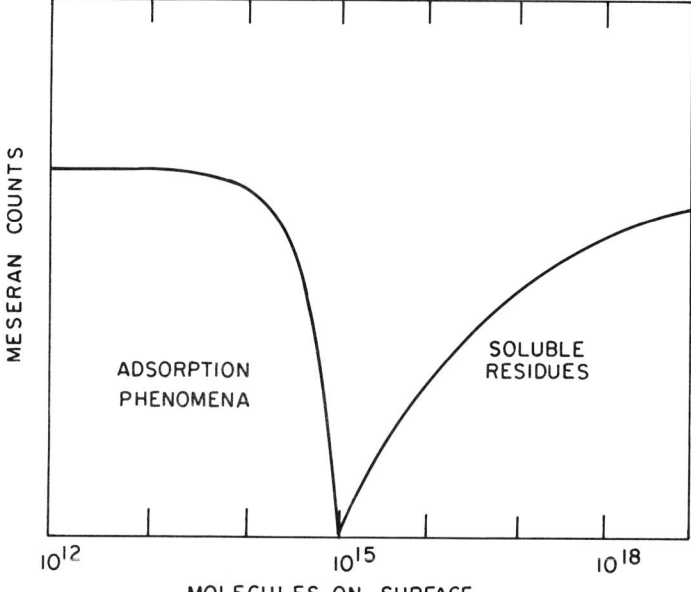

Fig. 6. Overall sensitivity to surface residues.

BIBLIOGRAPHY

General and Descriptive

1. J. L. Anderson, D. E. Root, Jr., and Gary Greene, "Measurement and Evaluation of Surfaces and Surface Phenomena by ERA", J. Paint Tech., 40, 321 (1968). A general discussion of ERA principles and applications.
2. J. L. Anderson, "Using Evaporative Rate Analysis to Solve Converting Problems", a talk delivered at the Packaging Institute Symposium, New Brunswick, N. J., March 20, 1969, and printed in part in Modern Converter (1969).
3. J. L. Anderson, "Application of Evaporative Rate Analysis to Coatings Evaluation", a paper delivered on June 2, 1969, at the 11th Annual Symposium on New Coatings and New Coatings Raw Materials, Fair Hills Resort, Detroit Lakes, Minn. Published (in German) as "Ermittlung der Verdampfungsgeschwindigkeit zur Bewertung von Anstrichstoffen", Farbe und Lack, 76 (8), 757 (1970). A general paper with emphasis on coatings.
4. J. L. Anderson, "Introduction to ERA", Coatings and Plastics Preprints, 29 (2), 250 (1969). ERA I.
5. J. L. Anderson, "New Surface Analysis of Materials", Modern Packaging, Nov. (1969). A general description of ERA.
6. J. M. Carter, "Physical Chemistry Aspects of ERA Involving Impermeable Surface", Coatings and Plastics Preprints, 30 (2), 299 (1970). ERA II.
7. A. Schaff, Jr., "Some Observations Related to ERA Data Reduction, Coatings and Plastics Preprints, 30 (2), 306 (1970). ERA II.

Residues and Lubricants (Including Cleaning Efficiency)

1. J. L. Anderson, "Quantitative Detection of Surface Contaminants", presented at the Second Annual Meeting of the Amer. Assoc. for Contamination Control, Boston, May 2, 1963; published in Contamination Control II (6), 9 (1963). The first published description of the principles of ERA; in this case applied to the detection of residues.
2. J. L. Anderson, "Quantitative Detection of Surface Contaminants: II. Experimental Verification", presented at Fall Meeting of Committee F1-X, ASTM, Skytop, Pa., Nov. 7, 1963.
3. J. L. Anderson, "Quantitative Detection of Surface Contaminants: III. Scope and Limitations of the Evaporation Rate Method", presented at the Third Annual Technical Conference, Amer. Assoc. for Contamination Control, Los Angeles, (1964).
4. J. L. Anderson, "The Quantitative Detection of Surface Contaminants by Evaporative Rate Measurements", presented at a Symposium on Surface Contamination, Gatlinburg, TN 1964; abstract published in Surface Contamination, p. 33, Pergamon Press Ltd., 1967, B. R. Fish, Editor.

5. D. A. Brewster and R. M. Cox (Kaiser Spokane), "The Determination of Dioctylsebacate (DOS) on Aluminum Stock by ERA", Coatings and Plastics Preprints, 29 (2), 277 (1969). ERA I. A comparison of ERA with the chloroform extraction method.
6. W. S. Koegel and A. C. Schachter (Grumman Aerospace), "The Application of ERA to the Determination of Non-Volatile Residue", Coatings and Plastics Preprints, 30 (2), 316 (1970). ERA II. ERA applied to the determination of NVR in trifluorotrichloroethane used in precision cleaning. The NVR is deposited by dropwise evaporation onto a clean surface and then determined as residue.
7. A. Schaff and J. F. Forbes, "An Analysis of ERA Measurements for NVR", Coatings and Plastics Preprints, 30 (2), 321 (1970). ERA II.
8. L. C. Jackson (Bendix, Kansas City), "How to Select a Substrate Cleaning Solvent. Contaminant Removal Using Solubility Parameter Technology", Adhesives Age, 17 (12), 23 (1974). The ERA method was used as the primary surface probe to evaluate cleanliness achieved.
9. "Non-Volatile Residue in Trichlorotrifluoroethane (MESERAN Procedure) Test for Amount of", *ASTM Book of Standards*, 25, 514 (1975).
10. "Lubricant on the Surface of Aluminum Foil (MESERAN Procedure) Test for Amount of", *ASTM Book of Standards*, 25, 519 (1975).
11. C. B. Hamilton (U.S. Steel, Monroeville), "Evaluation of Electrolytic Alkaline Cleaners by ERA", Plating and Surface Finishing, (June), 581 (1975).
12. L. C. Jackson (Bendix, Kansas City), "Solvent Cleaning Process Efficiency", Adhesives Age, 19, (7), 31 (1976).
13. L. C. Jackson (Bendix, Kansas City), "Surface Characterization Based on Solubility Parameters", Adhesives Age, 19, (10), 17 (1976).

Cure and Polymer Characterization

1. J. L. Anderson, "Coating Cure Determination by ERA", paper delivered in the National Meeting of the National Coil Coaters Association, Chicago, Ill., Oct. 31 and Nov. 1, 1967 and published in the Report of Meeting.
2. A. G. Rossi and A. Paolini, Jr. (Campbell Soup), "Study of the Cure Behavior of Can Coatings by ERA--Effect of Bake Time", J. Paint Tech., 40 (523), 328 (1968). Oleoresinous and epoxyamine coatings were studied.
3. E. Cerceo (Boeing Vertol, Philadelphia), "Determination of Degree of Cure of Polymeric Materials through ERA", Anal. Chem. 41 (1), 191 (1969). ERA applied to cure properties of adhesives and a resin-glass fiber composite.
4. R. Awe and C. Monroe (Dow Corning), "ERA for Measurement of Coating Weight and Degrees of Cure of Silicone Paper Coatings", a paper delivered at the TAPPI Coating Conference, May 1969.

5. E. J. Helwig (U.S. Steel, Murrysville), "Determination of Cure of Sanitary Can Lacquers", Coatings and Plastics Preprints, 29 (2), 300 (1969). ERA I. Excellent results with alkyd and polybutadiene lacquers. Effect of substrate on the curing behavior of the lacquer was noted. JPT, 42, 385 (1970).
6. A. G. Rossi, G. A. Charland, and W. C. Stammer (Campbell Container), "Determining the Cure Parameters of Can Coatings by ERA", Coatings and Plastics Preprints, 29 (2), 308 (1969). ERA I. Also published in J. Paint Tech. 42 (546), 391 (1970). Oleoresinous, acrylic and epoxy-amine coatings were studied with respect to effects of time and temperature on cure.
7. J. L. Anderson and E. P. O'Connell, Jr., "ERA Applied to Thermoset Automotive Acrylic Enamels", Coatings and Plastics Preprints, 29 (2), 316 (1969). ERA I.
8. S. E. Stromberg (Spencer Kellogg, Div. of Textron), "Investigation of the Cure Behavior of Water Thinnable Baking Resins by ERA", Coatings and Plastic Preprints, 29 (2), 321 (1969). ERA I. Also published in J. Paint Tech., 42 (546), 398 (1970). Test results correlate well with acetone resistance and hardness tests for optimum curing conditions.
9. C. I. Sanders (Armstrong Cork), "Polyurethane Cure Measurement; Comparison of ERA with Other Techniques", Coatings and Plastics Preprints, 29 (2), 334 (1969). ERA I. Also published in J. Paint Tech., 42 (546), 405 (1970). Cure of polyurethane films by ERA correlates well with other measurements such as Sward and Tukon hardness, DSC at glass transition temperatures, and measurement of an isocyanate peak in the IR.
10. R. R. Rees, R. E. Stahl, and A. Paolini, Jr. (Desoto), "Correlation of Can Coating Physical Properties with Cure by ERA", J. Paint Tech. 42 (546), 422 (1970). Presented at the 47th Annual Meeting of the Federation of Soc. for Paint Tech., Chicago, Nov. 7, 1969. Studies of cure of an epoxy-urea can coating. Optimum cure determined. Effects of time, temperature, and substrate shown.
11. J. M. Carter and E. P. O'Connell, "Some Theoretical Aspects of ERA Measurements Involving Polymeric Substrates", Coatings and Plastics Preprints, 30 (2), 328 (1970). ERA II. Suggests that diffusion processes of both the low-boiling solvent(s) and the radiochemical into and out of the polymer account for most if not all of the observed phenomena with respect to cure and other properties measured by ERA.
12. A. G. Rossi and G. A. Charland (Campbell Container), "Determining the Efficiency of Various Bake Methods for Can Coatings by ERA", Coatings and Plastics Preprints, 30 (2), 337 (1970). ERA II.
13. T. J. Miranda (Whirlpool), "Comparison of ERA with Thermomechanical Analysis for Thermosetting Acrylic Coatings", Coatings and Plastics Preprints, 30 (2), 344 (1970). ERA II. Also published in J. Paint Tech., 43 (553), 51 (1971).

14. A. Paolini, Jr., R. E. Stahl, and J. W. Sullivan (DeSoto), "The Characterization of an Epoxy/Urea-Formaldehyde Coating by ERA", Coatings and Plastics Preprints, 30 (2), 350 (1970). ERA II. Also published in J. Paint Tech. 42 (551), 722 (1970). Minimum and optimum bake temperatures defined by ERA.
15. G. G. Schurr and T. P. Kinsella (Sherwin-Williams), "Evaluation of Catalysts for Coatings by ERA", Coatings and Plastics Preprints, 30 (2), 359 (1970). Work reported on thermosetting acrylic and alkyd modified vinyl organosol. ERA results correlated well with Tukon hardness. Also a polyester-melamine coating was studied. ERA II.
16. H. E. Hill, C. S. Pietras, and D. J. Damico (Lord Corp.), "Evaluation of Unblocking of Urethane Coatings by ERA", Coatings and Plastics Preprints, 30 (2), 366 (1970). ERA II. Also published in J. Paint Tech., 43 (553), 55 (1971).
17. C. T. Wise (Sherwin-Williams), "Determination of Curing Behavior of Thermoset Powder Coatings", Coatings and Plastics Preprints, 33 (2), 527 (1973). ERA together with a number of other methods was used to indicate cure.
18. B. A. Sok and C. Noll (Inland Steel), "Evaluation of ERA for Prediction of Thermal, Chemical, and Mechanical Properties of Irradiated Polyethylene", Coatings and Plastics Preprints, 33 (1), 579 (1973). ERA III.
19. T. J. Miranda (Whirlpool), "Measurement of Stabilizer Losses in Polypropylene by Thermal and ERA", Coatings and Plastics Preprints, 33 (1), 586 (1973). ERA III. Also published in Mechanical Behavior of Materials, Vol. 3, p. 392-404, by the Society of Materials Science (Japan) 1972.
20. D. E. Scherpereel (Whirlpool), "Characterization of Continuous Cleaning Porcelain Enamel by ERA", Coatings and Plastics PrePrints, 33 (1), 597 (1973). ERA III.

Adsorption/Desorption

1. J. L. Anderson, "Measurement and Evaluation of Surfaces by Evaporative Rate Analysis. III. Quality of Seal on Anodized Aluminum", Presented at Mid-Year Meeting of the Society of Automotive Engineers, Detroit, June 6-10, 1966. Preprint available from the SAE.
2. J. L. Anderson, R. A. Baker and J. F. Forbes, "Application of ERA to the Surface Properties of Metals", J. Colloid and Interface Science, 31 (3), 372 (1969). A number of experiments are described involving functional group/surface site interactions.
3. R. R. Seiler, W. Morrison (Martin-Orlando), J. L. Anderson and D. E. Root, Jr., "ERA Applied to Characterization of Aluminum Surfaces used in Adhesive Bonding", Coatings and Plastics Preprints, 29 (2), 256 (1969). ERA I.
4. R. A. Baker and J. L. Anderson, "Kinetics of Desorption of Some Fatty Acids from Aluminum Foil Surfaces by Heating. I.

Butyl Stearate and Stearic Acid", Coatings and Plastics Preprints, 29 (2), 338 (1969). See also ref. 6 of this series. ERA I.

5. J. F. Forbes and J. L. Anderson, "Application of ERA to the Study of Finely Divided Solids. I. Al_2O_3 and TiO_2", Coatings and Plastics Preprints, 29 (2), 345 (1969). ERA I.

6. "Round Table Discussion of ERA during ERA I", Coatings and Plastics Preprints, 30 (1), 442 (1970). Includes comments by a number of participants and others. Also includes additional experimental work by R. A. Baker and J. L. Anderson with respect to paper #4 above. Also includes experimental results and discussion on the effect of moisture on the desorption of a number of radiochemicals which had been adsorbed on aluminum foil surfaces. Moisture in the nitrogen passing over the surface always significantly increased the rates of desorption.

7. J. L. Anderson, "Adsorptive/Desorptive Behavior of ERA Radiochemicals", Coatings and Plastics Preprints, 30 (2), 374 (1970). ERA II. A proposed model to explain the mechanism of adsorption under ERA conditions, a method for estimating surface areas for low surface area materials, some anomalous desorptions showing successive increases in rates with time and apparent temperature-independent desorptions.

8. C. S. Pietras (Lord), "Detection of Adsorbed Monolayers by ERA", Coatings and Plastics Preprints, 33 (1) 599 (1973). ERA III.

9. J. R. McDowell and W. H. Saunders (Lord Corp.), "Desorption Kinetics of Organic Monolayers from Aluminum by ERA", Coatings and Plastics Preprints, 33 (1), 607 (1973).

Miscellaneous

1. R. F. Wegman and M. Otzinger (Picatinny Arsenal), "A Study of Adhesive Joint Failures by ERA", Coatings and Plastics Preprints, 29 (2), 265 (1969). ERA I. Also published in SAMPE Quarterly, 1, 32 (1969). By examining both sides of an adhesive joint failure, the mode of failure can be deduced.

2. J. L. Anderson, "Non-Poisson Distributions Observed During Counting of Certain Carbon-14 Labeled Organic (Sub)Monolayers", J. of Phys. Chem., 76 (24), 3603 (1972). Using the technique of ERA in depositing the chemisorbed monolayer, anomalous, narrow statistical distributions as shown by the s^2/m index of dispersion were obtained during lengthy desorptions and during counting of adsorbed stearic acid-C14.

3. J. L. Anderson, "Statistical Abnormalities in Counting Carbon-14 Monolayers", Coatings and Plastics Preprints, 33 (1), 614 (1973). ERA III. Some additional comments concerning the work of #2 above.

Patents

(U.S. Patents only are listed although a number of foreign equivalents have also issued. The inventor in each case is J. L. Anderson and all are exclusively licensed to ERA Systems, Inc.)

1. USP 3,215,839, issued Nov. 2, 1965. Not directly applicable to present ERA techniques, this patent claims detection of contaminants using an added non-volatile radioactive compound.
2. USP 3,247,385, issued April 19, 1966. Not directly applicable to present ERA techniques, this patent claims a process for detecting contamination using gaseous radioactive labeled compounds which are selectively retained by contaminants.
3. USP 3,297,874, issued Jan 10, 1967. This is the parent ERA case and claims a process for detecting contamination involving exposing the surface to a volatile radioactive chemical and measuring the rate at which the radiochemical evaporates.
4. USP 3,412,247, issued Nov. 19, 1968. Apparatus for examining surfaces using the evaporation rate of an added material from the surface is claimed. Also this patent contains a number of composition-of-matter claims for the Test Solutions. A C-I-P of #3.
5. USP 3,445,657, issued May 20, 1969. A C-I-P of #3, contains process claims for examining a surface using a mixture of two different chemicals.
6. USP 3,560,157, issued Feb. 2, 1971. A C-I-P of #3, the patent contains process claims covering the measurement of crosslinking of polymer systems.
7. USP 2,582,657, issued June 1, 1971. A C-I-P of #3, claims liquid dispensing apparatus and particularly the dispenser used in the MESERAN systems.

Radiation Absorption for Polymers

John R. Hallman

Nashville State Technical Institute
Nashville, Tennessee 37209

and

J. Reed Welker and C.M. Sliepcevich

University of Oklahoma
Norman, Oklahoma 73069

Polymeric type materials are being used in ever increasing amounts in both commercial and residential structures. Since many of the materials are capable of evolving noxious and flammable gases and vapors when heated, there is much concern to those involved in fire prevention and control. A major area involving fire safety but where information is lacking is the polymer surface absorptance of radiant heat with the subsequent ignition and burning of the materials. During a research study into the ignition characteristics of polymers, it was found that the polymer surface absorptance of radiant heat was a necessary parameter required in the energy equation. Subsequently, absorptance tests were made on a variety of commercial polymers and the information used for the ignition study. Average absorptances were derived from the equation

$$\alpha_{AV} = \frac{\Sigma\, \alpha_\lambda\, e_\lambda\, \Delta\lambda}{\Sigma\, e_\lambda\, \Delta\lambda}$$

over the monochromatic wavelength span of the heat sources. This paper will graphically present the monochromatic absorptance with respect to wave length of several commercial polymers. A table of polymer average absorptance will be listed for several heat sources.

Polymeric type materials are being used in ever increasing amounts in both commercial and residential structures. These materials, although generally decorative in application are capable of producing noxious gases and vapors when heated or burned. Companies and individuals engaged in fire insurance, fire safety and health safety are concerned not only with the type and amount of evolving gases, but also the flammability of the materials used in the construction.

One such flammability study was made and reported by Hallman, et al., to determine the ignition characteristics of polymers and rubber materials.[1] This investigation was primarily concerned with the susceptibility of the polymers to radiant heat decomposition and the subsequent burning. During this study, it was found that the polymer surface absorption of the heat radiation was a factor in determining the ignition time. In addition, it was found that the same polymer exposed to two different heat sources had a different time of ignition for the same rate of heating; again a factor of the polymer surface in absorbing the radiant heat. Figure 1 shows the ignition times of white polystyrene samples of varying thickness plotted as a function of incident irradiance.[2]

As is illustrated, both flames and tungsten lamps radiation were used as energy sources. It should be noted that there is a difference in ignition time for the same amount of irradiance. In contrast, Figure 2 shows the ignition time for the same white polystyrene samples when the absorption coefficient for the particular heat source is used as a correcting parameter, time again plotted as a function of incident irradiance. Three important factors are apparent. First, ignition time is strongly dependent on incident irradiance. Second, ignition time is dependent on the nature of the radiation source. Third, when the polymer absorptance is known, the values can be used to correlate the ignition time regardless of the heat source. A plot of the monochromatic absorptance with wavelength of white polystyrene is shown in Figure 3.

During the ignition test[3], it was determined that absorptance data was necessary to complete any heat transfer studies. A second investigation was performed to obtain such information. Using a Cary Spectrophotometer with an integrating sphere refractometer and Gier-Dunkle Hohlaum unit with a Beckman IR4, monochromatic reflectance data (r_λ) was obtained over a range of 0.3 to 7 micron wavelength. With the use of Kirchoff's Law, the law of reciprocity and the equivalency factor that emittance equals absorptance (α_λ), the reflectance-absorptance relationship becomes

$$\alpha_\lambda = 1 - r_\lambda \qquad (1)$$

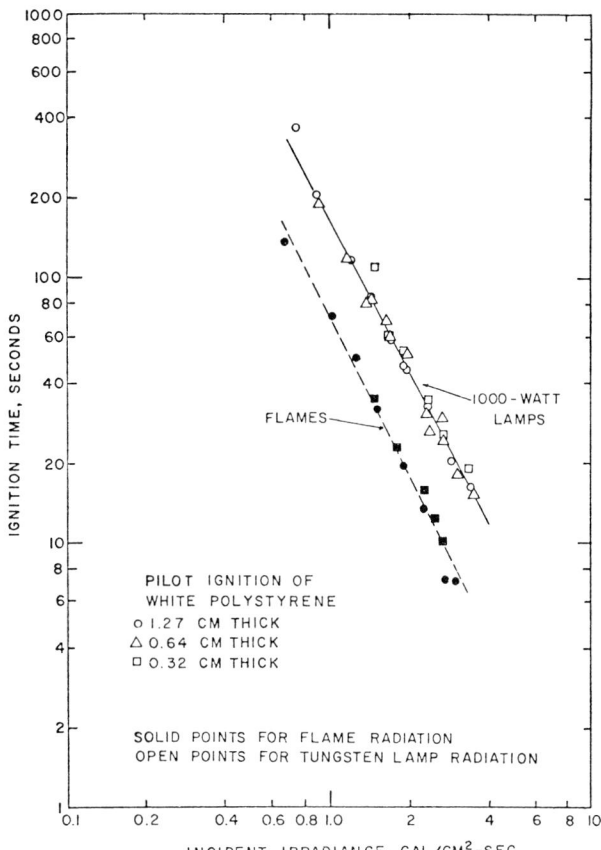

Fig. 1. Pilot Ignition of White Polystyrene.

where the values of absorptance listed were obtained from the reflectance data. This research and study has been reported in the work of Hallman, et al.[4].

While use of the absorptance information assisted in the heat transfer analysis, the particular heat source data were necessary also in the mathematical development. Monochromatic radiant intensity studies illustrate the spectral distribution of certain heat sources and are presented in Figure 4.[5,6,7,8]

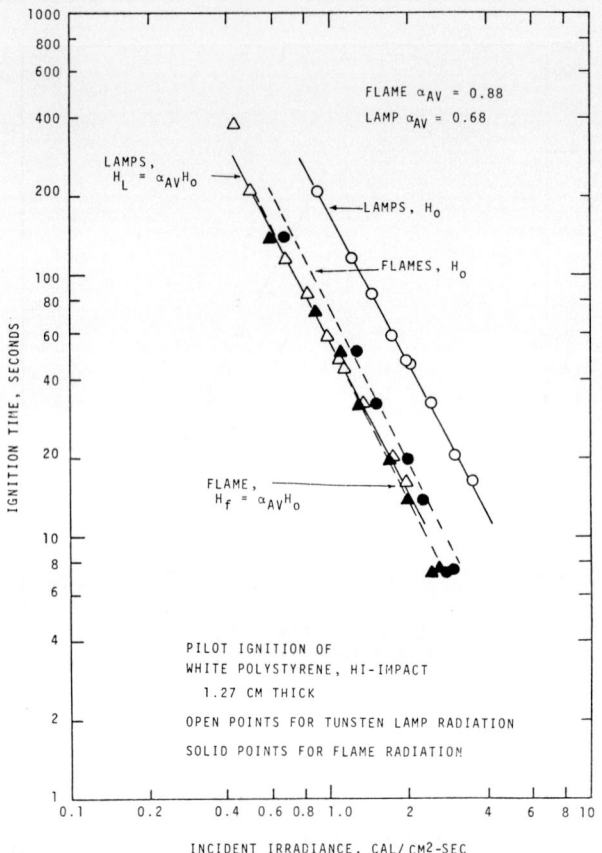

Fig. 2. Ignition Time for White Polystyrene with Varying Irradiance.

To account for the variation in absorbed irradiance, it was necessary to determine the magnitude of the monochromatic absorptance and the magnitude of radiation intensity at the same wavelength and to integrate over the total wavelength span involved. Using α_{AV} as the average absorptance, the equation becomes

$$\alpha_{AV} = \frac{\int \alpha_\lambda \, e_\lambda \, d\lambda}{\int e_\lambda \, d\lambda} \qquad (2)$$

where α_λ is the monochromatic absorptance of the polymer and e_λ is the monochromatic radiant intensity of the heat source.

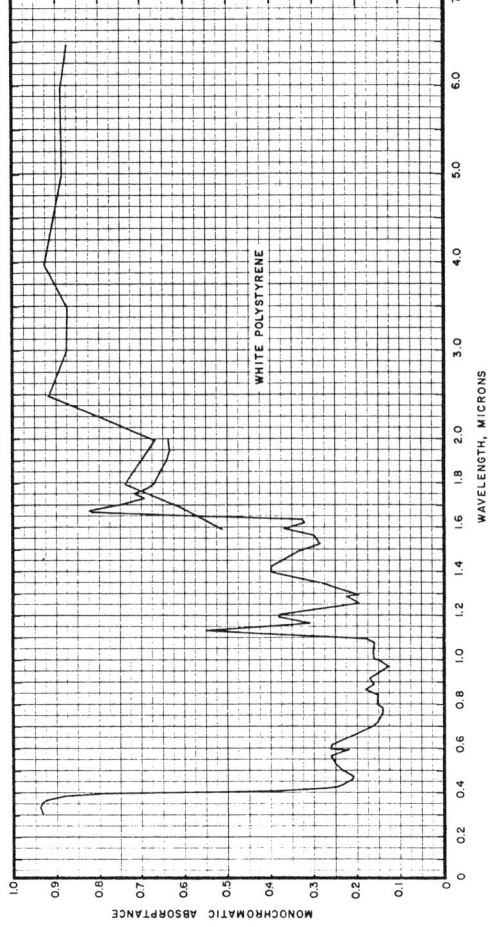

Fig. 3. Spectral Absorptance of White Polystyrene. (High-impact polystyrene - Dow).

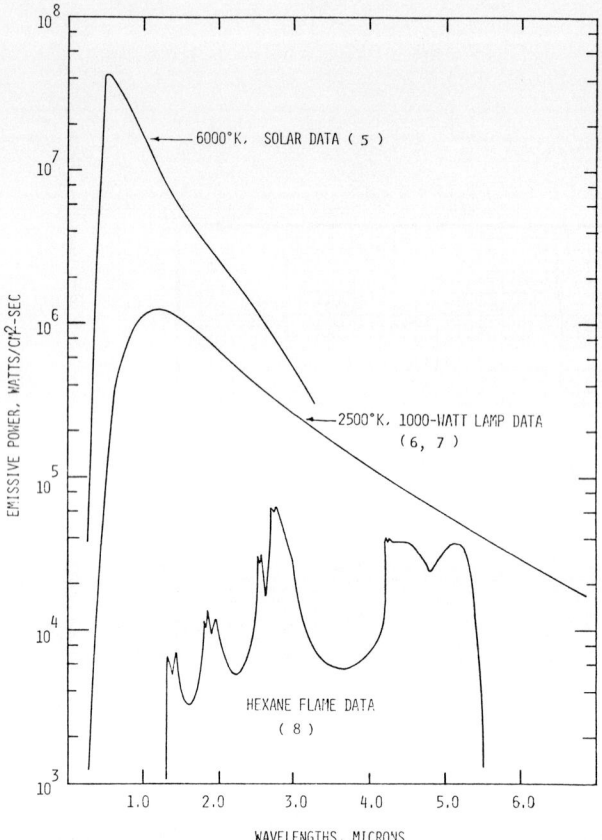

Fig. 4. Monochromatic Emissive Power of Several Heat Sources.

Equation (2) was converted into a trapezoidal rule summation for use in a standard computer program as shown by equation (3)

$$\alpha_{AV} = \frac{\Sigma\, \alpha_\lambda\, e_\lambda\, \Delta\lambda}{\Sigma\, e_\lambda\, \Delta\lambda} \qquad (3)$$

Absorptance values obtained by this computer program are listed in Table 1. The limits of integration are the wavelengths which include significant amounts of energy.

Since other forms of heat sources are known (wood and gasoline fires, ASTM tests and others), the absorptance values included here are not applicable for use in energy analysis involving these other sources of irradiance. Therefore, the following plots, Figure 5

TABLE 1

AVERAGE ABSORPTANCE FOR SEVERAL RADIATION SOURCES

Material	Radiation Source Blackbody Temperature, °K							Heat Sources	
	1000	1500	2000	2500	3000	3500		Flames	Solar
Gum Rubber	0.88	0.82	0.76	0.72	0.69	0.68		0.89	0.69
Cork Gasket Material	0.70	0.59	0.50	0.45	0.43	0.44		0.67	0.35
Neoprene Rubber, Solid	0.91	0.92	0.93	0.93	0.93	0.93		0.91	0.94
Chloroprene DC-100 Gasket	0.72	0.63	0.55	0.52	0.51	0.52		0.71	0.62
Formica	0.91	0.88	0.85	0.82	0.80	0.79		0.91	0.80
Polyphenylene Oxide (PPO)	0.86	0.78	0.70	0.63	0.57	0.53		0.88	0.48
Kydex, Red, Cast	0.91	0.90	0.89	0.88	0.87	0.86		0.92	0.86
Kydex, Gray, Rolled (vinyl chloride-acrylic copolymer (Rohm and Haas))	0.88	0.87	0.86	0.85	0.84	0.83		0.88	0.81
Texin (Urethane-ester copolymer)	0.92	0.89	0.83	0.77	0.72	0.68		0.93	0.62
Delrin	0.92	0.86	0.78	0.71	0.64	0.59		0.93	0.48
Hi-Impact White Styrene (Dow)	0.86	0.75	0.63	0.53	0.45	0.40		0.88	0.29
Plexiglas, White	0.91	0.86	0.78	0.70	0.62	0.56		0.92	0.42
Plexiglas, Black	0.94	0.94	0.95	0.95	0.95	0.95		0.94	0.96
Plexiglas, Clear	0.85	0.69	0.54	0.41	0.31	0.25		0.89	0.097
Lexan, Polycarbonate (Rough Surface)	0.87	0.83	0.78	0.75	0.72	0.71		0.88	0.69
Polypropylene	0.87	0.83	0.78	0.74	0.70	0.68		0.86	0.62
Polyethylene, Low Density	0.92	0.88	0.82	0.77	0.72	0.68		0.93	0.57
Polyvinyl Chloride, Gray	0.90	0.90	0.89	0.89	0.89	0.89		0.91	0.89
Polyvinyl Chloride, Clear 0.33 cm	0.81	0.65	0.49	0.38	0.30	0.24		0.85	0.15
Silicone Rubber	0.79	0.66	0.58	0.54	0.52	0.53		0.79	0.62

TABLE 1 -- Continued

Material	Radiation Source Blackbody Temperature, °K						Heat Sources	
	1000	1500	2000	2500	3000	3500	Flames	Solar
Buna-N Rubber	0.92	0.93	0.93	0.93	0.93	0.93	0.92	0.94
Butyl IIR Rubber	0.92	0.93	0.94	0.94	0.95	0.95	0.92	0.95
Nylon 6/6	0.93	0.90	0.86	0.82	0.75	0.71	0.93	0.62
Polystyrene, Clear (Styrolux - West Lake Plastics)	0.75	0.60	0.46	0.35	0.28	0.22	0.78	0.095
Cellulose Acetate Butyrate (Uvex)	0.84	0.71	0.56	0.43	0.34	0.27	0.88	0.12
Cycolac	0.91	0.86	0.77	0.71	0.65	0.61	0.92	0.55
Phenolic (Bakelite)	0.90	0.86	0.81	0.77	0.75	0.75	0.91	0.78
Cork	0.64	0.56	0.49	0.46	0.44	0.44	0.60	0.52
ACCOPAC Gasket CS-301 (Buna-S-Cork)	0.71	0.63	0.60	0.60	0.62	0.65	0.69	0.74
ACCOPAC Gasket CN-705 (Buna-N-Cork)	0.57	0.51	0.47	0.46	0.47	0.50	0.60	0.62
ACCOPAC Gasket AS-428 (Buna-N-Asbestoes)	0.92	0.92	0.92	0.92	0.92	0.92	0.91	0.92

through 22 are included for several polymers listed in Table 1. It should be noted that in the reflectance testing program, for the overlap domain of 1.6 to 2.0 microns wavelgnths, the Cary data were obtained in a continuous spectrum while the Gier-Dunkle/Beckman IR4 data were obtained at discrete wavelength settings. The error of closure is primarily due to the temperature of the test system and difference in the test apparatus. In the curves presented, absorptance values are dimensionless while the wavelength span is given in microns. A discussion of the theoretical-consideration of α_λ and the development of the reflectance-absorptance characteristics of polymer surfaces is given by Hallman, et al.[4].

This investigation has provided considerable data for the surface absorption of commercially available polymers. Also, it is indicated that additional work should be performed to obtain the absorption data for the many types and forms of the new polymeric materials available. Rewards of such research will enable those engaged in fire control and safety to predict more readily the behavior of the polymers when subjected to radiant heating and to provide information for the control of fires involving these materials.

REFERENCES

1. J.R. Hallman, J. Reed Welker and C.M. Sliepcevich, "Ignition of Polymers", SPE Journal, 28, 43, 1972.
2. J.R. Hallman, J. Reed Welker and C.M. Sliepcevich, "Ignition Times for Polymers", Polymer-Plastics Technology and Engineering, 6 (1), 1976.
3. J.R. Hallman, "Ignition Characteristics of Plastics and Rubber", PhD Dissertation, University of Oklahoma, 1971.
4. J.R. Hallman, J.R. Welker and C.M. Sliepcevich, "Polymers Surface Reflectance-Absorptance Characteristics", Polym. Eng. and Sci., 14, 10, 1974.
5. Smithsonian Miscellaneous Collection 74, No. 1. "Solar Function at Sea Level", Smithsonian Institute, Astrophys. Observatory, Ann. V, 108 (1932).
6. B.T. Lee and N.J. Alvares, "The Effects of Irradiance on the Ignition and the Rate of Fire Spread Along Composite Cellulosic Materials", 1967 Spring Meeting, Western States Section, The Combustion Institute, La Jolla, California (April 1967).
7. T.J. Love, *Radiative Heat Transfer*, Merrill Publishing Company, Columbus, Ohio (1968).
8. L.R. Ryan, G.J. Penzias, and R.H. Tourin, "An Atlas of Infrared Spectra of Hydrocarbon Flames in the 1-5µ Region", Scientific Report #1, Contract AF19 (604)-6106. Air Force Cambridge Research Laboratories, Bedford, Massachusetts (July 1961).

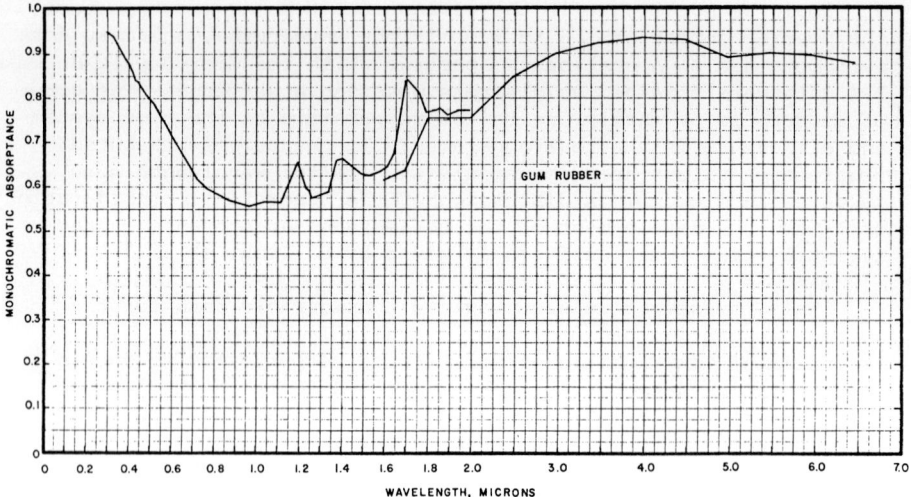

Fig. 5. Spectral Absorptance of Gum Rubber.

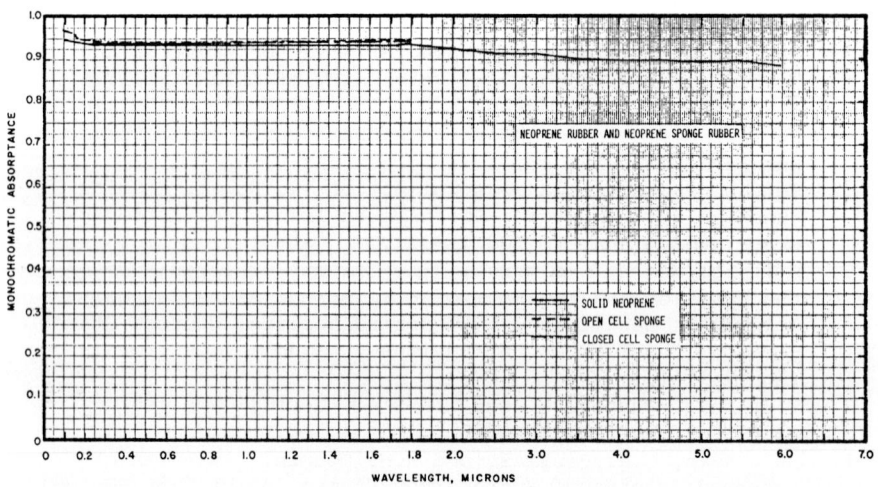

Fig. 6. Spectral Absorptance of Neoprene Rubber.

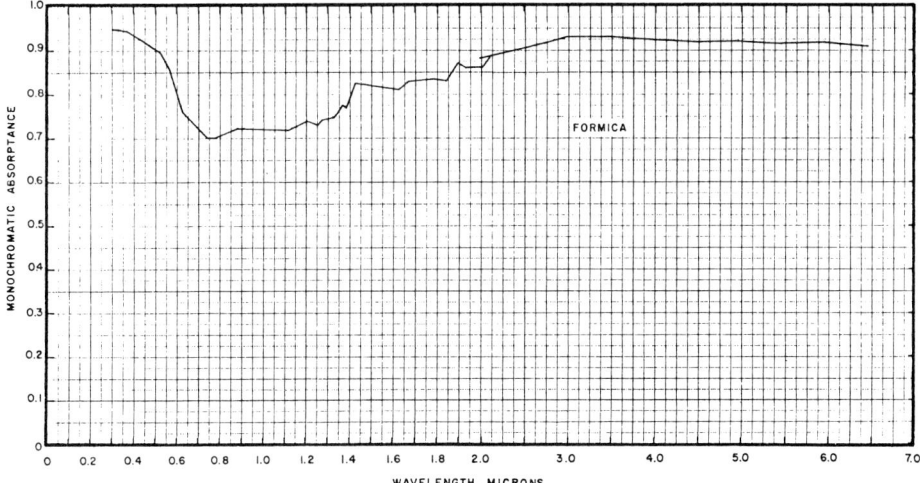

Fig. 7. Spectral Absorptance of Formica.

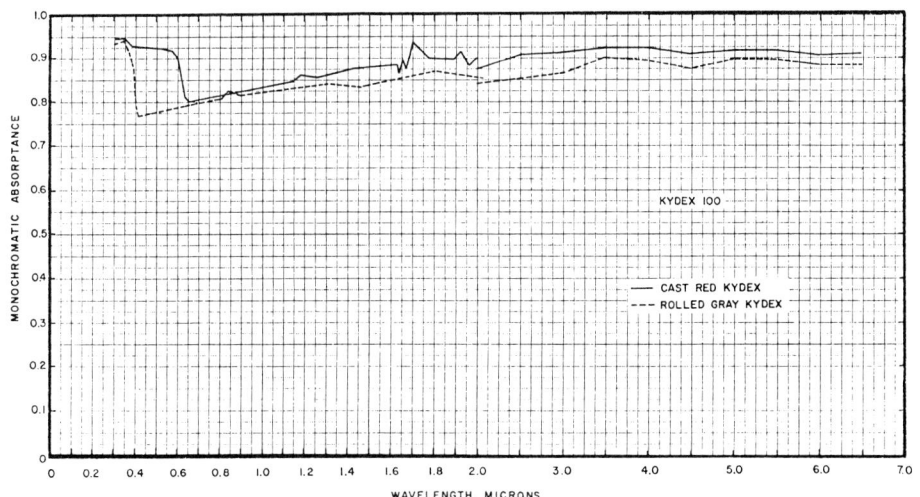

Fig. 8. Spectral Absorptance of Kydex 100 (vinyl chloride acrylic copolymer - Rohm & Haas).

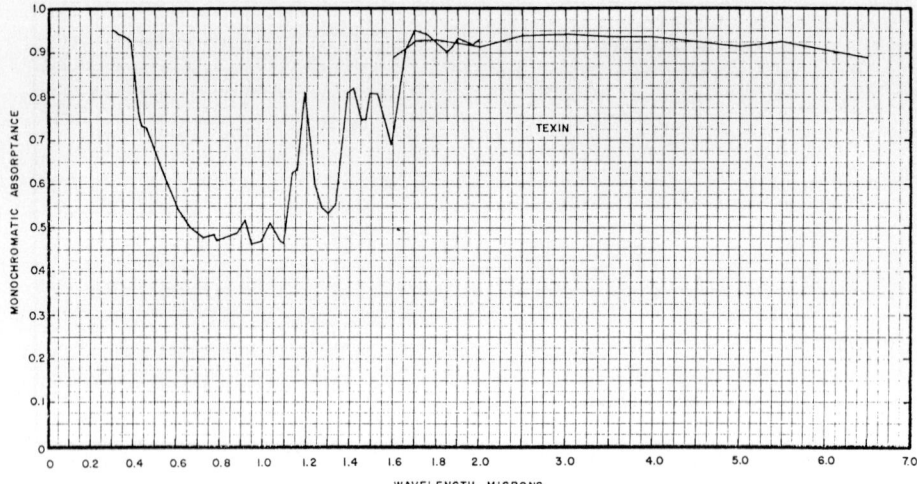

Fig. 9. Spectral Absorptance of Texin (Urethane-ester copolymer - A.L. Hyde Co.).

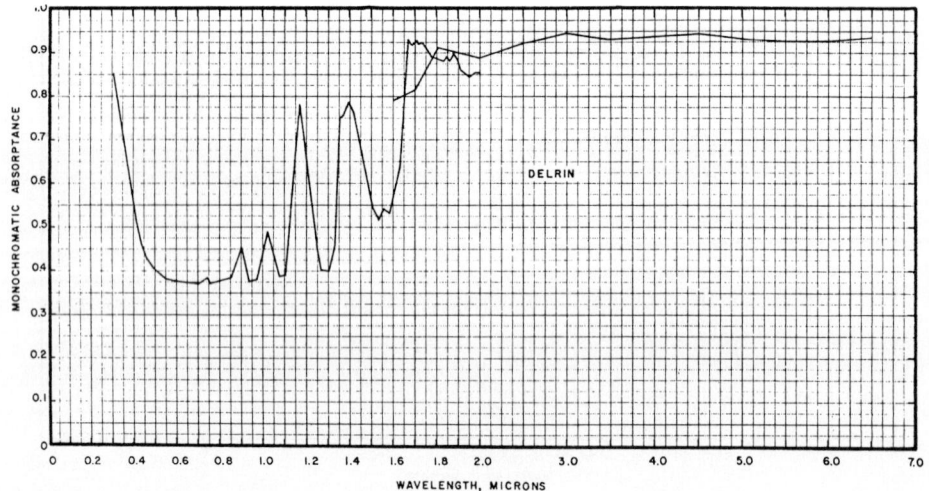

Fig. 10. Spectral Absorptance of Delrin.

Radiation Absorption   441

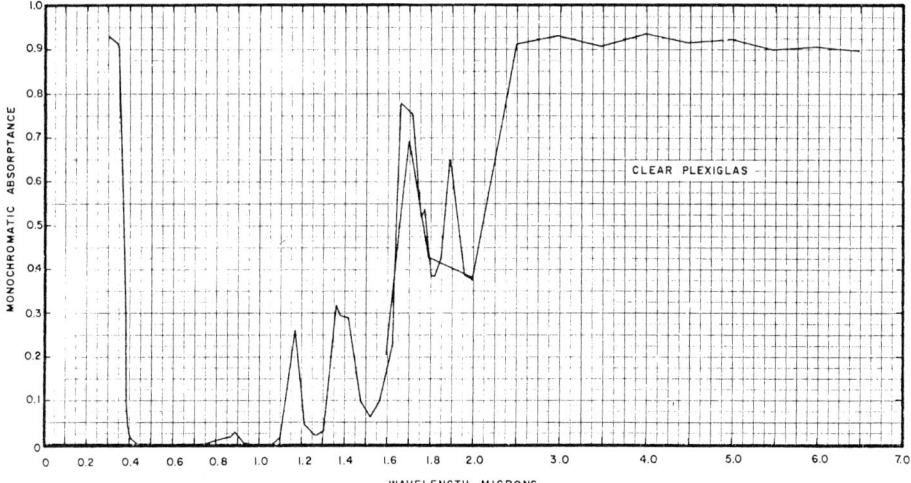

Fig. 11. Spectral Absorptance of Clear Plexiglas (Polymethylmethacrylate).

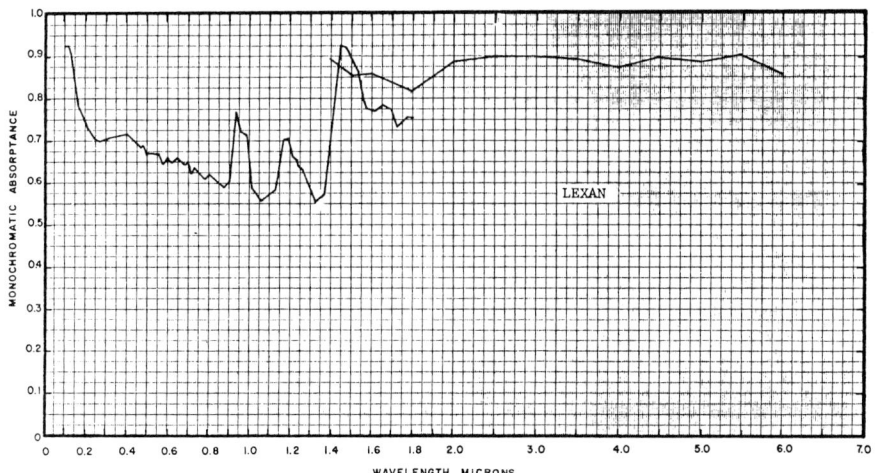

Fig. 12. Spectral Absorptance of Lexan Polycarbonate.

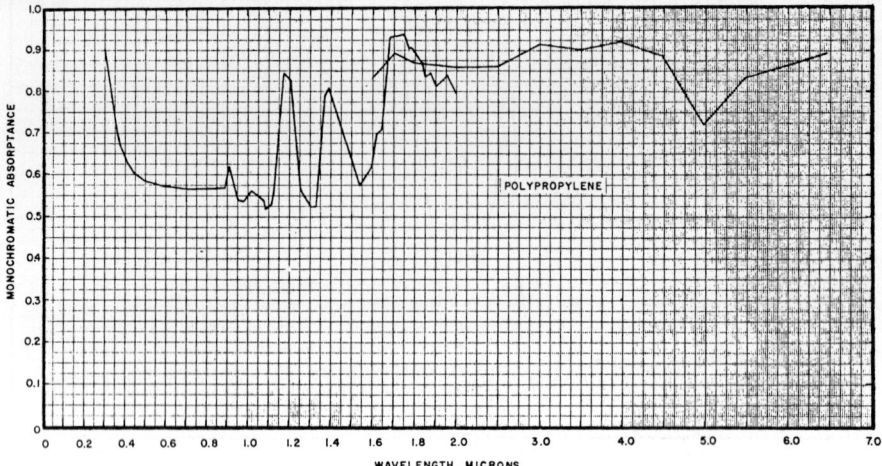

Fig. 13. Spectral Absorptance of Polypropylene.

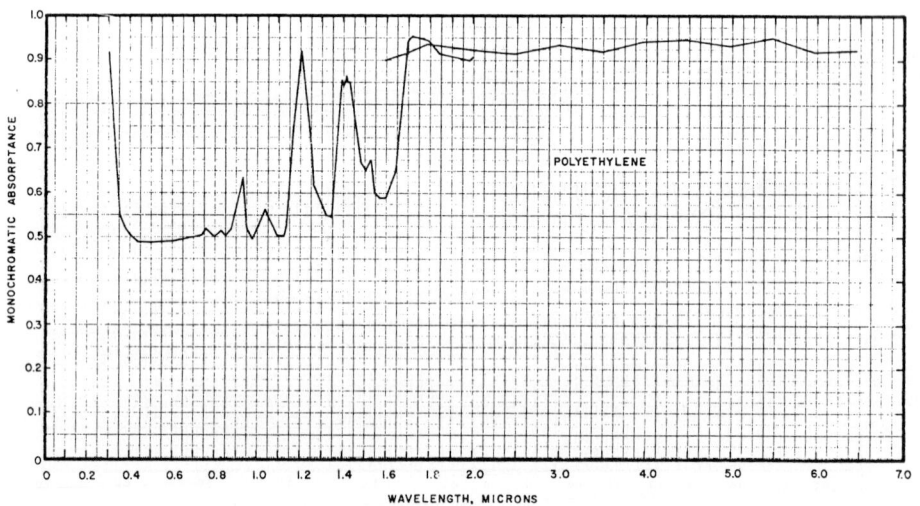

Fig. 14. Spectral Absorptance of Polyethylene.

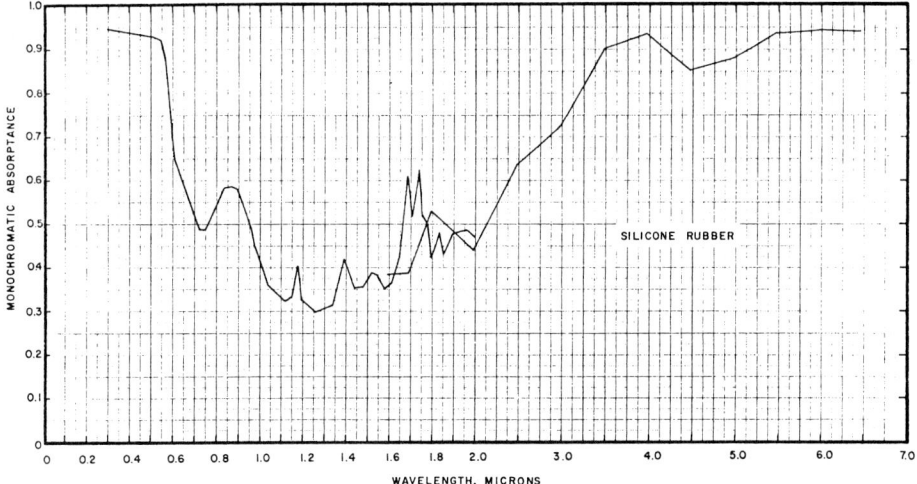

Fig. 15. Spectral Absorptance of Clear Polyvinyl Chloride.

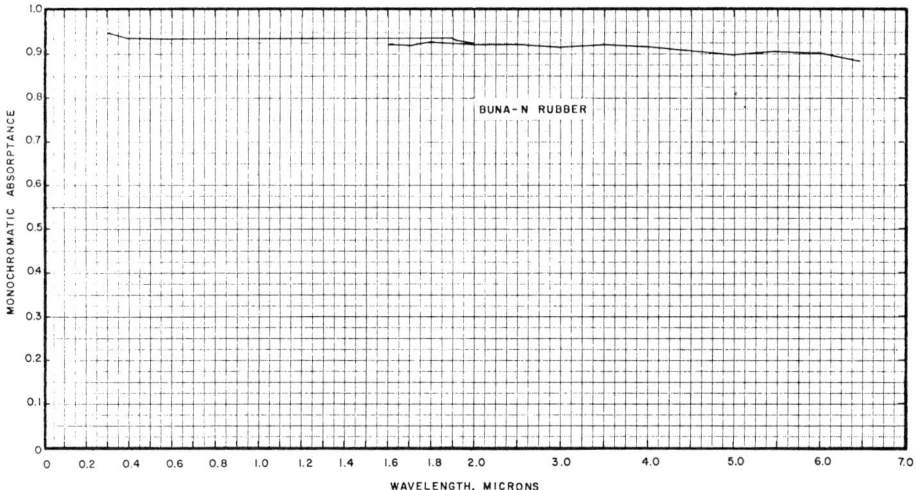

Fig. 16. Spectral Absorptance of Silicone Rubber.

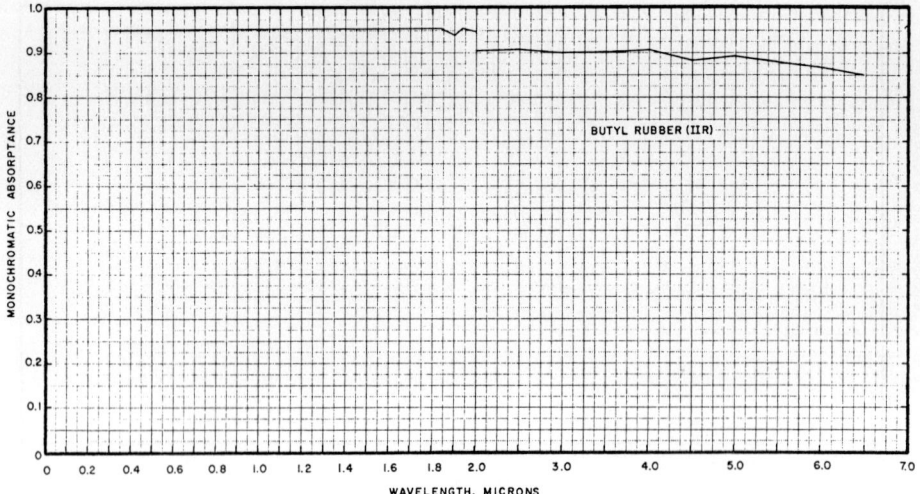

Fig. 17. Spectral Absorptance of BUNA-N Nitrile Rubber.

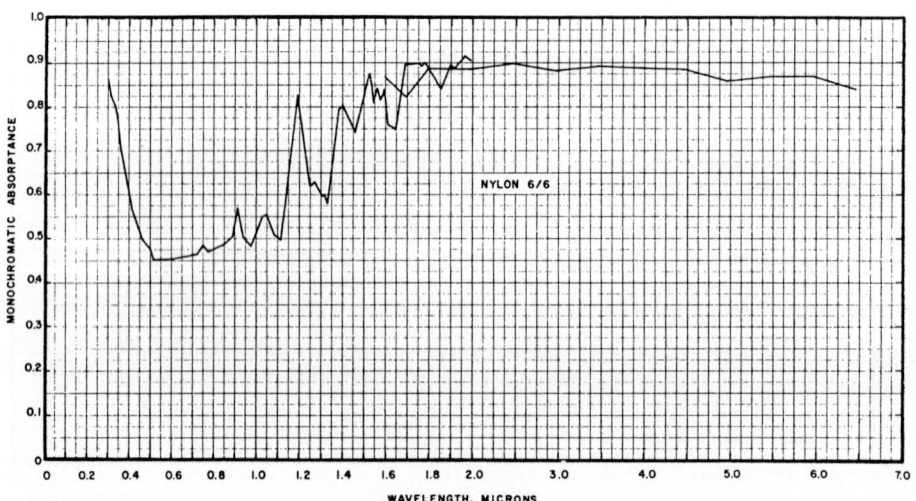

Fig. 18. Spectral Absorptance of Butyl IIR Rubber.

Radiation Absorption 445

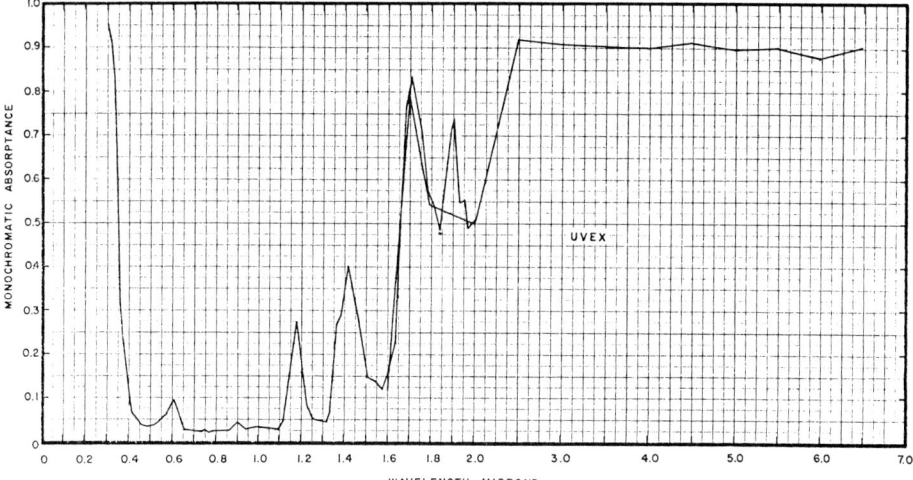

Fig. 19. Spectral Absorptance of Nylon 6/6.

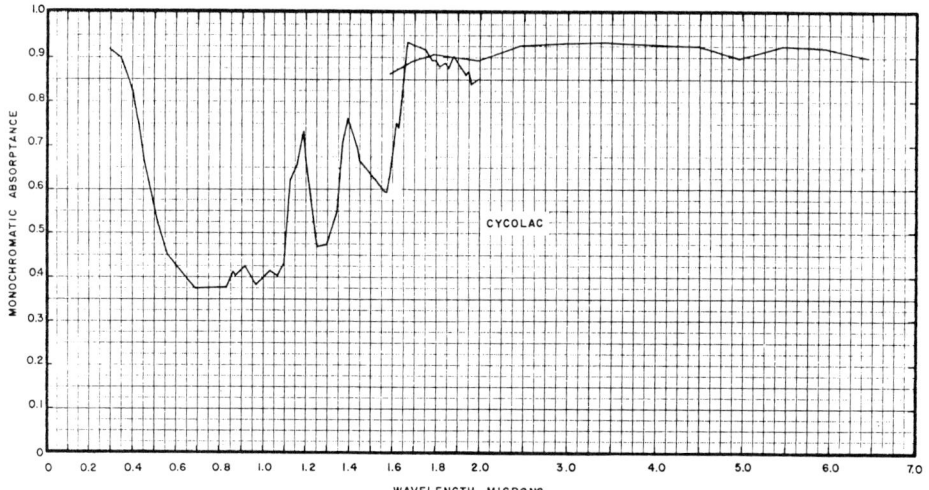

Fig. 20. Spectral Absorptance of Uvex (Cellulose Acetate Butyrate).

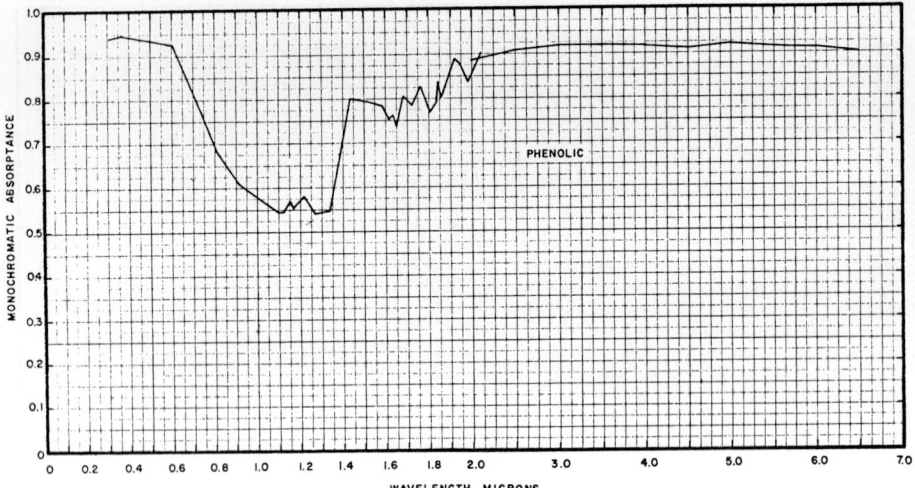

Fig. 21. Spectral Absorptance of Cycolac ABS Polymer.

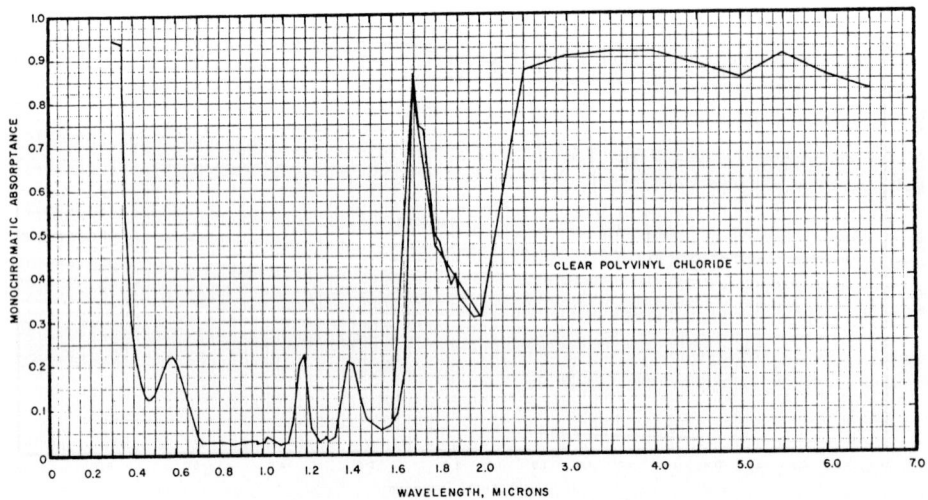

Fig. 22. Spectral Absorptance of Bakelite Phenolic Resin.

About Authors

David Clark is a Reader in Chemistry at Durham University, England. Dr. Clark has published over 100 articles related to ESCA, physical organic and theoretical chemistry. He was a Fulbright scholar at California Institute of Technology in 1972.

Dwight E. Williams is Laboratory Supervisor in the Analytical Department of Dow Corning Corporation. Dr. Williams' publications include papers on molecular spectroscopy, NMR and Mössbauer spectroscopy.

Donald M. Soignet is a Research Chemist, USDA, Southern Regional Research Center. Dr. Soignet's interests are in the area of electrochemistry and ESCA.

Merle M. Millard has been with the USDA Western Regional Research Center since 1971. Dr. Millard's research interests include fibers, ESCA and plasma chemistry.

Joe Andrade is Associate Professor of Materials Science and Engineering at the University of Utah. Dr. Andrade is the Editor of ACS Symposium volume, Hydrogels for Medical and Related Applications, 1976.

N. J. Harrick is the President of Harrick Scientific Corp. Dr. Harrick is the Author of the book <u>Internal Reflection Spectroscopy</u>. His research interests include magnetic resonance, semiconductor transport properties and surface physics.

David L. Allara is a member of the technical staff of Bell Laboratories. Dr. Allara has published over twenty articles.

Rodney D. Andrews is Professor of Chemistry at the Stevens Institute of Technology. Dr. Andrews' research interests are mechanical and optical properties of polymers and their relation to molecular and solid-state structure.

Timothy R. Hart is Associate Professor of Physics and Engineering Physics at the Stevens Institute of Technology. Dr. Hart's research interests include the use of lasers to study molecular structure of solids.

About Authors

W. R. Goynes is a Research Chemist at the Southern Regional Research Center, USDA. Mr. Goynes has been engaged in microscopical research since 1960.

Alfredo G. Causa is a senior member in the Tire Textile Department of the Goodyear Tire & Rubber Co. Dr. Causa holds patents and has published several articles in polymer science.

Norman Weeks is a Senior Materials Engineer with Pratt & Whitney Aircraft. Dr. Weeks has co-authored six papers on the properties of highly oriented polyethylene.

David W. Dwight is a Research Associate at the Virginia Polytechnic Institute and State University. Dr. Dwight's research interests include x-ray diffraction, electron microscopy and surface characterization in general.

Clifford H. Griffiths is a Senior Scientist in the Physical and Chemical Sciences Laboratory of the Xerox Webster Research Center. *Dr. Griffiths* did his graduate work at the University of London. His research interests include structure-property relationships in organic and inorganic polymers.

John W. Vanderhoff is Professor of Chemistry, Lehigh University. Dr. Vanderhoff is an Associate Director of the Center for Surface and Coatings Research and the Director of the National Printing Ink Institute. His research interests include emulsion polymerization and surface chemistry of polymers.

Robert L. Rowell is Associate Professor of Chemistry, University of Massachusetts. Dr. Rowell was the Chairman of the National Colloid Symposium of 1969. Currently, he is the co-editor of Colloid Science Series.

John L. Anderson is associated with ERA Systems. Dr. Anderson discovered the basic principles of Evaporative Rate Analysis in 1960. He received the A. K. Doolittle Award of ACS in 1970.

John R. Hallman is Professor and Head of Chemical Engineering Technology Department, Nashville State Technical Institute. Dr. Hallman worked as an Aerospace Engineer at General Dynamics, San Diego, for fourteen years. He has published several articles on polymer ignition.

Lieng-Huang Lee is a Senior Scientist of Xerox Corporation. Dr. Lee is the Editor of <u>Recent Advances in Adhesion, Advances in Polymer Friction and Wear,</u> and <u>Adhesion Science and Technology</u>. He was the Chairman (1975-76) of the Division of Organic Coatings and Plastics Chemistry of the American Chemical Society.

Author Index

A

Aberg, T., 51
Adams, D.B., 50, 51
Adams, D.H., 240
Adams, J.H., 206
Adams, S.B., 238
Ahlstrom, D.H., 240
Akana, Y., 357
Akelroyd, E.I., 394
Albridge, R.G., 100
Allan, A.J.G., 331
Allara, D.L., 151, 190, 191, 193, 206
Allen, G.C., 100
Alvares, N.J., 437
Ambrozh, K., 357
Amy, J.W., 51
Anderson, J.L., 409, 422, 423, 424, 425, 426, 427
Andrade, J.D., 133, 140, 141
Andrews, M.W., 100
Andrews, R.D., 151, 207, 238
Angell, C.L., 240
Antanaitis, C., 239
Archer, R.J., 191
Arnett, L.M., 395
Asano, S., 408
Aspnes, D.E., 190, 206
Awe, R., 423

B

Backer, S., 287, 288
Baer, E., 312
Baer, Y., 51
Baier, R.E., 140
Bailey, W.O., 331
Bair, H.E., 357
Baitinger, W.F., 51
Baker, R.A., 425, 426
Bamford, C.H., 395
Banderet, A., 395
Barber, M., 100
Barrales-Rienda, J.M., 357
Basch, H., 51
Bascom, W.D., 312
Basilier, E., 50
Bassett, D.C., 312

Bates, J.B., 151
Batra, S.K., 288
Bell, A.T., 100
Bell, R.J., 151
Bellamy, L.J., 206
Benderly, A.A., 312
Beninate, J.V., 83
Bergmark, T., 50, 51, 100
Bernett, M.K., 331
Bernstein, H.J., 238, 240
Berthou, H., 51
Bevington, J.C., 395
Bikales, N., 265
Bischoff, K.B., 140
Bitsch, B., 395
Blades, H., 287
Bless, P., 312
Bllica, H.R., 288
Bonis, L.J., 311
Booth, C., 239
Bosley, D.E., 287
Bower, D.I., 239
Boyd, G.E., 394
Boylston, E.K., 83
Bradbury, J.H., 100
Bradford, E.B., 394, 395
Brewster, D.A., 423
Briggs, T.H., 206
Brodnyan, J.G., 394
Brolin, S.E., 100
Bruck, S.D. 140
Buechler, E., 238
Bunsell, A.R., 287
Bunzl, K., 394
Burns, C.M., 311

C

Cannizzaro, A.M., 265
Carey, P.R., 240
Carlson, T.A., 51, 141
Carlsson, D.J., 206
Carra, J.H., 83, 251
Carter, G.B., 287
Carter, J.M., 422, 424
Casale, A., 357

Causa, A.G., 267
Cembrola, R.J., 408
Cerceo, E., 423
Chadha, S.C., 394
Chambers, R.D., 50
Chan, M.G., 151, 191, 206
Chapman, B.M., 287
Charland, G.A., 424
Chiu, H.Th., 140
Chow, S., 240
Chow, Y.L., 240
Cirlin, E.H., 331
Clark, D.T., 5, 50, 51, 101, 131, 132, 140, 331
Cobiana, N., 357
Coleman, I., 206
Colthup, N.B., 150
Comisarow, M.B., 151
Conley, R.T., 206
Cooke, D.D., 407
Cooley, J.W., 151
Cooney, R.P., 240
Cornell, S.W., 239
Cornish, E.H., 357
Cottington, R.L., 312
Cox, R.M., 423
Craver, C.D., 150
Cross, P.C., 238
Cross, P.M., 287
Crystal, R.G., 357
Cudby, M.E.A., 239
Culp, L.A., 141
Cumming, R.D., 140
Curthoys, G., 240
Curtis, J.W., 287
Czanderna, A.W., 206

Donnet, J.B., 100
Drake, G.L., Jr., 83
Drake, G.W., Jr., 265
Drmaj, D.J., 190
Ducann, D.E., 408
Dwight, D.W., 313, 331

E

Edelhauser, H.A., 394
Egerton, T.A., 240
El-Aasser, M.S., 395
Elder, M.E., 395
Ellison, A.H., 151, 190, 206
Erickson, U., 100
Escard, J., 100
Evanco, M.A., 312
Evans, E.L., 100
Evans, H., 240

F

Fadley, C.S., 100
Fahlman, A., 100
Fahrenfort, J., 151, 191
Fang, J.C., 331
Farinato, R.S., 397
Feast, W.J., 50, 51, 132, 140, 331
Feruse, A., 140
Field, J.E., 357
Figucia, F., 287
Filshie, B.K., 100
Finestone, A.B., 395
Fitch, R.M., 395
Fleischmann, M., 240
Flory, P.J., 357, 358
Folkes, M.J., 239
Foltz, C.R., 240
Forbes, J.F., 423, 425, 426
Force, C.G., 394
Ford, J.W., 140
Forhlich, H., 238
Forster, M.J., 287
Fox, T.G., 357, 358
Francis, G., 132
Francis, S.A., 151, 190, 206
Fraser, G.V., 239
Fredericks, R.J., 312
Frei, R.W., 192
Frisch, H.L., 311
Froese, A., 240
Frushour, B.G., 238
Furusawa, K., 395

D

Dabrio, M.I., 357
Dall'asta, G., 357
Dalton, J.C.S., 100
Daly, L.H., 150
Damico, D.J., 425
Danna, G.F., 265
Dauksch, H., 100
Davis, L.E., 53
Davitt, R., 312
De Bruyn, H., 394
De Palma, V.A., 140
Dechant, J.I., 151
Decius, J.C., 238
Decker, C., 206
Dekeyser, W., 50, 132
Dettre, R.H., 311, 331
Dilks, A., 50, 51, 101, 131
Dimonie, M., 357
Dolova, I.A., 312

G

Gall, M.J., 238, 239
Gamache, R., 238

Gandolfi, G., 357
Geigley, A.G., 240
Gelius, U., 50, 51, 100
Genzel, L., 238
Gerrard, D.L., 240
Ghiradella, H., 311
Ghosh, P., 394
Gill, D., 239, 240
Ginnard, C.R., 331
Glang, R., 51
Goggin, P.L., 239
Gonzales-Ramas, J., 357
Good, R.J., 331
Goodwin, J.W., 395
Goswami, B.C., 287
Gott, V.L., 140
Goynes, W.R., 83, 251, 265
Greenler, R.G., 151, 190, 206, 240
Gregg, R.A., 395
Gregonis, D.E., 140
Griffiths, C.H., 333
Grosberg, P., 287
Grunthaner, F.J., 100
Gucker, F.T., 407
Guest, M.F., 51
Gurnee, E.F., 395

H

Hagstrum, H.D., 132
Hair, M.L., 150
Hallman, J.R., 429, 437
Hamrin, K., 51, 100
Hannah, R.W., 190
Hansen, J.L., 394
Hansen, R.H., 131, 206
Hansen, W.N., 191
Harada, I., 240
Hardin, A.H., 240
Harrick, N.J., 140, 151, 153, 189, 206
Hart, T.R., 151, 207, 238
Hartley, A.J., 239
Hawkins, W.L., 151, 191, 206
Hayes, K.E., 190, 191
Hearle, J.W.S., 287
Hearn, J., 395
Hecht, H.G., 192
Heden, P.F., 51, 100
Hedman, J., 51, 100
Hedra, P.J., 238
Heller, J., 357
Heller, W., 407, 408
Hellmuth, E., 312
Helwig, E.J., 424
Hendra, P.J., 238, 239, 240
Hercules, D.M., 3
Herskovitz, T., 239
Heyde, M.E., 240
Hibbs, J., 141

Hibler, G.W., 239
Hill, D., 141
Hill, H.E., 425
Hillier, I.H., 51
Hirata, H., 357
Hoffman, J.D., 358
Hollahan, J.R., 100
Holland-Moritz, K., 150, 238, 239
Holm, R.H., 239
House, D.A., 395
Howe, M.L., 240
Hu, T.J., 238
Huang, Y.S., 239
Hubbard, D., 140
Huber, H.F., 395
Huddlestrone, R.H., 132
Hummel, D.O., 150, 239
Husner, H.H., 311

I

Inagaki, F., 240
Inglis, A.S., 100
Ishida, I., 151
Iwamoto, G.K., 133

J

Jackson, L.C., 423
Jaffe, H.H., 51
Jenkins, A.D., 395
Johannessen, B., 83
Johansson, G., 51, 100
Johnson, G., 100
Johnson, R.E., Jr., 311, 331
Jorgensen, C.K., 51

K

Kaelble, D.H., 331
Karasz, F.E., 357
Kardos, J.L., 312
Keilmann, F., 238
Keller, A., 239
Kelley, E.L., 394
Kendrick, J., 51
Kerker, M., 407
Kiefer, W., 238
Kilcast, D., 50
Kilponen, R.G., 239, 240
Kim, K.S., 51, 100
Kim, S.W., 140, 141
Kimura, A., 357
King, R.M., 140, 141
Kinsella, T.P., 425
Kirkwood, B.H., 287
Kiselev, A.V., 312
Kitagawa, T., 239
Kitchener, J.A., 394
Klein, M.P., 100

Kobayashi, M., 239, 240
Koegel, W.S., 423
Koenig, J.L., 151, 238, 239, 240
Kohlmayr, G.M., 289
Kolthoff, I.M., 394, 395
Konopasek, L., 287
Kortum, G., 192
Kotera, A., 395
Kramer, L.N., 100
Kratohvil, J.P., 394, 407, 408
Krause, M.O., 51
Kressman, R., 394
Krieger, I.M., 395
Krishnan, K.S., 151
Kucinischi, V., 357
Kusahoyahi, S., 357
Kvikstad, J.A., 331
Kwei, T.K., 311
Kwolek, S.L., 287

L

Laible, R.C., 287
Lando, J.B., 312
Lau, W.W.Y., 311
LeBlanc, R.B., 265
Lederman, D.M., 140
Lee, B.T., 437
Lee, E.S., 141
Lee, I.J., 140
Lee, K.S., 100
Lee, L.H., 132, 147
Lee, R.G., 140
Leeder, J.D., 100
Leininger, R.I., 140
Leonard, S.L., 132
Leung, Y.K., 239
Levit, A.B., 407
Lewis, A., 238
Liebman, S.A., 240
Lin, V.J.C., 238
Lindberg, B.J., 100
Linz, A., 238
Lippert, J.L., 238, 239
Litt, M.H., 151
Livit, A.B., 407
Livramento, J., 238
Loader, E.J., 240
Loader, L., 150
Lomas, B., 287
Lontz, J.F., 311
Love, B.T., 437
Low, M.J.D., 206
Lucas, G.L., 140
Luongo, J.P., 206
Lyman, D.J., 140, 141
Lyons, W.J., 287

M

Machi, S., 312
Macneil, J.D., 192
Maddams, W.F., 240
Madras, P.N., 141
Maissel, L.I., 51
Makagaki, M., 407
Makinen, M.W., 238
Mandal, B.N., 395
Manne, R., 51
Marshall, A.G., 151
Marshall, M.C., 394
Marshall, T.R., 407
Martin, T.P., 238
Matheson, M.S., 395
Matijevic, E., 394, 407
Matsuda, T., 151
Maune, R., 100
Maxfield, J., 239
Mayo, F.R., 206, 395
Mazzanti, G., 357
Mazzeno, L.W., 265
McCann, G.D., 395
McDowell, J.R., 426
McFaden, P., 407
McGraw, G.E., 239
McIntyre, J.D.E., 190, 206
McNeill, B., 133
McQuillan, A.J., 240
Meek, R.L., 206
Mellillo, L., 331
Meluskey, J.T., 240
Melveger, A.J., 239
Melville, H.W., 395
Mie, G., 408
Migliore, A.J., 408
Millard, M.M., 86, 100, 331
Miller, I.K., 395
Miranda, T.J., 424, 425
Miskowski, V., 239
Miyazawa, T., 239
Mizushima, S.I., 239
Moddeman, W.E., 51
Modric, I., 238, 239
Monroe, C., 423
Morris, G., 288
Morrison, W., 425
Mortensen, L.E., 239
Morton, W.A., 140, 141
Moss, T.H., 239
Muir, W.M., 140
Mukherjee, A.R., 394
Mukherjee, P., 395
Muller, J., 50, 51
Murrell, J.N., 51
Muschlitz, E.E., Jr., 132
Musgrave, W.K.R., 50, 132, 331

AUTHOR INDEX

N

Nakanaga, T., 240
Nakanishi, K., 51
Natta, G., 357
Newman, S., 311
Nicolaon, G., 407
Nielsen, J.R., 238
Niki, E., 206
Nilsson, A.G., 50
Noll, C., 425
Nordberg, R., 100
Nordling, C., 51, 100
Nyilas, E., 140, 141

O

O'Connel, E.P., Jr., 424
O'Connor, P.R., 394
O'Kane, D.F., 331
O'Malley, J.M., 51
O'Reilly, J.M., 357
Obremski, R.J., 240
Ogilvie, J.L., 100
Okabayashi, H., 239
Okuyama, M., 239
Olf, H., 239
Orchin, M., 51
Osher, J.E., 132
Otocka, E.P., 289
Ottewill, R.H., 394, 395
Otzinger, M., 426
Overberger, C.G., 395

P

Packer, A., 240
Painter, P.C., 238
Palit, S.R., 394, 395
Palmberg, P.W., 51
Paolini, A., Jr., 423, 424, 425
Parkinson, D.B., 357
Parmeland, G., 395
Partington, S., 50
Paton, A., 50
Patridge, R.H., 132
Pavlath, A.E., 100, 331
Peacock, C.J., 238, 239
Peeling, J., 50, 51
Penzias, G.J., 437
Peterlin, A., 51, 239
Peterson, J.H., 395
Peterson, R.V., 141
Peticolas, W.L., 238, 239
Pietras, C.S., 425, 426
Placzek, G., 238
Poling, G.W., 151, 190
Porto, S.P.S., 151
Powell, C.J., 51
Prescott, B., 238

Princen, L.H., 249
Pugh, T.L., 408
Purvis, J., 239

Q

Quinn, E.J., 240

R

Rabek, J.F., 132
Radigan, W., 311
Rahl, F.J., 312
Rahn, R.R., 206
Raman, C.V., 151
Ranby, B., 132
Rankin, J.M., 312
Reed Welker, J., 437
Rees, R.R., 424
Reeves, W.A., 83, 265
Reidel, D., 132
Reimschussel, A.C., 312
Reuwer, J.F., 240
Rice, D.W., 331
Richards, D.H., 51
Richardson, W.H., 206
Riess, G., 395
Riggs, W.M., 331
Rimai, L., 239, 240
Ritchie, I., 50, 132, 331
Roberts, R., 331
Robinson, J.W., 132
Rodwell, W., 51
Rogers, G.E., 100
Rollins, M.L., 265
Root, D.E., Jr., 422, 425
Rossi, A.G., 423, 424
Rothery, F.E., 100
Rowell, R.L., 397. 407, 408
Russ, J.C., 265
Ryan, L.R., 437

S

Salaneck, W., 50
Samson, J.A.R., 132
Sanders, C.I., 424
Saunders, W.H., 426
Sawyer, P.N., 140
Scanlan, I.W., 50
Schachter, A.C., 423
Schaff, A., Jr., 422, 423
Schaffert, R.M., 357
Schaufele, R.F., 151, 238, 239
Schenk, V.T.J., 287
Schenkey, J.H., 394
Scheraga, H.A., 238
Scherpereel, D.E., 425
Scherrer, P., 358
Schneider, H., 240

AUTHOR INDEX

Scholl, F., 150
Schonhorn, H., 131, 311, 358
Schoppee, M.M., 288
Schurr, G.G., 425
Scilly, N.F., 51
Scofield, J.H., 51
Scott, R.G., 288
Scott, W.E., 407
Segal, L., 265
Seguchi, T., 312
Seiler, R.R., 425
Semen, J., 312
Sevich, J., 151
Shapiro, E., 239
Shaw, J.N., 394, 395
Shepherd, I.W., 239
Sheppard, N., 240
Sherman, P.O., 83
Sherratt, S., 312
Shih, P.T.K., 151, 240
Shimanouchi, T., 239, 240
Siegbahn, K., 50, 51, 100
Siesler, H.W., 150
Signer, R., 238
Singleterry, C.R., 312
Skelton, J., 288
Slager, T.L., 240
Slater, R.F., 100
Sliepcevich, C.M., 429, 437
Small, E.W., 238
Smekal, A., 151
Smith, L., 141
Smith, S., 83
Soignet, D.M., 73
Sok, B.A., 425
Solomon, O.F., 357
Sparrow, J.T., 287
Speak, R., 239
Spiro, T.G., 238, 239, 240
Stacey, M., 240
Stahl, R.E., 424
Stammer, W.C., 424
Steiner, G., 311
Stejny, J., 239
Stoicheff, B.P., 151
Stolka, M., 357
Stratton, P.M., 238
Strekas, T.C., 239, 240
Stromberg, S.E., 424
Struker, L.J., 407
Sullivan, J.W., 425
Sutton, L.E., 51
Suwa, T., 312
Svensson, S., 50
Swingle, R.W., 51
Symons, N.K.J., 312

T

Tabb, D.L., 151, 239

Tadokoro, H., 239, 240
Takeda, Y., 395
Takehisa, M., 312
Takenaka, T., 240
Takokora, H., 150
Tam, N.T., 240
Tang, S.P.W., 239
Tashiro, K., 240
Tasumi, M., 240
Tausk, R.J.M., 394
Taylor, R.P., 395
Tess, R.W., 150
Thies, C., 395
Thomas, G.J., Jr., 238
Thomas, H.R., 50, 51
Thomas, J.M., 100
Tieszen, D.O., 357
Tobin, M.C., 150
Tomesku, M., 357
Tompkins, H.G., 151, 190, 206
Tourin, R.H., 437
Trask, B.J., 265
Treloar, L.R.G., 287
Tucker, P.M., 100
Tuinstra, F., 312
Tukey, J.W., 151
Tuma, J., 407
Turkevich, J., 238
Turner, I.D.M., 240
Tweedale, P., 51

U

Uesaka, T., 239

V

Van Den Hul, H.J., 394, 395
Van Os, G.A., 394
Van Wagenen, R., 141
Vanderhoff, J.W., 365, 394, 395
Vanveld, R.D., 288
Varela, C., 288
Viser, W.M., 191

W

Wagner, C.D., 331
Wallace, T.P., 407, 408
Wannberg, B., 50
Ward, R.S., 141
Watson, D.S., 238
Watters, K.L., 240
Weeks, N.E., 289
Wegman, R.F., 426
Wehner, G.K., 51
Weiler, J., 238
Welker, J.R., 429, 437
Wendlandt, W.W., 192
Werme, L.O., 51
White, C.W., 206

Wiberley, S.E., 150
Wiles, D.M., 206
Williams, D.E., 53
Williams, D.J., 357
Williams, F.J., 51
Williams, V.A., 100
Willis, H.A., 239
Wilson, E.B., 238
Windle, J.J., 100
Winograd, N., 51, 100
Winter, C., 100
Winterling, G., 238
Wise, C.T., 425
Witeczek, J., 407
Wolberg, A., 100
Wood, D.L., 151
Woodward, A.E., 240
Wozniak, W.T., 239
Wunderlich, B., 51, 132, 312, 331, 358

Y

Yacoby, Y., 238
Yamamato, G., 408
Yashin, I., 312
Yeadon, D.A., 265
Yeh, G.S.Y., 358
Yin, T.P., 311
Yoshimoto, A., 357

Z

Zaukelies, D.A., 287
Zbinden, R., 150
Zell, K.D., 287
Zimmerer, N., 238
Zimmerman, J., 287
Zisman, W.A., 331

Subject Index

A

Absorptance, 430
  of radiation, 429
Absorption
  cell, 214
  of carbonyl, 203
  of radiation, 429
Acetic acid
  on copper, 204
Acoustic mode, 224
Acrylic acid
  as emulsifier, 365
Adhesion, 147
  by ERA, 421
  work of, 134
Adhesive bonding, 101
Adsorbed
  impurity, 61
  species on metal, 166
Adsorption
  of polymer, 149
  of pyridine on oxide, 236
  of solvent, 418
Aerosol coagulation, 405
Alkane-styrene
  copolymer (ESCA), 45
Alpha helix, 217
Aluminum foil, 316
Amine
  nitrogen line, 98
Amino acid
  ESCA, 87
Ammonia gas, 74
Angle of incidence, 154, 195
Anisotropic
  distribution, 136
  solution, 271
Anisotropy, 222
Antistoke
  spectrum (Raman), 211
Annealing
  of fluoropolymer (SEM), 321
Aramid
  tire cord, 267
Argon
  ion bombardment, 102
  ion gun, 55
  ion laser, 213
  ion treatment, 114
  laser line, 224
Autoxidation
  mechanism, 193
Avcothane-51, 135
Axis spacing, 352
Azobisisobutyronitrile, 348

B

Ball-milling
  oxidation on, 78
Band-intensity
  (ERS), 171
Barium carbonate
  C^{14}-tagged, 410
Beam splitter, 174
Binding energy, 5, 88
Biological polymer
  Raman, 216
Biomedical application
  of polymer, 133
Birefringence
  degree of, 222
Black film, 168
Block copolymer,
  urethane-siloxane, 135
Blood
  clotting, 133
  coagulation, 133
  compatibility, 133
  plasma protein, 133
  thrombosis, 133
Bond polarizability, 209
Bovin serum
  albumin, 217
Bremsstrahlung, 6, 121
Brewster's angle, 153
Buna-N rubber
  (radiation), 436
Butyl rubber
  (radiation), 436

n-Butyraldehyde
  solution, 335

C

Carbon
  1s spectra, 88
  tetrachloride, 237
CASING, 102
Catheter
  polyvinyl chloride, 135
Cell
  exudate, 136
  substrate-interaction, 137
Cellophane bag
  for dialysis, 377
Cellulose
  acetate-butyrate, 436
  crosslinking, 78
Chain-fold
  nuclei, 333
  surface energy, 355
Chain length
  and crystallinity, 347
  of PVK, 343
Chain scission, 205
Chemical modification
  of polymer, 149
Chemisorption
  of polymer, 148
Chromophore process, 121
Coalescence
  of resin, 310
Coating
  drying property, 421
Colloidal dispersion, 366
Compression flex test, 269
Conductivity, 134
Conductometric
  titration, 365, 372
Configuration
  of molecule, 223
Conformation
  energy of helix, 223
  transition, 208, 217
Contact angle, 134
  advancing, 322
  hysteresis, 322
  receding, 313
  Teflon, 306
Contamination, 147
  of hydrocarbon, 119
Copper
  oxide, 197
  stearate, 204
Cord-rubber
  composite, 267
Core level, 8
  spectra, 123

Cotton, 74
  flannlette, 75
  modifiéd, 73
  polyester blend, 80
Coupling interaction, 149
Critical surface tension, 134
Crosslinking mechanism, 102, 129
Crystal growth
  of PVK, 344
Crystalline
  density, 352
  polymer, 223
Crystalling force, 322
Crystallization
  epitaxial, 289
  temperature, 356
Cyanoccrylate, 150
Cyclic stress, 268
  fracture, 275
Cysteine, 98

D

Debye-Sherer
  method, 336
Deconvolution
  of ESCA spectra, 87
Degradation
  of polymer, 148
Degree
  of crystallinity, 226
  of crystallization, 421
  of cure (ERA), 421
Depolarization
  of Raman, 220
  ratio, 404
Depth profiling
  ESCA, 8
Dialysis, 367, 379
Dielectric constant, 156
Differential scanning
  calorimeter, 292, 335
Diffraction peak, 350
Diffuse
  reflection spectroscopy, 179
1,1-Dihydroxy-perfluorooctylamine, 74
Dilution factor, 405
Dipeptide
  nitrogen line, 98
Dipole moment, 209
Dispersion
  compensation, 6
  of PTFE, 291
Distribution width, 405
Double helix, 216
Doublet state, 41
DSC, 292, 335

Dye
  laser, 211
  partition, 382, 383

E

Effective
  thickness, 168
  wavelength, 171
Electric field
  amplitude, 156, 196
  vector, 160
Electromagnetic
  radiation, 102, 103, 398
Electron
  affinity, 130
  density, 80
  energy distribution, 107
  flux, 110
  photoemitted, 19
Electron diffraction
  of PVK, 334
Electron micrograph
  transmission, 319
Electron microscopy
  of latex particle, 381
Electrostatic analyzer, 107
Element mapping, 8
Ellipsoid
  light scattering, 398
Ellipsometry, 172, 198
Emmisive power
  of heat source, 434
Emulsifier, 365, 377
  Aerosol-MA, 377
  Aerosol-OT, 378
  Dresinate—214, 378
  Sipon—WD, 378
Emulsion polymerization, 365
  seeded, 366
End group analysis, 13
Endoplasm, 98
Energy of crystallization, 354
Energy transfer, 101
Epoxy resin
  amine-cured, 327
ESCA, 5
  -Auger spectrometer, C-1s spectra, 318
  of biopolymer, 133
  of cotton, 73
  of fluoropolymer, 327
  of modified polymer, 101
  of PTFE, 313
  of wool, 85
  sputtering analysis, 54
Escape depth, 19, 116
Esterification, 229
Ethylene
  -tetrafluoroethylene
    copolymer, 102, 121
Evanescent wave, 172
Evaporation
  rate analysis (ERA), 409
  rate of, 414
External
  reflection spectroscopy, 164
  specular, 153
Extinction coefficient, 34, 153
  transmission, 195

F

Fabric stiffness, 256
Fatigue tester
  SCEF, 270
Fermi level, 104
Ferredoxin, 220
Fiber fracture
  (SEM), 277
Fiber glass cloth, 318
Fiber structure, 147
Fibril
  of PTFE, 309
Fibrillation
  of crystal, 275
Fire prevention, 429
Flame retardant, 73
  finish, 75
Flammable material, 429
Flocculation
  of latex, 385
Flow anisotropy, 297
Fluorination
  of polyethylene, 102
  of surface, 102
Fluorine atom
  (ESCA), 78
Fluorocarbon system, 22
Fluorocarbonate, 14
Fluoropolymer, 5
  FC-218, 80
  surface, 313, 321
Fold period
  of crystal, 354
Fourier transform
  infrared, 148
Fracture
  composition, 59
  topography, 275
Free energy
  of side-fold, 355
  of linear chain surface, 355
Fresnel equation, 174
Fringe pattern, 161
Full width
  at half maximum, 87

G

Gandolfi
  camera, 336
  scattering curve, 336
Gas-film
  interaction, 195
Geiger detector, 410
Gel permeation
  chromatography, 335
Germanium prism, 175
Glass
  surface (Raman), 150
  temperature
    of polystyrene, 349
    of PVK, 335, 349
Glow discharge, 101
  modification of
    fluoropolymer, 123
Glycine, 218
Gold
  film, 259
  palladium film, 259, 316
Goniometer, 316
  double beam, 185
Graphite (ESCA), 95
Grating spectrometer, 214
Grazing incidence, 160
Gum rubber
  (radiation), 435

H

Hapten, 234
  and antibody, 234
Head-to-head
  configuration, 25, 26
Helical
  conformation, 220
Helium
  neon laser, 213
Heme protein, 219
Hemoglobin, 219
Hexadecane, 316, 322
Hole state, 22
Hydrogel, 135
Hydrolysis
  of polysiloxane, 149
Hydroperoxide, 203
Hydroxybutyl
  glutamine, 218
Hydroxyl group
  on latex, 382

I

Immersion
  depth of, 316
Impact tester, 268
Impurity, 134
Incident
  irradiation, 430
Index
  of refraction, 153
Indium foil, 135
Inelastic scattering, 66, 87
Infrared
  absorption of PVK, 348
  spectrometer, 198, 207
  spectroscopy, 193
  spectroscopy (internal reflectance), 147
  spectroscopy, MIR, 134
  spectroscopy (reflection-absorption), 148
Instrumentation IR,
  IR, 183
Insulator
  Valence band, 8
Insulin
  ESCA, 98
Intensity
  integrated, 126
  of diffraction peak, 356
  Raman line, 212
Interchain
  spacing, 344
Interfacial
  energy, 134, 351
Interference fringe, 156
Interferogram, 148
Internal reflection, 156
  element, 187
  specular, 153
  spectroscopy, 172
Ion, 102
  interaction, 134
  neutralization spectroscopy, 104, 130
  pair formation, 386
Ion-exchange
  resin, 367
  titration, 365
Ionic strength, 217
Iron
  dithiolene complex, 95
  porphyrin ring complex, 219
Isomer
  gauche and trans, 222
Isomerization
  of double bond, 229
L-Isotucyl-l-alanine, 98

K

Kinetic energy
  of electron, 102
Kirchoff's law, 430
Krypton
  ion laser, 231

L

Lamellae
 folded chain, 351
 of PVK, 342
Lamellar structure, 225
Laser
 beam, 207
 frequency, 211
Laser Raman
 spectroscopy, 149
 for polymer, 207
Latex
 butadiene-styrene, 380
 characterization, 365
 particle size, 391
 stability, 388
Lattice
 pseudohexagonal, 334
Level
 of additive, 421
Ligand field, 219
Light
 modulator, 174
 scattering (angular), 397
 scattering photometer, 405
Linewidth
 (ESCA), 57
Lorenz
 -Mie theory, 398
Lorenzian function, 87
Lubricant
 (ERA), 421
Lubrication, 147
Lysozyme
 (Raman), 218

M

Magnesium
 $K\gamma$ x-ray, 87
Maleic anhydride
 in polyester, 228
Malloy tube test, 268, 269
Mass transfer, 10
Maxwell distribution, 111
Mean free path
 of core level, 5
 of electron, 102
Mechanical anisotropy, 275
Metal
 density of state, 8
 optics, 166
 oxide, 95
 oxide (ERS), 168
 probe, 107
Metastables, 102
Methyl orange
 (Raman), 237
N-Methylolphosphine, 75

N-Methylol urea, 78
Michelson interferometer, 148
Mirage, 173
Mirror attachment
 twin parallel, 187
Modal size, 405
Modification
 rate of, 124
Molecular packing 335
Molecular weight
 distribution, 335
 of latex, 381
 of PTFE, 291
 of PVK, 334
Monochromatization
 of x-ray, 6
Monodisperse
 fraction of PVK, 335
 latex, 365
Monopole
 excited state, 8
 selection rule, 30
Multiple
 collector, 6
 internal reflection, 174
 reflection, 195
Mylar film
 for reflection, 164

N

Neoprene rubber
 (radiation), 435
Nitrogen
 in polypeptide, 98
 N 1-s signal, 95, 136
Noncentrosymmetric
 chain, 352
Nuclear magnetic
 resonance broadline, 221
Nucleation
 efficiency, 359
 growth theory, 354
 site, 342
Nylon 6, 6
 cord, 271
 radiation, 436

O

Oil repellency
 of cotton, 73
 soil release agent, 78
Oligomeric
 radical, 366
Optical
 cavity, 161
 constant, 154
 density of solvent, 380
 rotary dispersion, 218

SUBJECT INDEX

Organic polymer
  sputtering of, 66
Organosilicone
  polymer, 53
Oxidation
  of poly (1-butene), 193
  of polymer, 9, 147
Oxygen
  plasma, 91
  sensitivity ratio, 18

P

Paracrystalline phase, 353
Peak
  intensity, 231
  width at half height, 329
Penning
  ionization process, 104
Perturbation, 30
  theory (second order), 211
Phase change, 168
Phenol
  substituted, 231
Phenol-formaldehyde, 230
Phenolic
  (radiation), 436
Phosphorous
  ESCA, 75
  Kγ-X-ray, 260
Photoconductive
  properties of PVK, 333
Photocopying device, 333
Photoelectron
  emissivity, 58
Photoemission, 58
  angular dependence, 61
  from PTFE, 102
Photo energy, 104
Photoionization, 19, 104, 121
Photon, 18
  energy, 8
  flux, 121
Pi
  eclipsed arrangement, 23
  electron acceptor, 34
  electron donor, 34
  pi* transition, 33
Piperidine
  on oxide, 236
Plane of
  incidence, 194
Plasma
  polymer modification, 101
Plasticization
  of polymer, 147
Plastocyanine, 219
Platinum
  PTFE on, 306

Platt's
  spectroscopic moment, 36
Plexiglas
  (radiation), 435
Polarization, 404
  control, 154
  parallel, 160
  selection rule, 212
Polarized
  incident light, 399
  p-, light, 153
Polarizer
  IR, 186
Polyacrylic acid
  ESCA, 16, 19
Polyadenylic acid
  Raman, 216
Poly (1-alkylethylene)
  Raman, 226
Polybutadiene rubber
  Raman, 223
Poly (1-butene)
  on gold, 148
  oxidation, 193
Polybutyl acrylate
  ESCA, 12
Polybutyl methacrylate, 258
Polycarbonate
  (radiation), 435
Poly (decamethylene sebacate), 226
Poly (n-decyl acrylate)
  ESCA, 27
Polydimethylsiloxane, 32
  ESCA, 133
  Raman, 222
Polydiphenylsiloxane, 32
Polydispersity, 398
Polyelectrolyte, 369
Polyester
  cord, 271
  ESCA, 73, 80
  unsaturated, 223
Polyethylene, 32
  ESCA, 104
  fluorination, 102
  irradiation, 149
  Raman, 223
  spreading, 310
Polyethylene oxide, 226
Polyethylene terphthalate, 6, 10, 123
  ESCA, 80
  Raman, 226
Poly (2-ethylhexyl acrylate)
  ESCA, 27
Poly (n-hexyl acrylate)
  ESCA, 38
Polyisopropyl acrylate
  ESCA-12

Polymer
  backbone, 227
  degradation, 231
  endgroup, 382
  film on metal, 193
  latex, 397
  metal composite, 193
  plasma treated, 101
  -polymer interaction, 386
  structure, 5
Polymerization
  of vinylcarbazole, 335
Polymethyl acrylate, 209
  ESCA, 23
  Raman, 223
Polymethyl methacrylate
  ESCA, 16, 57
  Raman, 220
  spreading, 310
Polynucleotide
  Raman, 216
Polyoctadecyl acrylate
  ESCA, 27
Polypeptide, 98, 208
  Raman, 216
Polyphenyl acrylate
  ESCA, 38
Polyphenylene oxide
  (radiation), 435
Polypropylene, 209
  laser Raman, 225
  (radiation), 435
Polystyrene, 32, 123
  ignition time, 432
  latex, 365, 405
  para-substituted, 34
  Raman, 208, 222
Polytetrafluoroethylene,
  copolymer with
  hexafluoropropylene, 314
  dispersion, 313
  ESCA, 10
  fusion, 292
  SEM, 295
  wetting, 289, 295
Poly (trans-1, 4-butadiene), 228
Polyurethane
  ESCA, 55, 133
Polyuridylic acid, 216
Polyvinyl alcohol,
  embedment, 259
  iodide complex, 232
Polyvinyl bromide, 259
Polyvinyl carbazole,
  characterizations, 333
  ESCA, 40
  isotactic, 334
Polyvinyl chloride
  degradation, 231
  ESCA, 12
  polarization, 149
  radiation, 435
  vinylidene chloride copolymer, 55
Polyvinylidene fluoride, 117
Poly (1-vinylnaphthalene)
  ESCA, 50
Poly (4-vinylpyridine)
  ESCA, 34
Porosity
  by ERA, 421
Positive ion, 103
Potassium persulfate, 382, 386
Potentiometric
  titration, 375
Prism, 172
Protein
  ESCA, 98
Pyrolytic graphite, 289

R

Radiant intensity, 432
Radiation
  abosrption by polymer, 429
  Mg K$_\gamma$, 135
  short wave length, 121
Radiochemical
  molecule diffusion, 411
Radio frequency
  plasma, 102
Rayleigh
  linewidth spectroscopy, 397
  scattering, 211
Raman
  depolarization ratio, 222
  hypochroism, 216
  scattering, 209
  scattering tensor, 211
  spectroscopy, 147
  tensor, 220
Red
  kidney bean, 98
Reflectance
  absorptance relationship, 430
Reflection
  absorption spectroscopy, 193
  spectroscopy, 153
Reflectivity
  of interface, 155
Refractive index
  complex, 196
  of medium, 147
Relaxation energy, 30
Remol, 230
Resonance
  enhancement effect, 235
  method, 213

Retro-mirror
  accessory, 187
Re-wetting, 321
RNA, 216
Roughness
  of polymer surface, 324
  of PTFE, 314
Rubber
  oxidation of, 149
Rugosity, 324
Rydberg transition, 121

S

Saran
  ESCA, 55
Satellite
  shake-up, 8
Scanning electron microscopy
  of fiber, 251
  of PTFE, 289, 313
Scott peel, 281
Secondary electron, 109
Semiconductor
  space charge, 174
  surface, 174
  surface state, 174
Shake up, 9
  in polymer, 30
Side chain,
  perfluoromethyl, 324
Signal
  to noise ratio, 215
Silica gel
  bromine, 236
Silicon hydride, 176
Silicone
  polymer, 61
  resin, 62
  rubber, 55, 435
Silver chloride cell, 175
Single crystal, 225
Singlet state, 41
Sintering
  of PTFE, 290, 314
  oven, 314
Site
  -adsorbate interaction, 418
Skived film, 321
Slit
  filtering of monochromator, 215
Sodium
  lauryl sulfate (C^{14}), 380
  naphthalene, 289
Soil-release
  of cotton, 73
Solvent
  high boiling, 410
Specific
  surface, 399
  volume of PVK, 350
Spectral
  constant (ERS), 169
  resolution, 210
  sensitivity, 148
  subtraction technique, 135
Spectrometer
  ESCA, 5, 75, 87, 106, 135
  IR, 183
Sphere, 398
  breadth parameter, 398
  particle size, 398
  refractive index, 398
  scattering angle, 398
Spreading drop, 296
Sputtering
  of polymer, 53, 66
Standing wave, 156, 166
  ratio, 110
Starch
  complex with iodide, 232
Stokes spectrum, 211
Stratification
  of fluorocarbon, 330
Streaming potential, 134
Structural isomerism, 23
Substituent effect, 10
Sulfate ion-radical, 366
2-Sulfoethyl
  methacrylate, 365
Sulfonic acid, 91
Sulfur
  ESCA, 86
  oxidized, 91
Surface
  ablation, 119
  absorptance, 429
  analyzer, 411
  area of latex, 397
  charge density, 372
  density of element, 57
  free energy, 134, 289
  oxidation of polymer, 114
  reaction, 103
Surface tension
  in sintering, 290
  of hexadecane, 316
  of wash water, 369
Surfactant,
  and azo dye, 237
  in PTFE dispersion, 318

T

Tacticity
  of polymer, 27
Tail-to-tail
  configuration, 25, 26

SUBJECT INDEX 465

Teflon
  -FEP-epoxy, 313, 326
  -S, 315, 329
Telephone cable
  coating, 168
Tensile stress, 268
Test solution
  for ERA, 411
Testing machine
  MTS, 270
Tetrabromoethane, 412
Tetrakis (hydroxylmethyl)
  phosphonium chloride, 74, 252
Textile
  fiber-modified, 251
  finish, 74
Thermal degradation
  of epoxy, 327
Thick film
  ERS, 166
  IRS, 176
Thin film
  ERS, 166
  IRS, 176
THPC/urea, 74
THPOH
  -amide, 74
  -NH_3, 252
Tinius-Olsen
  stiffness test, 256
Total reflection, 173
Transmission
  spectroscopy, 153
Triboelectric
  phenomena, 9
Triplet state, 41
Tris-2, 3-dibromopropyl phosphate, 259
Tyndall scattering, 224

U

Ultraviolet
  absorption of resin, 368
  irradiation of metal, 109
Urethane-ester
  copolymer, 435

V

Valence level, 8
Versatile
  reflection attachment, 187
Vibration
  antisymmetric, 204

N-Vinylcarbazole, 335
Viscoelastic
  property in sintering, 290
Viscosity
  in sintering, 290
Vitamin B_{12}
  Raman, 220

W

Wagner
  sensitivity factor, 98, 326
Warp stiffness, 256
Wavelength
  vacuum, 399
Weighting coefficient
  of transition, 31
Wettability
  anisotropic, 314
  of wool, 85
  spectrum, 134
Wetting
  balance, 316
  PTFE, 290
Wool fiber, 85
Worm structure
  PTFE, 306

X

X-ray
  analysis, 251
  diffraction of PVK, 333
  fluorescence, 381
  mapping, 259
  pattern of PVK, 351
  radiation of polymer, 121

Y

Yarn strength
  of wool, 85

Z

Zeolite
  Raman, 236
Zero tension
  cyclic loading, 281
Zeta potential, 134
Ziegler-Natta
  catalyst for PVK, 334
Zig-Zag
  configuration, 226

A
B 7
C 8
D 9
E 0
F 1
G 2
H 3
I 4
J 5